Land-Cover and Land-Use Changes in Eastern Europe after the Collapse of the Soviet Union in 1991

Garik Gutman · Volker Radeloff
Editors

Land-Cover and Land-Use Changes in Eastern Europe after the Collapse of the Soviet Union in 1991

Springer

Editors
Garik Gutman
NASA Headquarters
Washington, DC
USA

Volker Radeloff
Department of Forest and Wildlife Ecology
University of Wisconsin
Madison, WI
USA

ISBN 978-3-319-42636-5 ISBN 978-3-319-42638-9 (eBook)
DOI 10.1007/978-3-319-42638-9

Library of Congress Control Number: 2016947198

Cover illustration: The border area of Belarus and Russia as depicted in a Landsat satellite image from the summer of 2000. Belarus lays to the west, and this is where agriculture remains active, field boundaries are clearly visible, and many fields appear in white and beige, representing freshly plowed soils, or bright orange, representing young crops. The situation in Belarus is very different from that in Russia, to the east, where almost all agriculture was abandoned: field boundaries are diminished, and many areas appearing in blue-green colors represent early-successional vegetation. Settlements on the Belarussian side include Drybin and Mscislau, on the Russian side Monastyrshchina and Khislavichi. The satellite image was recorded by Landsat 7 ETM+ on June 6th of 2000, and provided by the US Geological Survey Earth Resources Observation and Science Center.

Printed on acid-free paper

This Springer imprint is published by Springer Nature
The registered company is Springer International Publishing AG Switzerland

Preface

The breakup of the Soviet Union and collapse of the socialism in Eastern Europe in the late 1980s and early 1990s had major ramifications for the structure of societies, economies, and the way people used their land. Collectivized farms have transitioned to privately owned fields. Many arable lands have been abandoned and urban centers experienced substantial changes, both growth and decay, depending on the country, as manifested by space observations of nighttime lights. Forests damaged by industrial pollution in Soviet era recovered towards the end of the twentieth century. At the same time, satellite observations have been accumulated during the last 20–25 years to provide data-rich time series over land surfaces, allowing systematic analysis of the changes in land cover and land use observed from space.

This book describes and analyzes the effects of the collapse of socialist governance and management systems on land cover and land use in various parts of Eastern Europe including the countries of the former Soviet block, former Soviet Union republics, and European Russia. This book is a compilation of results from studies on land-cover and land-use changes and their interactions with carbon cycle and environment, effects of institutional changes on urban centers and agriculture, as well as changes in wildlife populations. The book is a truly international interdisciplinary effort written by an international team consisting of scientists from the USA, Europe, and Russia under the auspices of the Northern Eurasia Earth Science Partnership Initiative (NEESPI) supported by the NASA Land-Cover/Land-Use Change Program.

This book is of interest and directed to a broad range of scientists within natural and social science, involved in studying recent and ongoing changes in Europe, be they senior scientists, early career scientists, or students. The chapters of this book summarize analyses of the dramatic changes in land uses triggered by the abrupt change in the economies of the region and in land management. The satellite data used for these studies were mostly from optical sensors, including night lights observations, with both coarse and medium spatial resolution. The volume includes analysis of the drivers of agricultural land abandonment, forest changes in Black

Sea region, an extreme drought event of 2010, impacts of fires on air quality and other land-cover/land-use issues in Eastern Europe.

We warmly thank all the contributors of this book and acknowledge NASA support.

Washington, DC, USA — Garik Gutman
Madison, WI, USA — Volker Radeloff

Contents

Introduction

V.C. Radeloff and G. Gutman

1 Background

If lucky, each generation of land use scientists can study one or two major natural experiments, those historic events that rapidly change land use over large areas. The collapse of socialism throughout Eastern Europe in the late 1980s and early 1990s was such an event. The collapse was a natural experiment of rare magnitude that affected every aspect of societies, economies and land use practices. This book focuses on Eastern Europe, including the European part of Russia, because of the opportunity that the collapse of socialism entails for land science.

For almost half a century, starting with the descent of the Iron Curtain in the aftermath of the Second World War, Eastern Europe was part of the Soviet Bloc. Socialist governments had taken power in Russia and parts of Ukraine and Belarus after the First World War. However, other parts of Eastern Europe became socialist only after Stalin's Red Army defeated Nazi Germany and conquered all of Eastern Europe. What followed was almost half a century of socialist rule, and with it the collectivization of agricultural fields, conversion of forests into public ownership and construction of industrial centers (Lerman et al. 2004).

Economies throughout Eastern Europe were centrally planned in each nation's capital, and communist parties ruled without much open opposition. Few goods were traded in global markets, but the Soviet Bloc created its own internal market, and various subsidies kept inefficient enterprises afloat. For land use, this meant that farming was largely organized in collectivized farms, which were modeled after the

V.C. Radeloff (✉)
SILVIS Lab, Department of Forest and Wildlife Ecology, University of Wisconsin-Madison, 1630 Linden Drive, Madison, WI 53706, USA
e-mail: radeloff@wisc.edu

G. Gutman
NASA Land-Cover/Land-Use Change Program, 300 E Street SW, Washington DC 20546, USA

G. Gutman and V. Radeloff (eds.), *Land-Cover and Land-Use Changes in Eastern Europe after the Collapse of the Soviet Union in 1991*,
DOI 10.1007/978-3-319-42638-9_1

Russian 'kolchozes' with large field sizes, heavy reliance on fertilizers and pesticides, and the draining of wetlands (Lerman et al. 2004).

The collapse of socialism started in August 1989, when the very first opening of the Iron Curtain occurred near the small town of Sopron, Hungary. What was intended to be a 3-hour long opening of the border to allow Austrian and Hungarian locals to cross and meet turned into an opening of floodgates that Eastern Germans vacationing nearby used to escape. At this point, reform was in the air in many Eastern European countries, starting with the 'perestroika' (restructuring) of the Russian government and the 'glasnost' (openness) policy reforms that Mikhail Gorbachev, the last leader of the Soviet Union, initiated in the late 1980s.

Once the Iron Curtain started to rip, changes came swiftly, and it was a heady time. In November 1989, the dreaded wall between East and West Germany fell. Pictures of people in Berlin chiseling holes into the concrete made global news. Remarkably, no shots were fired in Berlin and most other capitals as the protests multiplied. The same socialist governments that had ruled with iron hands for so many decades collapsed largely without bloodshed. The notable exception was Romania, where Nicolae Ceaucescu ordered that protesters be gunned down in the city of Timisoara on December 17th, only to be tried and executed eight days later, on Christmas Day 1989. Exactly two years later on December 25th 1991, Mikhail Gorbachev resigned, and the mighty Soviet Union, a global power throughout the cold war, formally disbanded. The remaining fifteen socialist republics emerged as post-Soviet countries.

In the stale language of science, this collapse was the treatment in an epic natural experiment. The collapse triggered massive changes in the way economies throughout Eastern Europe operated and how land was managed. Many of these changes followed the guise of 'shock therapy', advocated by J. Sachs and other economists from Harvard University. Their basic premise was that a liberal market economy would be the best cure for the ailing socialist state economies. Their advice was to privatize what was owned by the state, remove subsidies and open Eastern European markets to global trade. What followed was certainly a shock. Russia's GDP, for example, was cut in half in a single year (http://data.worldbank.org). The human toll associated with these changes was enormous, as Eastern Europe experienced a phase of chaos in the early 1990s. Suicide rates spiked and birth rates dropped as people lost faith in the future. Additionally, health care systems collapsed and pensioners, who had trusted that the state would provide for them in old age, suddenly had no money left to live on. Simultaneously, there were those who seized the moment and became enormously rich by accruing far more than their fair share of previously state-owned wealth. What occurred after the collapse of the Soviet Union was not a controlled experiment but a rather messy one. Differences in the cultures and histories of many Eastern Europe countries dictated how each would transition from a state-controlled economy to a market economy.

Indeed, there were important differences in the ways that socialism had been implemented in each country. For example, most countries did not allow private land ownership other than small garden plots; however, much of Poland's agricultural land remained in small, private holdings, which affected agricultural changes after the collapse (Lerman et al. 2004). Another important difference was based on the times that different countries became socialistic. In those countries where socialism was established after the Second World War, land was often restituted after the collapse to the heirs of prior owners. However, this was not the case in Ukraine, Belarus and Russia, where socialism took hold earlier. Agricultural policies also differed. For example, Belarus retained many of its agricultural subsidies after the collapse, while most other countries did not (Prishchepov et al. 2012, 2017). Differences among countries is the second reason why Eastern Europe in the 1990s is such a great natural experiment. The fact that so many countries were involved and transitioned differently provides opportunities to compare countries and identify how specific policies and histories affect land-use change.

Finally, the dramatic changes in Eastern Europe occurred during a time span for which frequent and consistent satellite images are available. Therefore, land-use changes that occurred during that span can be accurately mapped over large areas at high spatial resolution. Landsat data with 30-m resolution became available when Landsat 4 was launched in July 1982. Landsat 5 has an incredible record of continuous image capture from March 1984 to January of 2013 making Landsat data the most important data source for land-use change monitoring (Kuemmerle et al. 2008; Knorn et al. 2009; Prishchepov et al. 2012; Potapov et al. 2015). The fact that consistent satellite data are available for the last phase of socialism, and ever since, and that all of this data is now readily and freely available thanks to USGS's data policies (Wulder et al. 2012) means that the natural experiment that Eastern Europe provides was very well recorded. In addition to Landsat, MODIS sensors on both Aqua and Terra satellites have been a key data source for land monitoring in Eastern Europe after 2001, sometimes in combination with AVHRR data to capture long time periods (de Beurs and Henebry 2004; de Beurs et al. 2017).

In summary, why does this book focus on land-use change in Eastern Europe after the collapse of socialism? It is because of the unique natural experiment that this collapse represents, the magnitude of the shock to Eastern European societies and the differences among countries that allow the different effects of policy changes to be studied. Why compile a book on land-use change in Eastern Europe now? A quarter century has passed since the collapse of socialism, which means that enough time has passed to identify the changes that were temporary and those that are more permanent.

2 Major Land-Use Trends After the Collapse

Each chapter of this book provides a detailed look at the land-use changes that occurred after the collapse of socialism, highlighting the richness of the patterns and their underlying processes. This introduction only provides a general overview of the major land-use change trends in agricultural areas, forests and urban areas, providing a framework within which the specific changes that occurred can be interpreted. Further details are described in the subsequent chapters.

2.1 Agricultural Changes

Agricultural land use was strongly affected by the collapse of socialism. One of most notable results of the collapse was the widespread abandonment of agricultural lands in most Eastern European countries (Alcantara et al. 2013; Estel et al. 2015; de Beurs et al. 2017; Feranec et al. 2017; Lyalko et al. 2017) (Fig. 1). The reasons for this abandonment were manifold. One important factor throughout the region was the abolishment of agricultural subsidies that farms had relied on previously (Prishchepov et al. 2013, 2017). Lacking subsidies made it exceptionally difficult to transition to a market economy. The lack of subsidies was compounded by a lack of access to credit, which limited the ability to invest in new technologies and machines.

Market changes and trade partnerships also fostered agricultural abandonment. During socialist times, markets for agricultural goods were guaranteed and sheltered from competition from the West; however, that changed quickly after the collapse. At the same time, it became more costly to obtain necessary inputs, such as fertilizers and pesticides, making it difficult to continue to farm.

A third major reason for the abandonment of agricultural lands was uncertainty in land ownership (Lerman et al. 2004; Baumann et al. 2011). In some countries, such as Slovakia, land was restituted to pre-socialist owners, but the parcel sizes

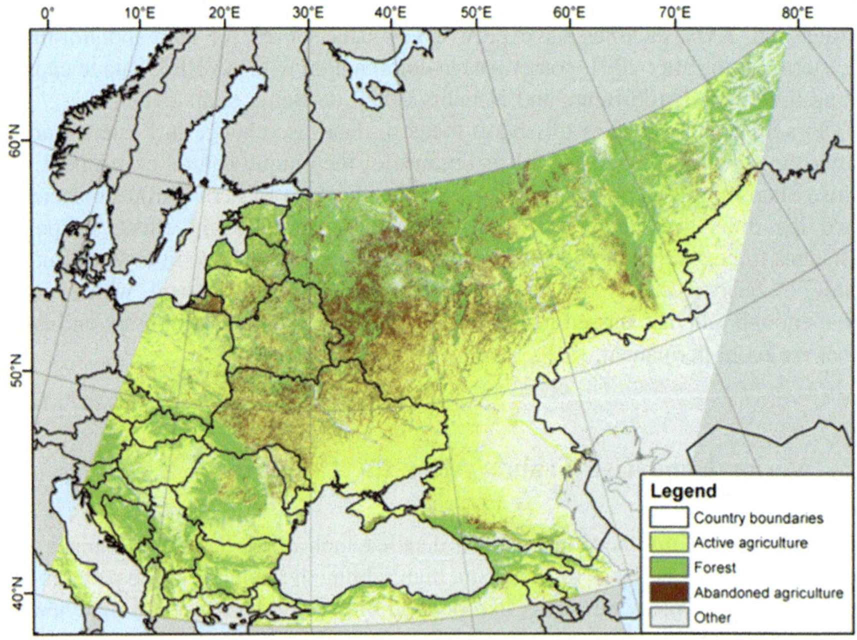

Fig. 1 Abandoned agricultural areas in Eastern Europe in 2005 mapped using MODIS satellite imagery (Alcantara et al. 2013)

were too small to make modern farming viable. In addition, many of the heirs of prior owners lived in cities and had little interest in farming (Kuemmerle et al. 2006). In other countries, such as Russia, ownership shares in communal farms were distributed to their former workers, making it difficult for those who wished to continue farming to obtain the rights to do so. This created uncertainty as to whether or not those attempting to farm would be able to reap the benefits of any investments that they were making.

However, agricultural abandonment exhibited strong spatial patterns. The rates of abandonment differed greatly throughout Eastern Europe. Temperate European Russia is one region where abandonment rates were high, abandonment was permanent and fields were converted to forests (de Beurs et al. 2017; Prishchepov et al. 2017). These changes were due to the area's marginal farming conditions, harsh climate, poor soils and limited market access. Another major reason for abandonment in northern Russia was the depopulation of many rural areas following the collapse, which reduced the local labor force (Ioffe and Nefedova 1997, 2000). In other words, abandonment was the response to new economic conditions that made farming no longer profitable.

In the central part of Eastern Europe, a somewhat different picture emerged, where abandonment was common during the 1990s, but re-cultivation has occurred on large areas since (Estel et al. 2015; de Beurs et al. 2017). The reasons why abandonment was less permanent here included better environmental conditions and access to internal markets. In addition, countries west of Belarus and Ukraine joined the EU in 2003, providing access to subsidies and easier access to West-European markets.

In southern European Russia, where soils are very fertile, abandonment was generally less common, and yet a different trend emerged (de Beurs et al. 2017). Here, industrial agriculture has taken hold, and yields are much higher now than they were during socialist times. Yield increases in this black-earth region are the reason why Russia is now a wheat-exporting country in most years. This situation is vastly different from times when the Soviet Union had to purchase wheat from the United States to meet its demand. However, the drought risk is high in parts of southern Russia (Loboda et al. 2017). Thus, fields may not provide harvests each year (de Beurs et al. 2017), and agricultural burning is widespread (McCarty et al. 2017).

In addition to these general patterns, it is important to note that agricultural abandonment can take many different forms (Estel et al. 2015). In the most extreme case, fields that were formerly ploughed and used for row crops have been completely abandoned. Forests have regrown in these areas, which are typically dominated by early-successional species such as birch and pine. However, various forms of partial abandonment and reduced agricultural land use-intensity also occurred. For example, areas that were formerly used for row crops have been converted to pastures or are mowed for hay. In addition, some fields are now only plowed every second or third year. Fallow years allow for the regeneration of soils, and ploughing prevents forest growth. Similarly, hay fields may be cut less frequently within a given year or are no longer fertilized. Finally, in countries that are part of the European Union, grasslands may only be cut to become eligible for subsidy payments, without actually utilizing the hay. These forms of land-use changes can represent the initial steps toward total abandonment, but they can also represent a temporary measure until economic conditions become suitable for more intensive farming.

2.2 *Forest Changes*

When socialism collapsed, so did timber markets in Eastern Europe. Logging rates generally declined throughout the region in the 1990s and have only partially recovered thereafter (Potapov et al. 2015). However, there have been many important deviations from this general trend due to environmental and policy differences (Fig. 2).

The reasons that caused logging rates to decline immediately after the collapse were similar to those that caused agricultural abandonment. Guaranteed markets for timber vanished. The companies that formerly processed wood largely ceased to exist. In addition, customers were less likely to purchase furniture or build houses in the early 1990s. However, the collapse of socialism also diminished the level of

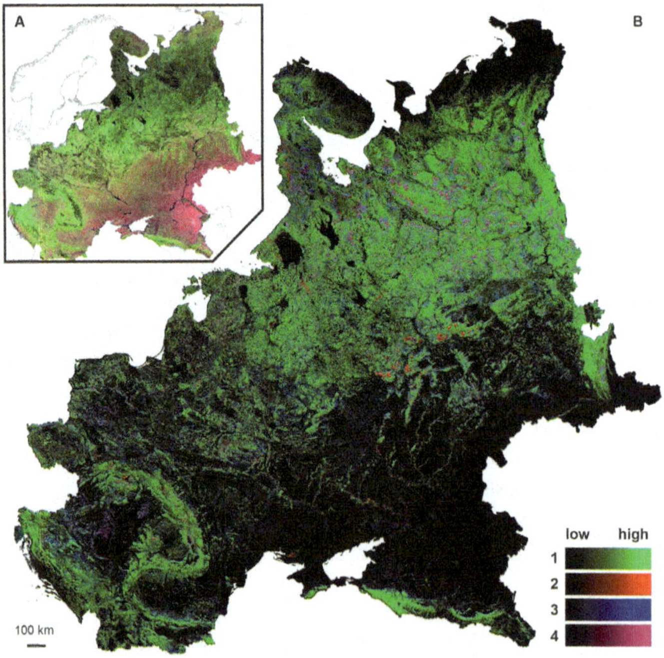

Fig. 2 Forest losses and gains in Eastern Europe from 1985 to 2012 (main map) and a Landsat satellite composite (*inset* map) (Potapov et al. 2015)

enforcement of the rules and regulations governing logging, causing a spike in logging in some areas at the time of the collapse (Baumann et al. 2011). Many of these logging activities were illegal or semi-legal, i.e., logging that exploited loopholes and stretched the definition of what was allowable. For example, one important loophole was associated with 'sanitary cuts', i.e., harvests after insect defoliation, which became widespread in parts of the Carpathians (Kuemmerle et al. 2011).

Another important factor that caused a rapid logging increase in some countries was the restitution of forest lands to former owners. In the Baltic countries, this occurred relatively early. Harvesting rates have been exceptionally high there and are likely not sustainable, i.e., harvesting levels are exceeding timber growth rate (Potapov et al. 2015). Logging spiked in Baltic countries due to the nearby, well-developed Scandinavian forest industry. The restitution of forest lands was not always immediate though. For example, restitution did not take place in Romania until the 2000s. However, Romanian forests that have been restituted have also experienced high logging rates associated with high demand, as international companies have built large sawmills in the country (Knorn et al. 2013; Munteanu et al. 2015, 2017).

The logging examples in the Baltic countries and Romania highlight that access to global markets has largely caused logging rates to increase in various countries, particularly after 2000. Global demand for timber was high until the economic crisis of 2008. Eastern Europe's vast forests, many of which are relatively old, constitute a valuable resource. Notably, international demand is for not only roundwood and pulp but also firewood. Firewood production is typically the result of local demand in various countries, such as Georgia (Ozdogan et al. 2017). However, German firewood demand is a major reason why beech forests are being harvested in Bulgaria and Romania. These practices threaten forests with high conservation values to support what is considered a more 'green' method of heating homes in the West.

2.3 Urbanization

The third major land-use trend is the growth of urban areas throughout Eastern Europe (Elvidge et al. 2017; Feranec et al. 2017). Major cities have experienced large population increases, including in countries where overall population numbers have declined. These trends are due to a combination of market forces and policies. Concomitant with the shift of Eastern European economies from state-controlled to open markets, a shift also occurred among sectors, as both the agricultural and industrial sectors declined and the service sector grew. However, many of these new service jobs were concentrated in cities to which young people moved.

In terms of policies, the growth of cities was possible because land-use regulations were relatively weak, or weakly enforced, putting few obstacles in the way of new housing development. Another important policy change was the abolishment of restrictions associated with moving from the countryside to cities, which limited internal migration in the former Soviet Union. The result was the collapse of many rural areas and the abandonment of villages due to the rapid growth in Moscow (Fig. 3), St. Petersburg and many other metropolitan areas throughout Eastern Europe.

However, some rural areas have witnessed considerable housing growth, and those are areas with high natural amenities, such as mountains, or that are relatively

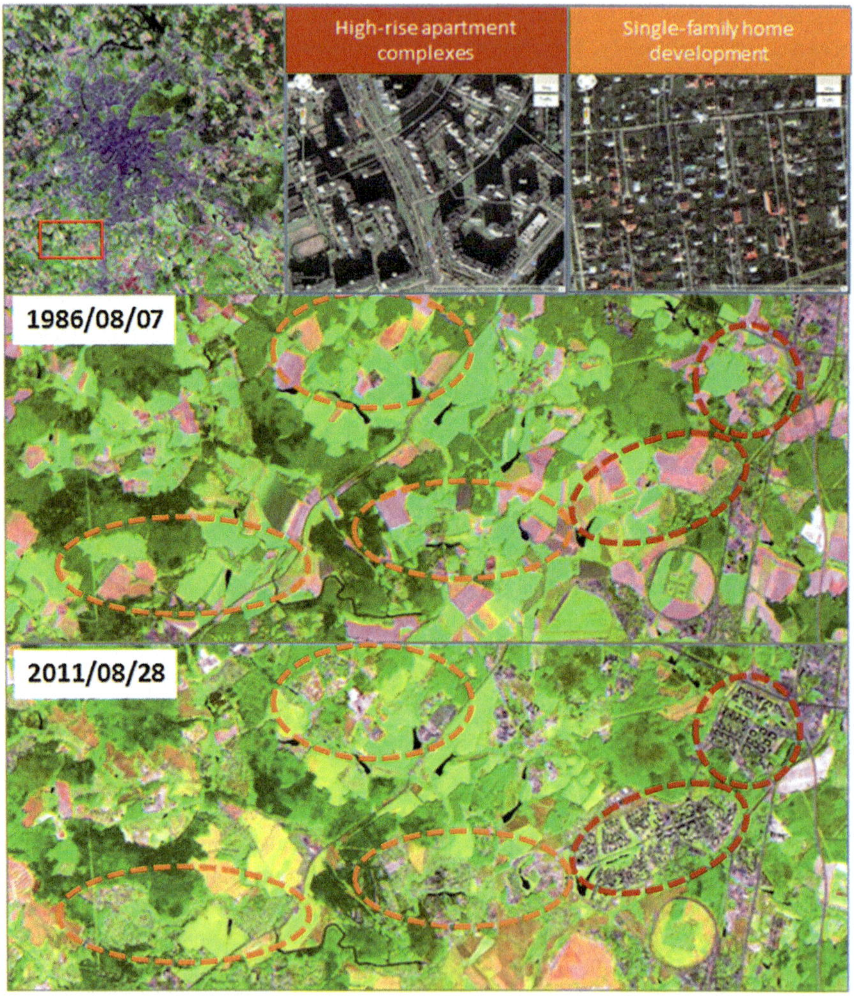

Fig. 3 Rapid urban growth in the areas around Moscow. The city of Moscow is shown in the *upper left* image. The *red box* highlights the areas that are depicted in detail below. The *center image* in the first row shows typical high-rise apartment buildings, which are highlighted by *red ellipses* in the images below. The *right* image in the first row shows a typical single-family home development, as highlighted by the *orange* ellipses in the images below. The two *bottom* Landsat images from August 1986 and August 2011 illustrate areas where extensive urban growth is clearly visible. All images are courtesy of C. Huang, University of Maryland College Park

close to urban centers. In many of these areas, seasonal homes that are primarily used for recreation have been built at rapid rates.

3 Summary

A quarter century has passed since socialism collapsed in Eastern Europe. The natural experiment that started in the late 1980s and early 1990s is not over, but important phases have concluded, and major trends in land use are clear at this point. The following chapters examine those major trends in more detail and provide a rich picture of the changes that occurred, and their underlying drivers. Satellite image archives that captured the entire period from socialism to today, especially the images recorded by the series of Landsat satellites provide the key data source for most of the research findings reported here. Collaborations between scientists from the region, and from Western countries have been key to monitor the patterns of change, and identify the processes causing these changes. We, the editors of this book, are grateful to the authors of the chapters, and all those who contributed to research described therein. It is thanks to their efforts that so much can be learned from what happened in Eastern Europe.

References

Alcantara C, Kuemmerle T, Baumann M, Bragina EV, Griffiths P, Hostert P, Knorn J, Mueller D, Prishchepov AV, Schierhorn F, Sieber A, Radeloff VC (2013) Mapping the extent of abandoned farmland in Central and Eastern Europe using MODIS time series satellite data. Environ Res Letters 8(3):035035

Baumann M, Kuemmerle T, Elbakidze M, Ozdogan M, Radeloff VC, Keuler NS, Prishchepov AV, Kruhlov I, Hostert P (2011) Patterns and drivers of post-socialist farmland abandonment in Western Ukraine. Land Use Policy 28:552–562

de Beurs KM, Henebry GM (2004) Land surface phenology, climatic variation, and institutional change: analyzing agricultural land cover change in Kazakhstan. Remote Sens Environ 89:497–509

de Beurs K, Ioffe G, Nefedova T, Henebry G (2017) Land change in European Russia: 1982–2011. In: Gutman G, Radeloff VC (eds) Land-cover and land-use changes in Eastern Europe after the collapse of the Soviet Union in 1991. Springer (this volume)

Elvidge C, Hsu F-C, Baugh KE, Ghosh T (2017) Lighting tracks transition in Eastern Europe. In: Gutman G, Radeloff VC (eds). Land-cover and land-use changes in Eastern Europe after the collapse of the Soviet Union in 1991. Springer (this volume)

Estel S, Kuemmerle T, Alcantara C, Levers C, Prishchepov A, Hostert P (2015) Mapping farmland abandonment and recultivation across Europe using MODIS NDVI time series. Remote Sens Environ 163:312–325

Feranec J, Soukup T, Taff GN, Stych P, Bicik I (2017) Overview of changes in land cover and land use in Eastern Europe. In: Gutman G, Radeloff VC (eds) Land-cover and land-use changes in Eastern Europe after the collapse of the Soviet Union in 1991. Springer (this volume)

Ioffe GV, Nefedova T (1997) Continuity and change in rural Russia: a geographical perspective. Westview Press, Boulder, Colorado

Ioffe G, Nefedova T (2000) Areas of crisis in Russian agriculture: a geographic perspective. Post-Sov Geogr Econ 41:288–305
Knorn J, Rabe A, Radeloff VC, Kuemmerle T, Kozak J, Hostert P (2009) Land cover mapping of large areas using chain classification of neighboring Landsat satellite images. Remote Sens Environ 113:957–964
Knorn J, Kuemmerle T, Radeloff VC, Keeton WS, Gancz V, Biris IA, Svoboda M, Griffiths P, Hagatis A, Hostert P (2013) Continued loss of temperate old-growth forests in the Romanian Carpathians despite an increasing protected area network. Environ Conserv 40:182–193
Kuemmerle T, Radeloff VC, Perzanowski K, Hostert P (2006) Cross-border comparison of land cover and landscape pattern in Eastern Europe using a hybrid classification technique. Remote Sens Environ 103:449–464
Kuemmerle T, Hostert P, Radeloff VC, van der Linden S, Perzanowski K, Kruhlov I (2008) Cross-border comparison of post-socialist farmland abandonment in the Carpathians. Ecosystems 11:614–628
Kuemmerle T, Olofsson P, Chaskovskyy O, Baumann M, Ostapowicz K, Woodcock CE, Houghton RA, Hostert P, Keeton WS, Radeloff VC (2011) Post-Soviet farmland abandonment, forest recovery, and carbon sequestration in western Ukraine. Glob Change Biol 17:1335–1349
Lerman Z, Csáki C, Feder G (2004) Agriculture in transition: land policies and evolving farm structures in post-Soviet countries. Lexington Books, Lanham, MD
Loboda T, Krankina O, Savin I. Kurbanov E, Hall J (2017) Land management and impact of the 2010 extreme drought event on agricultural and ecological systems of European Russia.In: Gutman G, Radeloff VC, (eds) Land-cover and land-use changes in Eastern Europe after the collapse of the Soviet Union in 1991. Springer (this volume)
Lyalko V, Ivanov S, Starodubtsev V, Palamarchuk J (2017) The effects of institutional changes on landscapes in Ukraine. In: Gutman G, Radeloff VC (eds) Land-cover and land-use changes in Eastern Europe after the collapse of the Soviet Union in 1991. Springer (this volume)
McCarty J, Krylov LA, Prishchepov AV, Banach DM., Tyukavina A, Potapov P, Turubanova S (2017) Agricultural fires in European Russia, Belarus, and Lithuania, and their impact on air quality, 2002–2012. In: Gutman G, Radeloff VC., (eds) Land-cover and land-use changes in Eastern Europe after the collapse of the Soviet Union in 1991. Springer (this volume)
Munteanu C, Kuemmerle T, Keuler NS, Mueller D, Balazs P, Dobosz M, Griffiths P, Halada L, Kaim D, Kiraly G, Konkoly-Gyuro E, Kozak J, Lieskovsky J, Ostafin K, Ostapowicz K, Shandra O, Radeloff VC (2015) Legacies of 19th century land use shape contemporary forest cover. Global Environ Change-Hum and Policy Dimensions 34:83–94
Munteanu C, Radeloff VC, Griffiths P. Halada L, KaimD, Knorn J, Kozak J, Kuemmerle T, Lieskovsky J, Mueller D, Ostapowicz K, Shandra O (2017) Land change in the Carpathian Region before and after major institutional changes. In: Gutman G, Radeloff VC (eds) Land-cover and land-use changes in Eastern Europe after the collapse of the Soviet Union in 1991. Springer (this volume)
Ozdogan M, Olofsson P, Woodcock CE, Baccini A (2017) Forest change and carbon budgets in the Black Sea region. In: Gutman G, Radeloff VC (eds) Land-cover and land-use change in Eastern Europe after the collapse of the Soviet Union in 1991. Springer (this volume)
Potapov PV, Turubanova SA, Tyukavina A, Krylov AM, McCarty JL, Radeloff VC, Hansen MC (2015) Eastern Europe's forest cover dynamics from 1985 to 2012 quantified from the full Landsat archive. Remote Sens Environ 159:28–43
Prishchepov AV, Radeloff VC, Baumann M, Kuemmerle T, Mueller D (2012) Effects of institutional changes on land use: agricultural land abandonment during the transition from state-command to market-driven economies in post-Soviet Eastern Europe. Environ Res Letters 7:024021–024022
Prishchepov AV, Mueller D, Dubinin M, Baumann M, Radeloff VC (2013) Determinants of agricultural land abandonment in post-Soviet European Russia. Land Use Policy 30:873–884
Prishchepov A, Mueller VD, Baumann M, Kuemmerle T, Alcantara C, Radeloff VC (2017) Underlying drivers and spatial determinants of post-Soviet agricultural land abandonment in

temperate Eastern Europe. In: Gutman G, Radeloff VC (eds) Land-cover and land-use changes in Eastern Europe after the collapse of the Soviet Union in 1991. Springer (this volume)
Wulder MA, Masek JG, Cohen WB, Loveland TR, Woodcock CE (2012) Opening the archive: How free data has enabled the science and monitoring promise of Landsat. Remote Sens Environ 122:2–10

Overview of Changes in Land Use and Land Cover in Eastern Europe

Jan Feranec, Tomas Soukup, Gregory N. Taff, Premysl Stych and Ivan Bicik

Abstract This chapter presents an analysis of land cover changes in Eastern Europe between 1990 and 2006, assessed using CORINE (Co-ORdination of INformation on the Environment) Land Cover (CLC) datasets. The plethora of potential land cover change categories were condensed into seven categories of major land use change processes: urbanization, agricultural intensification, agricultural extensification, afforestation, deforestation, construction and management of water bodies, and other changes. The amounts of each change category and their spatial distributions are summarized, and the change categories were also mapped to show the relative amounts of change (per 3 × 3 km^2) between 1990 and 2000 and between 2000 and 2006. The results showed that while more afforestation than deforestation was observed in the first period, the reverse was true in the second period, when deforestation outpaced afforestation. Urbanization and suburbanization were major processes in Eastern Europe, particularly around existing major cities, and the speed of this process generally increased from the first to the second period. Both the intensification and extensification of agriculture were common

J. Feranec (✉)
Slovak Academy of Sciences, Institute of Geography, Bratislava, Slovakia
e-mail: feranec@savba.sk

T. Soukup
GISAT, Prague, Czech Republic
e-mail: tomas.soukup@gisat.cz

G.N. Taff
Norwegian Institute of Bioeconomy Research, Ås, Norway
e-mail: gregory.taff@nibio.no

P. Stych
Department of Applied Geoinformatics and Cartography, Faculty of Science, Charles University in Prague, Prague, Czech Republic
e-mail: stych@natur.cuni.cz

I. Bicik
Department of Social Geography and Regional Development, Faculty of Science, Charles University in Prague, Prague, Czech Republic
e-mail: bicik@natur.cuni.cz

G. Gutman and V. Radeloff (eds.), *Land-Cover and Land-Use Changes in Eastern Europe after the Collapse of the Soviet Union in 1991*,
DOI 10.1007/978-3-319-42638-9_2

during both periods, but a larger effect was observed in the first period. Overall, land use changes were highest in central Europe and the Baltic countries and lowest in southeast Europe.

1 Introduction

The collapse of socialism in 1989 caused massive socio-economic and institutional changes (Prishchepov et al. 2013). This event has affected landscapes throughout Russia and Central and East European satellite states. This chapter addresses land cover (LC) changes in Eastern Europe between 1990 and 2006. First, a brief overview of previous studies on LC changes in Eastern Europe is provided here.

The landscape of Eastern Europe has experienced significant changes in grassland cover. Cremene et al. (2005) reported a significant recent reduction in Steppe-like grasslands in Eastern Europe. In Central Eastern Europe, including the Czech Republic, Hungary, Romania and the Slovak Republic, the main types of landscape changes were urbanization/industrialization, intensification of agriculture, extensification of agriculture, afforestation, deforestation, enlargement of areas of extraction (exhaustion) of natural resources, and other anthropogenic changes according to an analysis overlaying CORINE Land Cover (CLC) data with Landsat MSS (Multispectral Scanner) imagery corresponding to the late 1970s and visually observing the differences between CLC1990 and the 1970s imagery (Feranec et al. 2000). However, at the local scale, certain processes dominated. For example, along the Czech Republic-Austria border, afforestation prevailed from 1990 to 2000 (Kupkova et al. 2013). In the Czech Republic, the increase in forest areas is relatively permanent (an increase of 5 % during the 20th century), occurring in areas that are not as well suited for agriculture, i.e., in sub-mountain and mountain areas (Bičík et al. 2010).

While forest degradation, i.e., the partial extraction of timber and other goods from forests, was a common process in East Europe during the post-Soviet period, afforestation was generally widespread throughout the majority of the region since the fall of socialism (Taff et al. 2010). One major reason for afforestation was the abandonment of farmland, which was common throughout Central and Eastern Europe, according to an analysis of 250-m MODIS 8-day NDVI composites (Alcantara et al. 2013). Similarly, high rates of land abandonment were observed in northeastern Europe between 1990 and 2000, according to an analysis of Landsat TM data (Prishchepov et al. 2012). Baumann et al. (2012) used a combination of summer and winter Landsat images to analyze forest changes in European Russia and observed substantial regional variation, with an overall forest loss between 1990 and 1995 and an overall forest gain between 2005 and 2010. However, there were significant differences between countries in Eastern Europe in terms of their abandonment rates after the collapse of the Soviet Union (Prishchepov et al. 2012; Alcantara et al. 2013), indicating a strong influence of state-level institutional factors. Kuemmerle et al. (2006) also observed significant differences in land use (LU) changes between some countries within the Carpathian Mountains (Poland, Slovakia, and Ukraine). Kuemmerle et al. (2008) observed that drivers of LU changes also considerably varied between neighboring countries in the Carpathian region. Kozak (2003); Kozak et al. (2007a, b) investigated the main trends and drivers of LU/LC changes in the Carpathian Basin; afforestation and a decrease of arable land were the main long-term trends observed in these studies.

In addition to documenting the patterns of LU changes in general, it is particularly important to understand the drivers and consequences of recent LU and LC changes in this region resulting from the massive socio-economic perturbations that occurred after the fall of socialism (Prishchepov et al. 2010; Hostert et al. 2011). Key drivers of changes in LU patterns in former socialist countries included (1) country-specific land policies relating to the legal attitude towards private land ownership, (2) transferability of land, and (3) land allocation/redistribution strategies (Lerman et al. 2004). Macours and Swinnen (2000) investigated the changes in agriculture production and associated causes in Central and Eastern Europe during the transition period and observed that the primary involves deterioration in the terms of agriculture trade, transition uncertainties, and extreme weather events. During the period from 1990 to 2004, the Czech Republic experienced a significant increase in grassland areas, a process deeply influenced by the termination of significant subsidies given by the socialist state, which ended at the beginning of the 1990s (Bičík et al. 2010).

This chapter presents an overview of landscape changes (spatial distribution and intensity) in Eastern European countries based on CLC during the periods from 1990–2000 to 2000–2006 and describes a useful a mapping methodology for presenting landscape changes on a macro-scale.

2 Methodology

We quantified basic input information concerning recent LC changes in Central and Eastern Europe for 17 countries: Albania (AL), Bosnia/Herzegovina (BA), Bulgaria (BG), Croatia (HR), Czech Republic (CZ), Estonia (EE), Hungary (HU), Kosovo (KV), Latvia (LV), Lithuania (LT), Macedonia FYR (MK), Monte Negro (ME), Poland (PL), Romania (RO), Serbia (RS), Slovakia (SK) and Slovenia (SI). We considered two time periods, 1990–2000 and 2000–2006, and analyzed the changes based on the CLC database. The change data layers CLC1990–2000 and 2000–2006 are available at http://www.eionet.europa.eu/gis/, and details concerning these data are provided in Nunes de Lima (2005) and Feranec et al. (2007). A summary of the LC classes included in the change databases CLC1990/2000 and 2000/2006 is shown in Table 1.

We derived the main landscape changes for the second level of CLC classes after applying the conversion table (Table 2). This conversion table, i.e., the "matrix of changes", groups LC changes of the same type. There are $15 \times 14 = 210$ possible combinations of one-to-one changes between the 15 CLC classes at the second level (Feranec et al. 2010).

Table 1 CLC nomenclature (Heymann et al. 1994; Bossard et al. 2000)

1 Artificial surfaces

11 Urban fabric
- 111 Continuous urban fabric
- 112 Discontinuous urban fabric

12 Industrial, commercial and transport units
- 121 Industrial or commercial units
- 122 Road and rail networks and associated land
- 123 Port areas
- 124 Airports

13 Mine, dump and constructions sites
- 131 Mineral extraction sites
- 132 Dump sites
- 133 Construction sites

14 Artificial, non-agricultural vegetated areas
- 141 Green urban areas
- 142 Sport and leisure facilities

2 Agricultural areas

21 Arable land
- 211 Non-irrigated arable land
- 212 Permanently irrigated land
- 213 Rice fields

22 Permanent crops
- 221 Vineyards
- 222 Fruit trees and berry plantations
- 223 Olive groves

23 Pastures
- 231 Pastures

24 Heterogeneous agricultural areas
- 241 Annual crops associated with permanent crops
- 242 Complex cultivation patterns
- 243 Land principally occupied by agriculture, with significant areas of natural vegetation
- 244 Agro-forestry areas

3 Forest and semi-natural areas

31 Forests
- 311 Broad-leaved forests
- 312 Coniferous forests
- 313 Mixed forests

32 Scrub and/or herbaceous vegetation associations
- 321 Natural grasslands
- 322 Moors and heathland
- 323 Sclerophyllous vegetation
- 324 Transitional woodland-scrub

33 Open spaces with little or no vegetation
- 331 Beaches, dunes, sands
- 332 Bare rocks
- 333 Sparsely vegetated areas
- 334 Burnt areas
- 335 Glaciers and perpetual snow

4 Wetlands

41 Inland wetlands
- 411 Inland marshes
- 412 Peat bogs

42 Maritime wetlands
- 421 Salt marshes
- 422 Salines
- 423 Intertidal flats

5 Water bodies

51 Inland waters
- 511 Water courses
- 512 Water bodies

52 Marine waters
- 521 Coastal lagoons
- 522 Estuaries
- 523 Sea and ocean

Table 2 Conversion table (Feranec et al. 2010) for all of Eastern and Central Europe

	2000 classes															
1990 classes		**11**	**12**	**13**	**14**	**21**	**22**	**23**	**24**	**31**	**32**	**33**	**41**	**42**	**51**	**52**
	11	**0**	7	7	7	7	7	7	7	7	7	7	7	7	7	7
	12	7	**0**	7	7	7	7	7	7	7	7	7	7	7	7	7
	13	7	7	**0**	7	7	7	7	7	7	7	7	7	7	6	7
	14	7	7	7	**0**	7	7	7	7	7	7	7	7	7	6	7
	21	1	1	1	1	**0**	2	3	3	4	4	7	7	7	6	7
	22	1	1	1	1	3	**0**	3	3	4	4	7	7	7	6	7
	23	1	1	1	1	2	2	**0**	2	4	4	7	7	7	6	7
	24	1	1	1	1	2	2	3	**0**	4	4	7	7	7	6	7
	31	1	1	1	1	5	5	5	5	**0**	5	5	5	7	6	7
	32	1	1	1	1	2	2	2	2	4	**0**	5	7	7	6	7
	33	1	1	1	1	2	2	2	2	4	4	**0**	7	7	6	7
	41	1	1	1	1	2	2	2	2	4	4	7	**0**	7	6	7
	42	1	1	1	1	2	2	2	2	4	4	7	7	**0**	6	7
	51	1	1	1	1	7	7	7	7	4	4	7	7	7	**0**	7
	52	1	1	1	1	7	7	7	7	4	4	7	7	7	7	**0**

1—urbanization (industrialisation), *2*—intensification of agriculture, *3*—extensification of agriculture, *4*—afforestation, *5*—deforestation, *6*—water bodies construction and management, *7*—other changes (recultivation, dump sites, unclassified changes, etc.)

We grouped these 210 potential land use/cover change (LUCC) classes into seven major LU change processes:

- Urbanization: represents changes in agriculture (CLC classes 21, 22 and 23; codes are explained in Table 1), forest lands (classes 31, 32, and 33), wetlands (classes 41 and 42) and water bodies (51 and 52) into urbanized land (the construction of buildings designed for living, education, health care, recreation and sport) and industrialized land (the construction of facilities for production, all forms of transportation and electric power generation).
- Intensification of agriculture: represents the transition of LC types associated with lower intensity use (e.g., from natural areas—classes 32, 33, except forest class 31 and wetland class 4) into higher intensity agricultural use (classes 21 and 22).
- Extensification of agriculture: represents the transition of LC types from a higher intensity agricultural use (classes 21 and 22) to a lower intensity agricultural use (classes 23 and 24).
- Afforestation: represents forest regeneration, i.e., the establishment of forests by planting and/or natural regeneration in other natural areas or agricultural lands (change of classes 21, 22, 23, 24, 33, 41, and 42 into classes 31 and 32).
- Deforestation: involving forestland (class 31) changes into another LC or damaged forest (classes 21, 22, 23, 24, 32, 33 and 41).

- Construction and management of water bodies (abbreviated: Water bodies construction): involving the change of mainly agricultural (classes 21, 22, 23 and 24) and forest land (classes 31 and 32) into water bodies and the consequences of the management of water resources and the water surface area of reservoirs (Haines-Young and Weber 2006).
- Other changes: includes changes resulting from various anthropogenic activities, such as the recultivation of former mining areas, dump sites, unclassified changes, etc. More detailed characteristics of these changes are listed in Feranec et al. (2010).

The size of the changed area is generally too small to present on a map showing all of Central Europe (e.g., the smallest identified change area in the frame of the CLC mapping is 5 ha.). A practical solution for how to "visualize" such small areas of change is the presentation of the intensity/rate using a regular grid pattern. Consistent with Feranec et al. (2010), we used a 3×3 km grid as a compromise between the actual spatial distribution of the seven above-mentioned changes and their presentations on the Central European level at a meaningful scale. To this end, we first defined the mean LUCC value for the region for each of the seven LUCC types. For each LUCC type, the *mean LUCC* value represents the mean percent area of each 3×3 km^2 square covered by that LUCC type, taken only among 3×3 km^2 squares in which the LUCC type occurred. For each LUCC type, the mean LUCC value utilized in the map represents a ratio of the area of that LUCC type (in the whole study area) to the summed area of all squares of the 3×3 km grid in which such changes occurred (i.e., the denominator is 9 km^2 $\times$ the number of squares in which that LUCC type exists). The mean value of each LUCC type in both periods is listed in Table 3.

We mapped each of the seven types of changes in each of the two time periods (between 1990–2000 and 2000–2006) to show how these changes differed. We compared the LUCC types between the two time periods for each square. We assigned each square a red color when the percentage of the changed region remained steady or increased between the two time periods or a blue color when the percentage of the changed region decreased between the two time periods (Figs. 1, 2, 3, 4, 5, 6, and 7) as follows:

Table 3 Mean values of each LUCC type in both periods

	1990–2000 (%)	2000–2006 (%)
Urbanization	1.7	2.0
Intensification	3.5	2.9
Extensification	5.0	3.7
Afforestation	1.7	2.0
Deforestation	3.5	2.5
Water bodies construction	2.2	1.6
Other changes	3.1	2.4

Note these means are defined as the mean percent area of each 3×3 km^2 square covered by that LUCC type, taken only among 3×3 km^2 squares in which that LUCC type occurred

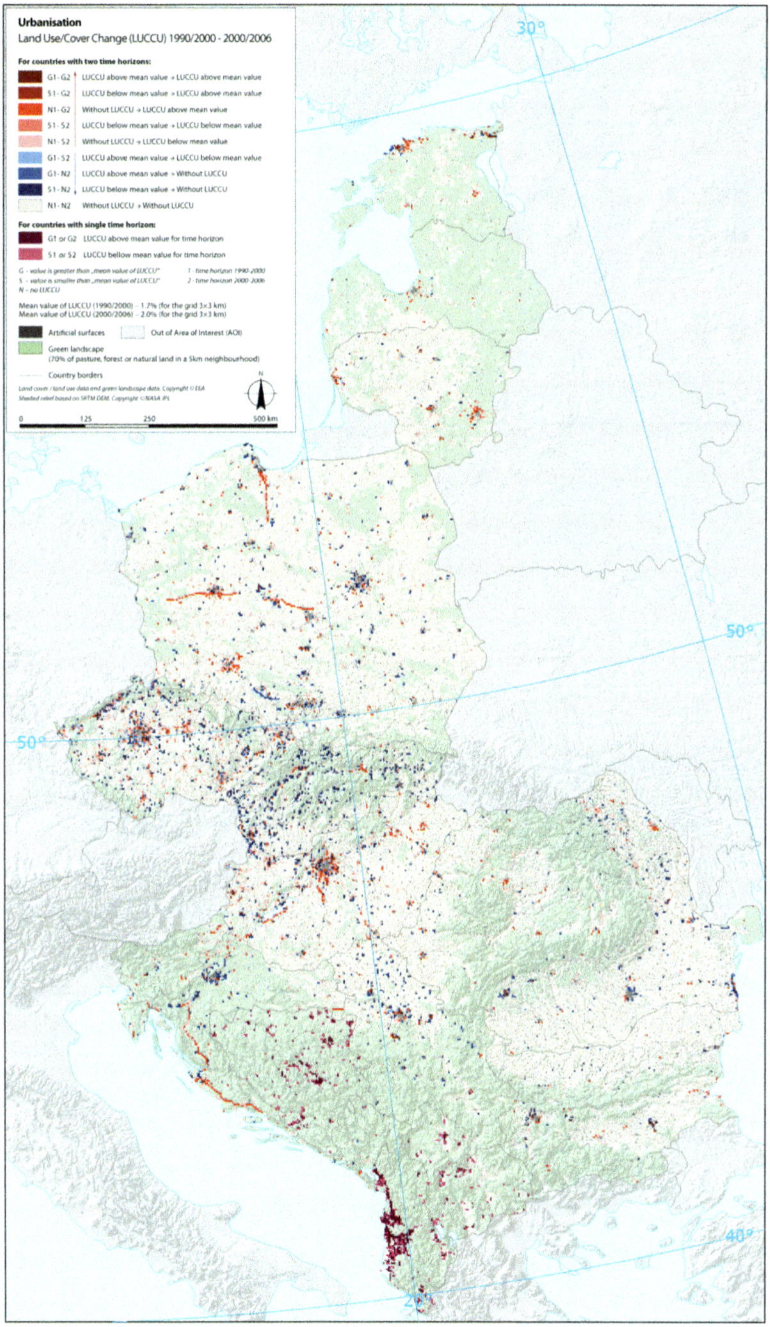

Fig. 1 Spatial distribution of urbanization in Central Europe for the periods 1990–2000 and 2000–2006

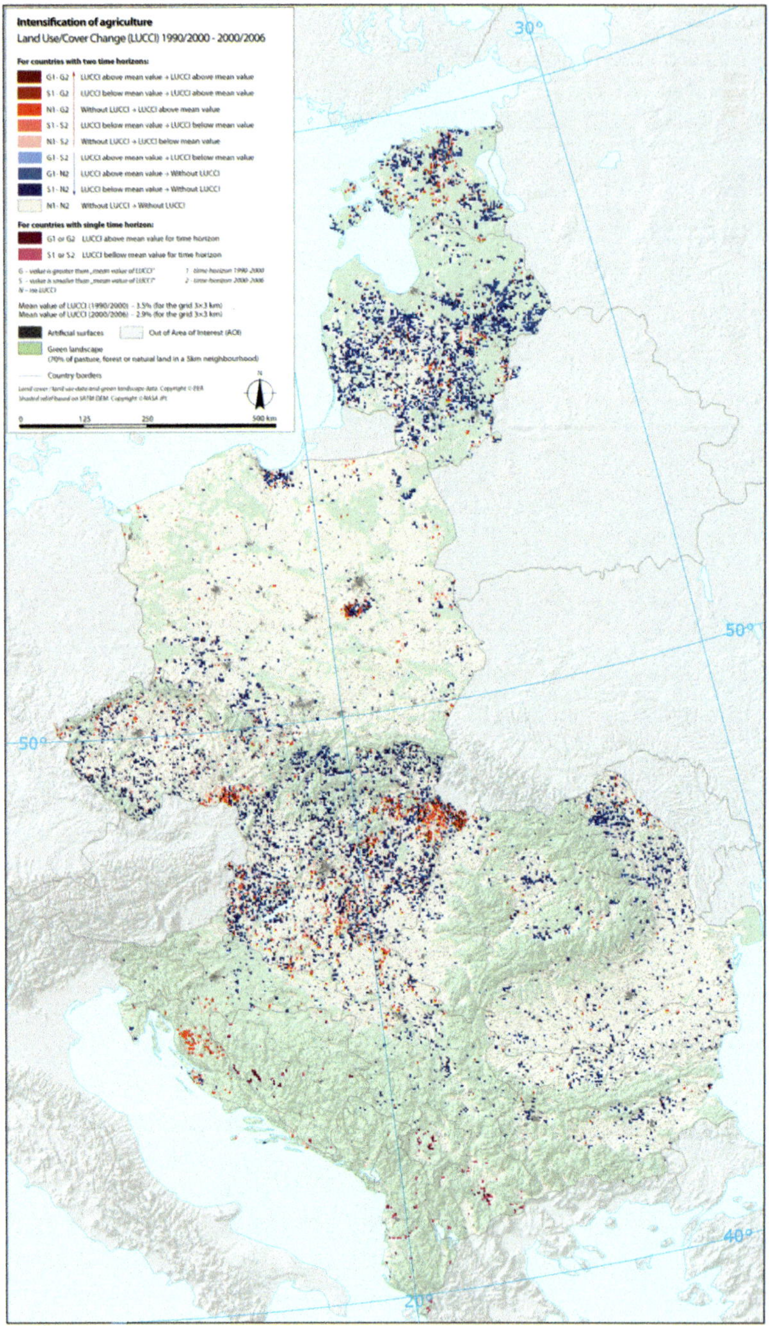

Fig. 2 Spatial distribution of intensification in Central Europe for the periods 1990–2000 and 2000–2006

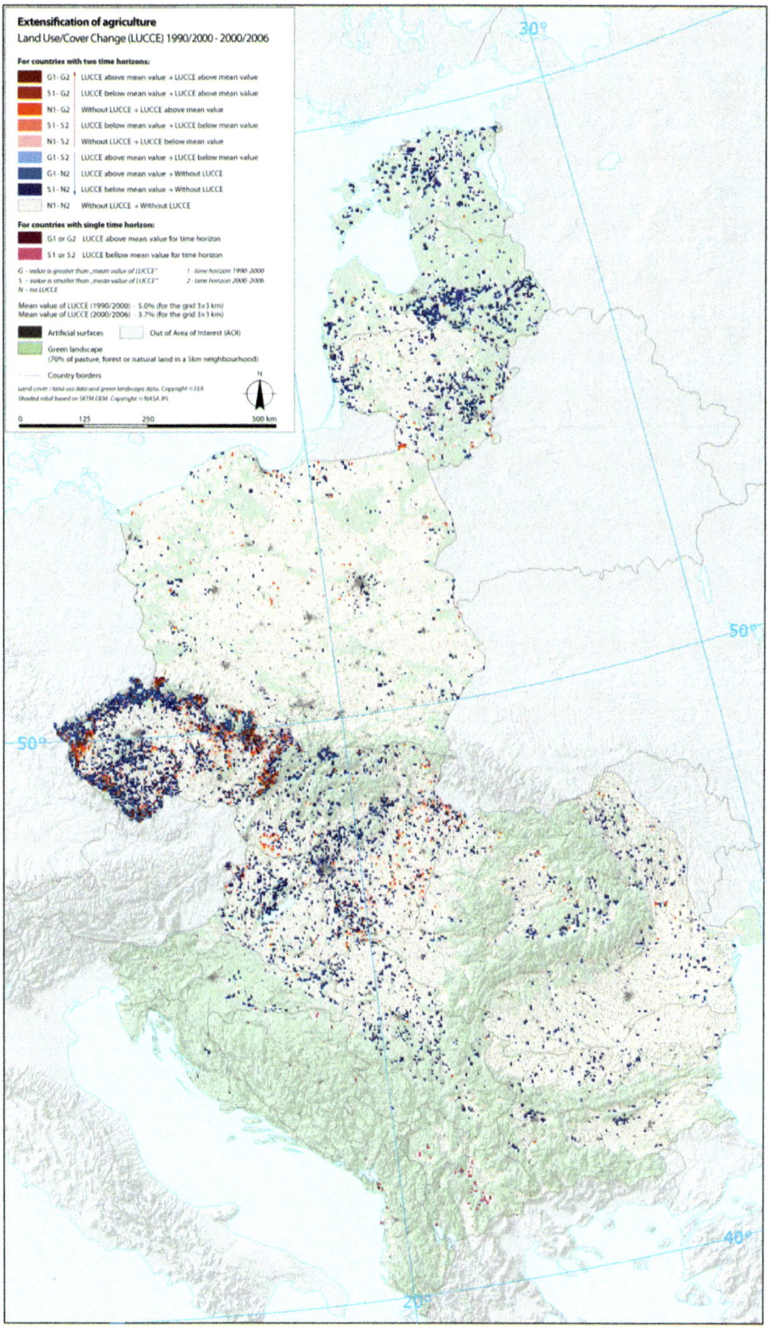

Fig. 3 Spatial distribution of extensification in Central Europe for the periods 1990–2000 and 2000–2006

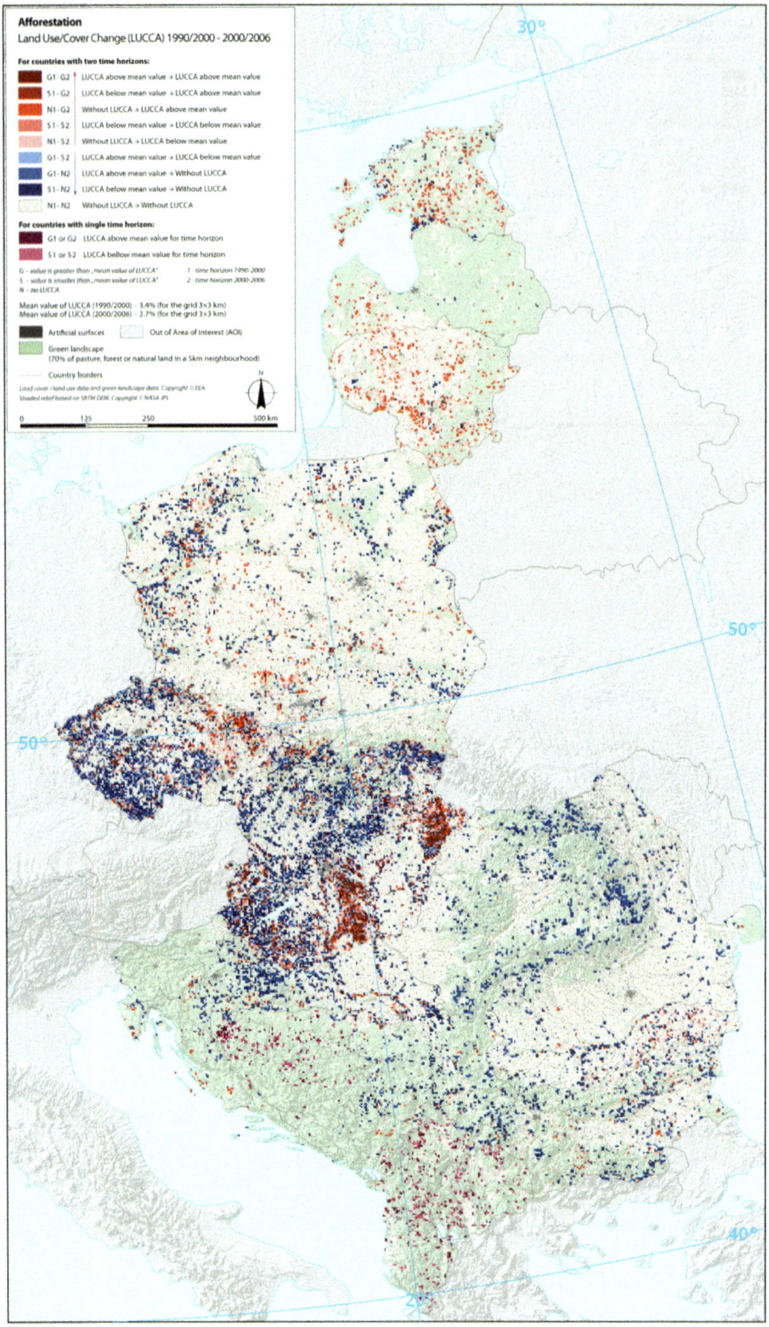

Fig. 4 Spatial distribution of afforestation in Central Europe for the periods 1990–2000 and 2000–2006

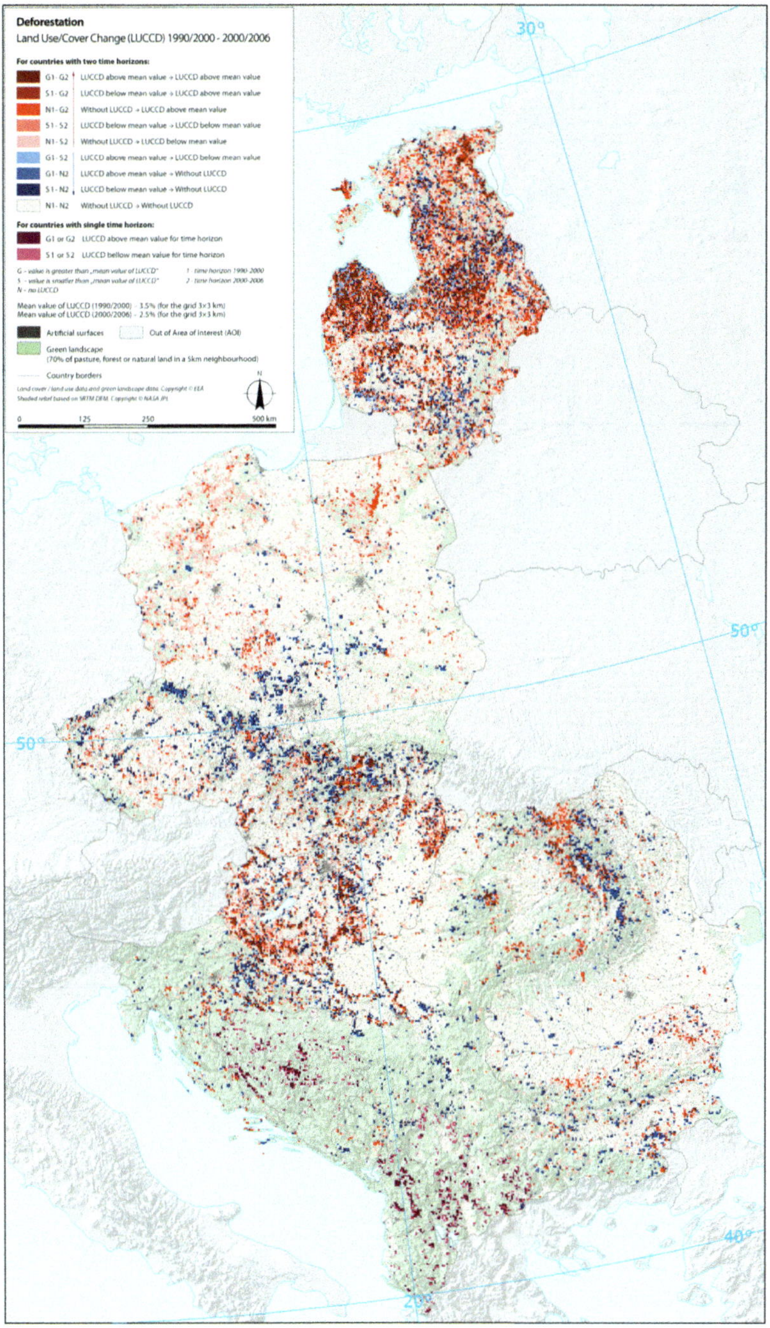

Fig. 5 Spatial distribution of deforestation in Central Europe for the periods 1990–2000 and 2000–2006

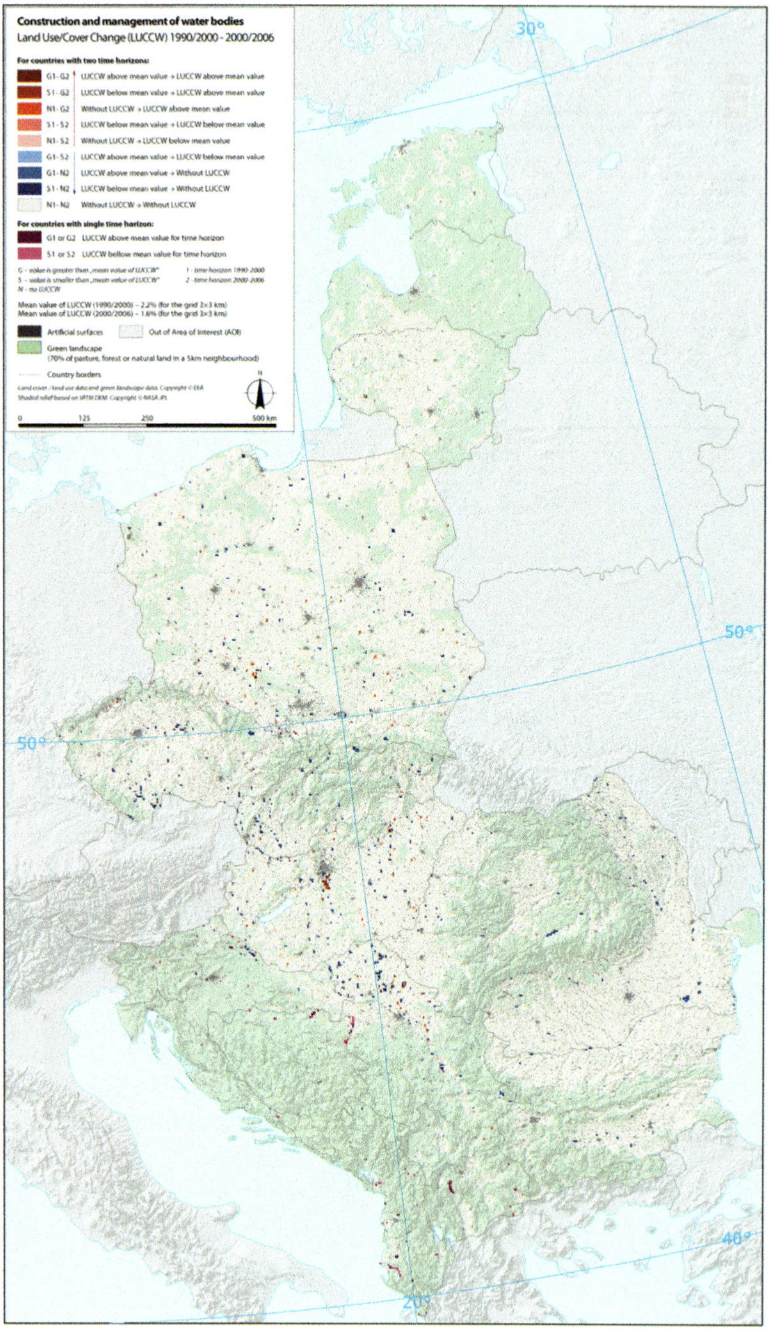

Fig. 6 Spatial distribution of the construction and management of water bodies in Central Europe for the periods 1990–2000 and 2000–2006

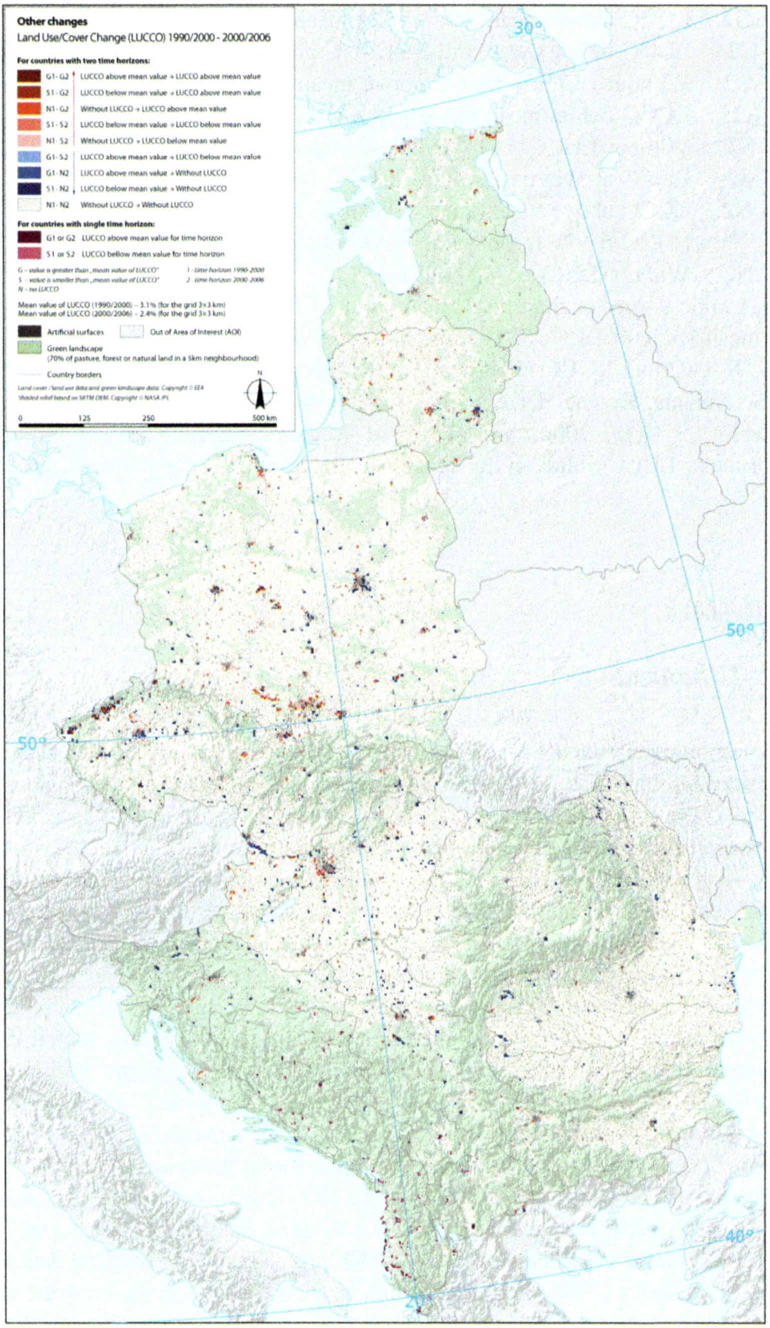

Fig. 7 Spatial distribution of other changes in Central Europe for the periods 1990–2000 and 2000–2006

G1—G2: LUCC above mean value—LUCC above mean value
S1—G2: LUCC below mean value—LUCC above mean value
N1—G2: Without LUCC—LUCC above mean value
S1—S2: LUCC below mean value—LUCC below mean value
N1—S2: Without LUCC—LUCC below mean value
G1—S2: LUCC above mean value—LUCC below mean value
G1—N2: LUCC above mean value—Without LUCC
S1—N2: LUCC below mean value—Without LUCC
N1—N2 Without LUCC—Without LUCC

G—value is greater than the "mean value of LUCC", **S**—value is smaller than the "mean value of LUCC", **1**—time horizon 1990–2000, **2**—time horizon 2000–2006, **N**—without LUCC identification (Feranec and Soukup 2013).

For Albania, Bosnia/Herzegovina, Kosovo and Macedonia, LC data are only available for 2000–2006, and we used a dark magenta color (G2) for an above-mean LUCC value and a light magenta color (S2) for below-mean LUCC value.

3 Results

3.1 Urbanization

Urbanization represents the expansion of artificial surfaces, including the construction of residential buildings, industrial areas, road and railway communications, etc. (Table 2). The results presented in Table 4 and Fig. 1 suggest increasing amounts of construction between the two time periods, particularly in the suburban areas of large cities, such as Budapest (Fig. 1), the northern and northeastern parts of Prague, northeastern Tallinn, northern and western Vilnius, western Warsaw, western Bucharest, and northeastern Bratislava. Major cities in the northern, southern and eastern parts of the study area were not as affected by the intensive urban and suburban processes compared with the changes that occurred in the cities in the central region. The construction of motorways dominated the western part of Croatia, central Poland, southwest of Hungary, and north of Slovakia. Additionally, a minor decline in the construction rate in 2000–2006 (relative to 1990–2000) occurred in several parts of Slovakia (Fig. 1), east and southeast of Warsaw, in the surrounding areas of Zagreb, western Prague (where some suburbs were created on abandoned agricultural lands, and the population increased 30–50 % during 2000–2010), west and east of Krakow, north of Serbia, and east of Bucharest (for a detailed analysis of the LU changes in Southern Romania see Kuemmerle et al. 2009b). In places where only the data for the period from 2000 to 2006 are available, urbanization occurred in the northern, central and southern regions of Bosnia/Herzegovina, Kosovo and the northern and south-western parts of Macedonia. A high rate of construction/urbanization occurred in the western part of Albania from 2000 to 2006.

Table 4 LUCC types in Central Europe for the periods 1990–2000 and 2000–2006

	1990–2000			2000–2006		
	Total area (ha)	Mean yearly increase in the period (ha)	Mean yearly change of total LUCC area (%)	Total area (ha)	Mean yearly increase in the period (ha)	Mean yearly change of total LUCC area (%)
Urbanisation	70,377	7037.7	3.2	131,143	21,857.2	9.5
Intensification	381,648	38,164.8	17.4	114,785	19,130.8	8.3
Extensification	486,275	48,627.5	22.1	93,115	15,519.2	6.7
Afforestation	619,346	61,934.6	28.1	344,569	57,428.2	24.9
Deforestation	580,318	58,031.8	26.4	652,129	108,688.2	47.1
Water bodies construction	17,204	1720.4	0.8	10,283	1713.8	0.7
Other changes	41,855	4185.5	1.9	39,715	6619.2	2.9
Total LUCC area	2,197,023	219,702.3	–	1,385,739	230,956.5	–
Total study area	122,375,321	–		134,022,612	–	

Countries where LUCC data available for 1990–2000 period: BG, CZ, EE, HR, HU, LT, LV, ME, PL, RO, RS, SI, SK
Countries where LUCC data available for 2000–2006 period: AL, BA, BG, CZ, EE, HR, HU, KV, LT, LV, ME, MK, PL, RO, RS, SI, SK

In total, from 1990 to 2000, an average of 7037.7 ha (3.2 %) of the total area experiencing LC changes (2,197,023 ha) occurred annually as urbanization (Table 4). During the six-year period between 2000 and 2006, 21,857.2 ha (9.5 %) of the total mean annual changes (230,956.5 ha) corresponded to urbanization. A comparison of the sizes of all types of changes showed that urbanization ranks fifth in the first period and third in the second period.

3.2 *Intensification of Agriculture*

The intensification of agriculture was widespread from 1990 to 2000; but from 2000 to 2006, it declined in all countries (Fig. 2). We observed a particularly strong decline in the intensification of agriculture in Lithuania, Latvia, Estonia, Slovakia, the Czech Republic and Hungary and a lesser effect in southern Poland, northeastern Romania, northern Bulgaria and central Serbia. Some degree of intensification of agriculture (changes of arable land into vineyards and orchards) was observed in northeastern and central Hungary, western Croatia and southeastern Czech Republic. A common occurrence of the transfer of agricultural lands into non-agricultural use was observed in Less Favored Areas (LFA) with poor soils and in the areas surrounding larger towns (for more details see Jelecek et al. 2012).

The share of intensification of agriculture (change of grassland into arable land and arable land into orchards and vineyards, etc.) was 38,164.8 ha (17.4 %), i.e., the fourth most extensive change (Table 4) in the first period. In the second period, the intensification of agriculture declined to 19,130.8 ha (8.3 %). The extent of these changes makes agricultural intensification the 4th most common LC change occurring in both time periods.

3.3 *Extensification of Agriculture*

The extensification of agriculture was primarily observed in areas the northeast of Croatia; the central regions of Serbia and the central part of Bulgaria; the northern, western and southern regions of the Czech Republic; areas north of Slovakia; areas north and center of Hungary; the eastern region of Lithuania; the southeastern region of Latvia; and the northern and central regions of Estonia. Muller et al. (2009) also documented an increase in agriculture land abandonment in central and northeastern regions of Romania.

The mean annual extent of the extensification of agriculture was 48,627.5 ha, (22.1 % of total changes, i.e., the third most extensive change in the first period; Table 4). This type of change decreased in the second period to 15,519.2 ha per year (6.7 %), representing the 5th most widespread change (Table 4).

3.4 *Afforestation*

In the first period, the mean annual changes totaled 61,934.6 ha (28.1 %) of afforestation, representing the most extensive change in the first period. Afforestation ranked second in the second period (Table 4), with a mean annual size of 57,428.2 ha (24.9 %; Table 4). The largest afforestation during both time periods was observed in the northeastern and central regions of Hungary, the northeastern region of the Czech Republic, and in Lithuania and Estonia. A lack of afforestation was detected in western, southern and central Czech Republic; central Slovakia; western, northwestern and northern Hungary; central and eastern Romania; central Bulgaria and southeastern Serbia. Below average values of afforestation were also evident in Bosnia/Herzegovina, Kosovo and Macedonia (Fig. 4).

3.5 *Deforestation*

The most extensive areas of deforestation occurred in Latvia, Estonia and Lithuania; in western, central and northeastern Hungary; the northeastern Romania and northern Slovakia. Less conspicuous signs of deforestation were also detected in the western, central and northern regions of Poland, northern Bulgaria and northeastern Croatia. A decline in the deforestation rate in 2000–2006 (relative to 1990–2000) was evident in the north, northeastern and eastern regions of the Czech Republic (Fig. 5). A high deforestation rate from 2000 to 2006 occurred in northeastern Albania and central Bosnia-Herzegovina.

Deforestation was the second most common LC change in the 1990s, reaching an average of 58,031.8 ha per year (26.4 %; Table 4). In the second period, deforestation was the biggest land cover change in Eastern Europe, reaching 108,688.2 ha per year (47.1 %; Table 4).

3.6 *New Construction of Water Bodies*

Changes of agricultural and forest landscapes into water bodies were sporadic, occurring in southern Poland, northeastern Hungary, and northern Serbia (Table 2; Fig. 6).

The increase in water bodies was the least widespread land use change in both time periods, comprising only 1720.4 ha (0.8 %) and 1713.8 ha (0.7 %; Table 4) in the first and second periods, respectively.

3.7 *Other Land Use/Cover Changes*

The re-cultivation of areas following the extraction of raw materials, landfills, and unclassified changes (Table 2) primarily occurred in northern Estonia, showing the widespread re-cultivation of areas where combustible shale had previously been mined, including southeastern Latvia, southern and southwestern Poland, north-eastern Czech Republic (where the former "Black Triangle" mining landscape is being re-cultivated), and in the surrounding areas of Budapest and northwestern Croatia (Fig. 7).

These "other" changes were second-to-last in terms of size. In the first and the second periods, these changes amounted annually to 4185.5 ha (1.9 %) and 6619.2 ha (2.9 %; Table 4), respectively.

4 Conclusions

The area of LUCC that we identified was approximately 21,970 km^2 in 1990–2000 and approximately 13,860 km^2 in 2000–2006 among 17 Central European countries, comprising a total area of approximately 1,340,000 km^2. The greatest changes were afforestation and deforestation, totaling 54.5 % of the total LUCC area in 1990–2000 and 72.0 % of the total LUCC area in 2000–2006 (Table 4).

This study showed that while the areas of afforestation were slightly larger than deforestation in the first period (1990–2000), deforestation far outpaced afforestation in the second period (2000–2006). This finding interestingly contrasts with Taff et al. (2010), who showed that the overall forest area increased in almost all Central and Eastern European countries between 2000 and 2005 using an analysis based on data obtained from national statistics and summarized by the UN-FAO (FAO 2006). This assessment of LU/LC changes, based on CLC data (which is based mostly upon satellite image analyses), questions the validity of national datasets on forest areas, suggesting that differing country definitions of forest land (or possibly the distinction between forest land and standing forest) might significantly affect these analyses. The causes of afforestation in the region likely reflect agriculture abandonment (Taff et al. 2010), while primary causes of deforestation are more numerous. Significant forest disturbances occurred in the Šumava Mountains in Germany and the Czech Republic, reflecting the calamitous outbreak of spruce bark beetle in the Šumava Mountains resulting from windbreak renovations of forest stands (Hais et al. 2008). Land ownership types influenced differences in forest changes in Poland after socialism (state vs. private), according to Kuemmerle et al. (2009a). Landowners occasionally practiced unsustainable clear-cutting on lands in Latvia in the early years (late 1990s and early 2000s) after post-Soviet land restitution (Taff 2005).

Significant changes occurred on the agricultural lands in the study region. Two antagonistic trends, extensification and intensification, were documented. Extensification (primarily over-grassing) was a prominent trend, particularly in Central Europe (the north, western, southwestern and northeastern region of the Czech Republic, and the northwestern region of Slovakia, Fig. 3) and in the Baltic states in the first period (1990–2000). These changes are also shown in Fig. 3. A particularly significant increase in grasslands was observed on arable lands in the Czech Republic in the period since 1990, which is the first period since the middle of the 19th century where grasslands have rapidly increased in the Czech Republic (Bičík et al. 2010). The abandonment of arable land accompanied grassing over during the transition of the agricultural sector in the Czech Republic in the early 2000s. Official government estimations reported that ca. 300 thousand hectares (10 % of the total area of arable land in the Czech Republic) were fallowed in the year 2003 (due to the Agricultural Policy Strategy for the period after accession to the EU 2004–2013). Part of this fallow arable land was re-cultivated after the accession of the Czech Republic into the EU when farmers obtained EU subsidies for agricultural production; however, a significant part of the fallow arable land has successively changed into grasslands or forests. This process particularly occurs in unfavorable areas with low-quality soils (Bičík and Jančák 2003). The intensification of agriculture was a more dominant trend on agricultural land during the period from 2000 to 2006, with a concentration in Central European areas with favorable conditions for arable land, including new vineyards and orchards. Generally both extensification and intensification decreased in area in the second period. Approximately 40 % of the total LUCC area led to agricultural land (either intensification or extensification) in 1990–2000, and 15 % of the total LUCC area led to agricultural land in 2000–2006.

The intensity of urbanization was three times higher in 2000–2006 than in the first period. Urbanization was concentrated in the largest core population areas (big cities) into which the main flows of investment were aimed. Urbanization was the third most common trend, in terms of area, in the second period with approximately 10 % of the total LUCC area. The smallest LUCC category (construction and management of water bodies) covered only 0.8 and 0.7 % of the total LUCC area in 1990–2000 and 2000–2006, respectively.

Territorial differences in LUCC trends were observed in the study area. Countries in the central part of the study area were affected by more intense changes and a wider spectrum of changes (often antagonistic: intensification and urbanization and land abandonment and afforestation). The second most intensive changes occurred in the Baltic states, particularly on agricultural and forestlands. However, the southern countries (e.g., Bulgaria, Bosnia-Herzegovina, Monte Negro), Slovenia and the central parts of Poland and Romania experienced an overall lower intensity of changes. This phenomenon is likely associated with the different conditions of these countries during the communist period. The Baltic states and former Czechoslovakia were more affected by the collectivization of agriculture, with radical changes on agricultural land. These processes were not introduced with respect to natural and market conditions. Areas with unfavorable conditions were often covered by agricultural land because the agricultural sector was highly subsidized. The Balkan countries and Poland maintained traditional agriculture, based on small private farms. After the collapse of socialism, LU in the Czech Republic, Slovakia and the Baltic states reverted back to more sustainable structures from the point of view of environmental and market factors. Transitional processes evoked intensive changes. Countries in the central region of the study area and the Baltic states joined the EU more quickly than other countries, in the first accession wave in 2004. The economy of these countries had to adopt EU markets and agricultural policies in a short period. These factors were important drivers of changes that increased the polarization between north-central countries and southern regions of the study area.

The observed trends showed a long-term tendency of transition from a local scale of societal organization into a regional, state and most recently, a global scale. Similar structures and trends of LUCC were observed in large regions with specific functions (residential, agricultural, recreation…). Thus, typological regions with specific function(s) and LUCC trends can be delimited in Central Europe.

Thus, the main purpose of the present study was to document major LUCC trends in Central Europe during the periods 1990–2000 and 2000–2006. Further analyses of territorial differentiation and the evaluation of driving forces could be developed based on the obtained results and LUCC maps. A detailed review of smaller LUCC case studies and comparisons of changes and drivers would also be useful for future research.

Acknowledgments This paper is one output of the project VEGA Grant Agency No 2/0006/13, Changes of cultural landscape: analysis of extension of urban fabric and farmland abandonment processes applying land cover databases, "pursued at the Institute of Geography of the Slovak Academy of Sciences" and "The Historical Geography Research Centre", Excellence grant project of Czech Science Foundation, GACR P410/12/G113." We also want to thank Hana Contrerasova for thorough reading of this paper in English.

References

Agricultural Policy Strategy for the period after accession to the EU (2004–2013) Ministry of agriculture of the Czech Republic. http://eagri.cz/public/web/file/26240/Koncepce_agrar_politiky_CR_pro_obdobi_po_vstupu_do_EU.pdf. Accessed 20 Jun 2014

Alcantara C, Kuemmerle T, Baumann M, Bragina E, Griffiths P, Hostert P, Knorn J, Mueller D, Prishchepovs A, Schierhorn F, Sieber A, Radeloff V (2013) Mapping the extent of abandoned farmland in central and eastern europe using MODIS time series satellite data. Environ Res Lett 8(3):38–47

Baumann M, Ozdogan M, Kuemmerle T, Wendland K, Esipova E, Radeloff V (2012) Using the Landsat record to detect forest-cover changes during and after the collapse of the Soviet Union in the temperate zone of European Russia. Remote Sens Environ 124:174–184

Bičík I, Jančák V (2003) The changes of rural space in Czechia in the period of transformation 1990–2015. Acta Universitatis Carolinae 2003. Geographica 1:11–20

Bičík I, Jeleček L, Kabrda J, Kupková L, Lipský Z, Mareš P, Šefrna L, Štych P, Winklerová J (2010) Vývoj využití ploch v Česku. Nakladatelství ČGS, Prague

Bossard M, Feranec J, Oťaheľ J (2000) CORINE land cover technical guide—Addendum 2000 Technical report, 40. European Environment Agency. http://www.eea.europa.eu/publications/tech40add. Accessed 3 Oct 2013

Cremene C, Groza G, Rakosy L, Schileyko AA, Baur A, Erhardt A, Baur B (2005) Alterations of steppe-like grasslands in Eastern Europe: a threat to regional biodiversity hotspots. Conserv Biol 19:1606–1618

Feranec J, Soukup T (2013) Map presentation of changes in Europe's artificial surfaces for the periods 1990–2000 and 2000–2006. Cent Eur J Geosci 5(2):323–330

Feranec J, Šúri M, Oťaheľ J, Cebecauer T, Kolář J, Soukup T, Zdeňková D, Waszmuth J, Vajdea V, Vijdea A, Nitica C (2000) Inventory of major landscape changes in the Czech Republic, Hungary, Romania and Slovak Republic. Int J Appl Earth Obs Geoinf 2(2):129–139

Feranec J, Hazeu G, Christensen S, Jaffrain G (2007) Corine land cover change detection in Europe (case studies of the Netherlands and Slovakia). Land Use Policy 24:234–247

Feranec J, Jaffrain G, Soukup T, Hazeu G (2010) Determining changes and flows in European landscapes 1990–2000 using CORINE land cover data. Appl Geogr 30:19–35

Feranec J, Soukup T, Hazeu G, Jaffrain G (2012) Land cover and its change in Europe: 1990–2006. In Giri Ch (ed) Remote sensing of land use and land cover: principles and applications, CRC Press, Taylor & Francis Group, Boca Raton

Food and Agriculture Organization of the United Nations (2006) Global forest resources assessment 2005: progress towards sustainable forest management. FAO, Rome

Haines-Young R, Weber J-L (2006) land accounts for europe 1990–2000, EEA report 11. EEA, Copenhagen

Hais M, Langhammer J, Jirsová P, Dvořák L (2008) Dynamics of forest disturbance in central part of the Šumava mountains between 1985 and 2007 based on Landsat TM/ETM + satellite data. Acta Universitatis Carolinae Geographica 1–2:53–62

Heymann Y, Steenmans Ch, Croissille G, Bossard M (1994) CORINE land cover. Office for Official Publications European Communities, Technical guide, Luxembourg

Jeleček L, Bičík I, Štych P, Bláha JD (2012) Case study area kobylí: change of land use patterns 1827–2001. In: Bičík I (ed) Land use/cover changes in selected regions in the world VII. IGU-LUCC, Prague
Kozak J (2003) Forest cover change in the western carpathians in the past 180 years—a case study in the orawa region in poland. Mt Res Dev 23:369–375
Kozak J, Estreguil C, Troll M (2007a) Forest cover changes in the northern carpathians in the 20th century: a slow transition. J Land Use Sci 2:127–149
Kozak J, Estreguil C, Vogt P (2007b) Forest cover and pattern changes in the Carpathians over the last decades. Eur J Forest Res 126:77–90
Kuemmerle T, Radeloff VC, Perzanowski K, Hostert P (2006) Cross-border comparison of land cover and landscape pattern in Eastern Europe using a hybrid classification technique. Remote Sens Environ 103(4):449–464
Kuemmerle T, Hostert P, Radeloff VC, Perzanowski K, Kruhlov I (2007) Post—socialist forest disturbance in the Carpathian border region of Poland, Slovakia, and Ukraine. Ecol Appl 17:1279–1295
Kuemmerle T, Hostert P, Radeloff VC, van der Linden S, Perzanowski K, Kruhlov I (2008) Cross-border comparison of post-socialist farmland abandonment in the Carpathians. Ecosystems 11:614–628
Kuemmerle T, Kozak J, Radeloff VC, Hostert P (2009a) Differences in forest disturbance among land ownership types in Poland during and after socialism. J Land Use Sci 4:73–83
Kuemmerle T, Müller M, Rusu M, Griffiths P (2009b) Land use change in southern Romania after the collapse of socialism. Reg Environ Change 9:1–12
Kupkova L, Bicik I, Najman J (2013) Land cover changes along the iron curtain 1990–2006. Geografie 118:95–114
Lerman Z, Csaki C, Feder G (2004) Evolving Farm Structures and Land-use Patterns in Former Socialist Countries. Q J Int Agric 43:309–335
Macours K, Swinnen JFM (2000) Causes of output decline in economic transition: the case of central and eastern european agriculture. J Comp Econ 28:172–206
Müller D, Kuemmerle T, Rusu M, Griffiths P (2009) Lost in transition: determinants of post-socialist cropland abandonment in Romania. J Land Use Sci 4(1–2):109–129
Nunes de Lima MV (2005) IMAGE2000 and CLC2000. EC, JRC and EEA, Ispra
Prishchepov AV (2010) Land-use and land cover change in times of massive socio-economic changes: detecting agricultural abandonment and understanding its drivers in post-Soviet Eastern Europe. Dissertation, University of Wisconsin-Madison
Prishchepov AV, Radeloff V, Baumann M, Kuemmerle T, Müller D (2012) Effects of institutional changes on land use: agricultural land abandonment during the transition from state-command to market-driven economies in post-Soviet Eastern Europe. Environ Res Letters 7(2):68–85
Prishchepov AV, Müller D, Dubinin M, Baumann M, Radeloff VC (2013) Determinants of agricultural land abandonment in post-Soviet European Russia. Land Use Policy 30:873–884
Taff GN (2005) Conflict between global and local land-use values in latvia's gauja national park. Landscape Res 30(3):415–430
Taff GN, Müller D, Kuemmerle T, Ozdeneral E, Stephen J, Walsh J S (2010) Reforestation in Central and Eastern Europe after the breakdown of socialism. In: Nagendra H, Southworth J (eds) Reforesting landscapes: linking pattern and process. Springer Landscape Series Volume 10, Dordrecht

Lighting Tracks Transition in Eastern Europe

C.D. Elvidge, F.-C. Hsu, K.E. Baugh and T. Ghosh

Abstract Previous studies have revealed that satellite-observed lighting data from most countries are relatively well correlated with both population and gross domestic product (GDP). Eastern Europe contains the largest concentration of countries with lighting patterns that do not adhere to this trend by exhibiting higher correlations with either GDP or population. We examined a time series of DMSP nighttime light data spanning two decades and found that GDP-centric countries experienced an increase in nighttime light in the two decades following the collapse of the Soviet Union. Conversely, population-centric countries experienced widespread lighting losses during the first decade and urban lighting growth during the second decade. The fact that lighting was lost without the loss of infrastructure indicates that lighting is a poor proxy for mapping the extent of constructed infrastructure in some cases. These results indicate that the use of nighttime lights as an anthropogenic land cover proxy may require national or even subnational calibration in Eastern Europe.

1 Introduction

Artificial lighting is a hallmark of modern technological culture. At night, cities are bathed in light, particularly urban infrastructure. The fact that lighting is used for roads, highways, buildings and parking lots has led to the use of satellite-observed nighttime light data as proxies for land cover and land use variables related to the density of built infrastructure. Because of the global extent, standardized production, and relative ease with which Defense Meteorological Satellite Program

C.D. Elvidge (✉)
Earth Observation Group, NOAA National Geophysical Data Center,
325 Broadway, Boulder, CO 80305, USA
e-mail: chris.elvidge@noaa.gov

F.-C. Hsu · K.E. Baugh · T. Ghosh
Cooperative Institute for Research in the Environmental Sciences,
University of Colorado, Boulder, CO, USA

G. Gutman and V. Radeloff (eds.), *Land-Cover and Land-Use Changes in Eastern Europe after the Collapse of the Soviet Union in 1991*,
DOI 10.1007/978-3-319-42638-9_3

(DMSP) nighttime light data can be accessed, they are widely used as a proxy for other variables that are more difficult to measure such as population or GDP. The logic behind this is that urban processes are highly correlated. If one process or activity can be measured well, then it can be used to make reasonable estimates of others. For example, nighttime light data have been used to map economic activity (Ghosh et al. 2010), fossil fuel carbon emissions (Rayner et al. 2010), spatial population distributions (Doll 2008; Sutton 1997), poverty (Elvidge et al. 2009a, b), constructed surface densities (Elvidge et al. 2007), food demand (Matsumura et al. 2009), water use (Zhao et al. 2011) and stocks of steel and other metals (Hsu et al. 2013).

Nighttime light changes are also excellent indicators of economic or population changes. For example, the cumulative brightness of nighttime lights is highly correlated to both population and economic expansion in Vietnam and China, which are both growing rapidly. In developed countries, such as Japan and the USA, lighting growth due to the expansion of infrastructure is offset by lighting fixture and lighting type improvements that reduce the emission of light into the night sky. Thus, the cumulative brightness of satellite-observed lighting in developed countries is stable from year to year.

Under normal circumstances, lighting continues to operate after being installed. Lights may be upgraded, but it is highly unusual for highways, bridges or shopping districts to lose their lighting. However, loss of lighting can occur. There are several causes for losses in satellite-observed lighting. None of these causes are associated with positive events. Electrical infrastructure is commonly attacked and destroyed during war (Witmer and Loughlin 2011). Natural disasters may knock out power lines and power stations. Electric power grids may lack sufficient stability or capacity. Economic collapse can also lead to a lighting collapse. Conversely, investments in electric power infrastructure can increase the brightness and extent of satellite-observed lighting (Agnew et al. 2008; Hodler and Raschky 2010).

Eastern Europe provides the world's most notable example of unusual behavior associated with satellite-observed nighttime lights. Elvidge et al. examined the relationship between the total brightness of lights, population and GDP for the countries of the world. They found that the correlation coefficients between population, GDP and total detected lighting were quite similar for the vast majority of countries. These results are illustrated in Fig. 1, which charts the correlation coefficients for population and GDP with lighting. Countries whose lights are highly correlated with both population and GDP fall within the tip of the data set, which is located at the upper right of the plot. These countries have experienced rapid population and GDP growth, which have driven the expansion of lighting. At the opposite end of the diagonal are countries whose aggregate lighting is negatively correlated with both population and GDP. In these countries, population and GDP grow, but the aggregate brightness of lighting is unaffected. This includes highly developed countries, such as the USA and Japan, where lighting efficiency improvements offset lighting gains from urban expansion. The countries that plot near the middle of the diagonal exhibit little or no correlation between lighting, GDP and population. The data in Fig. 1 mainly plot along the diagonal because the

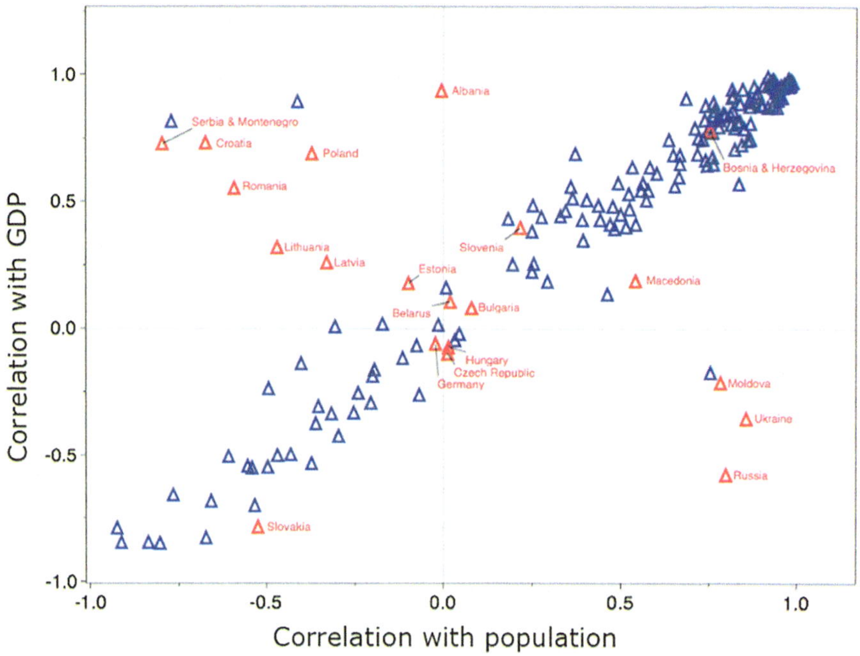

Fig. 1 Correlation coefficients (r values) for the DMSP sum of lights (1992–2012) versus GDP and population. Data points for Eastern Europe countries are colored *red*

populations and GDPs of most countries are synchronized. In other words, both population and GDP grow each year under normal circumstances.

When population and GDP are non-synchronized, countries plot further from the diagonal. One set of Eastern Europe countries plots above the diagonal, where the lighting correlation with GDP exceeds that with population (GDP centric). Another set of countries plots below the diagonal, where the lighting correlation with population exceeds that with GDP (population centric). A third set of countries largely plots along the diagonal. The objective of this chapter is to improve the understanding of the results shown in Fig. 1 in terms of country-level lighting behaviors.

2 Methods

A time series of cloud-free nighttime lighting annual composites was processed from the archived data collected by the U.S. Air Force Defense Meteorological Satellite Program (DMSP) Operational Linescan System (OLS) following the methods of Baugh et al. (2010). Because the OLS lacks a calibration system for the low-light imaging sensor, we employed an empirical intercalibration following Elvidge et al. (2009a, b).

The annual sum-of-lights (SOL) was extracted for each country by summing all of the DN values for lighting in each country. Lighting from gas flares was masked based on methods described by Elvidge et al. (2009a, b). As part of the extraction, we performed two important adjustments to the values for each grid cell. First, we applied an intercalibration based on the offsets and coefficients provided by Elvidge et al. The form of the calculation is $Y = C_0 + C_1X + C_2X^2$. Values higher than 63 were truncated at 63. Thus, the intercalibration increases the number of saturated pixels (DN = 63). The resulting values were then adjusted to compensate for the latitude-related changes in surface area in the 30 arc-second grid. The digital values were then summed to derive the "sum-of-lights" index value (or SOL) for each satellite year and country. To compensate for differences in the detection limits of the different products, only DNs of six or larger were added to the SOL.

In addition, we generated color composite images for Eastern Europe, ranging from 1992 to 2012. The color images are designed to show the spatial patterns of lighting changes over time. The composites use data from two years, with the difference between the two years shown in red. The earlier year difference is shown in green, while the later year difference is shown in blue. The green and blue color planes are then inverted, and the contrast is enhanced.

3 Results

3.1 Examination of Individual Countries

We were able to investigate the lighting behaviors in individual countries by examining the color composite images and temporal plots of the sum-of-lights index.

Albania: Figure 2 shows the nighttime light variations in Albania. While the temporal chart shows steady lighting growth across the two decades, the images show differences between urban and rural lighting development. The lighting brightness increased in and around the centers of major cities and in rural areas due to new lighting developments during the first decade. During the second decade, the lighting growth was concentrated in major urban areas, with no evident loss of rural lighting.

Belarus: Figure 3 shows the nighttime lights change in Belarus. Widespread lighting losses occur in both urban and rural areas during the first decade, with minor amounts of new rural lighting observed in the southeast. During the second decade, the brightness of urban lighting grew, and both new urban and rural lighting can be observed. The temporal chart is erratic; however, it generally indicates lighting losses in the 90s and gains from 2006 to 2012.

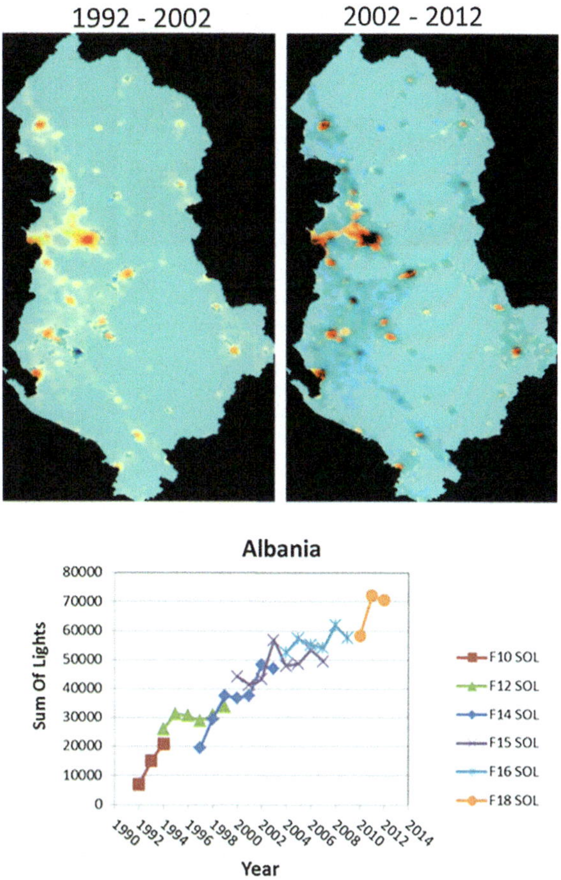

Fig. 2 Nighttime lights change images of Albania from 1992–2002 to 2002–2012 in addition to the sum-of-lights plot, which is based on data from each satellite

Bosnia-Herzegovina: Figure 4 shows the nighttime lights change in Bosnia-Herzegovina. Lighting growth was dramatic in the first decade, with large areas of new urban and rural lighting. In addition, the brightness of existing urban lighting increased in some urban centers. The brightness of urban lighting also increased during the second decade. The temporal chart mimics the image results quite well, with rapid lighting growth during the first decade and slow growth during the second decade.

Bulgaria: Figure 5 shows the nighttime lights change in Bulgaria. The temporal chart shows an erratic pattern from 1992 to 2006 and steady lighting growth from 2006 to 2012. The light image from the first decade indicates that lighting losses were common in most cities and many rural areas. The only urban area with lighting growth was the capital city of Sofia. New lighting developed in south-eastern, central and northeastern Bulgaria, including in some rural areas. Substantial urban lighting expansion occurred during the second decade. This trend is most pronounced around Sofia. New areas of rural lighting developed in the north-central region and northwest corner of the country.

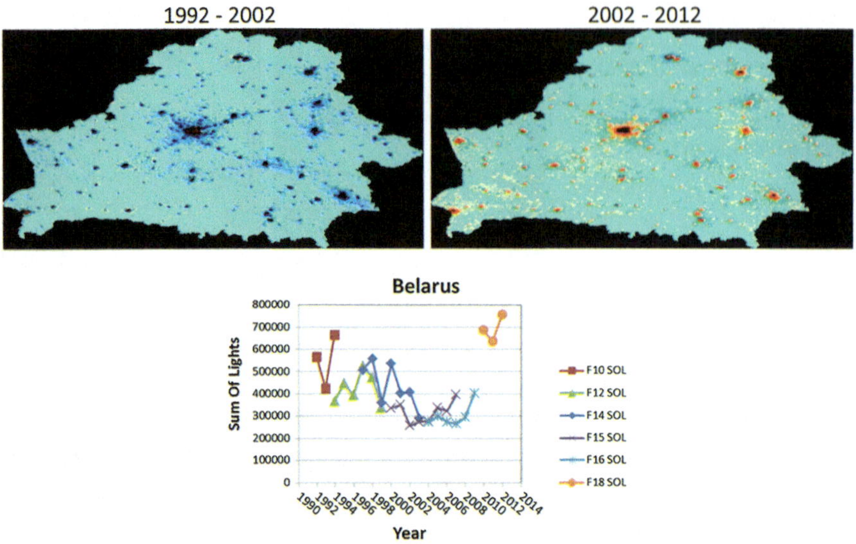

Fig. 3 Nighttime light change in Belarus from 1992–2002 to 2002–2012 in addition to the sum-of-lights plot, which is based on data from each satellite

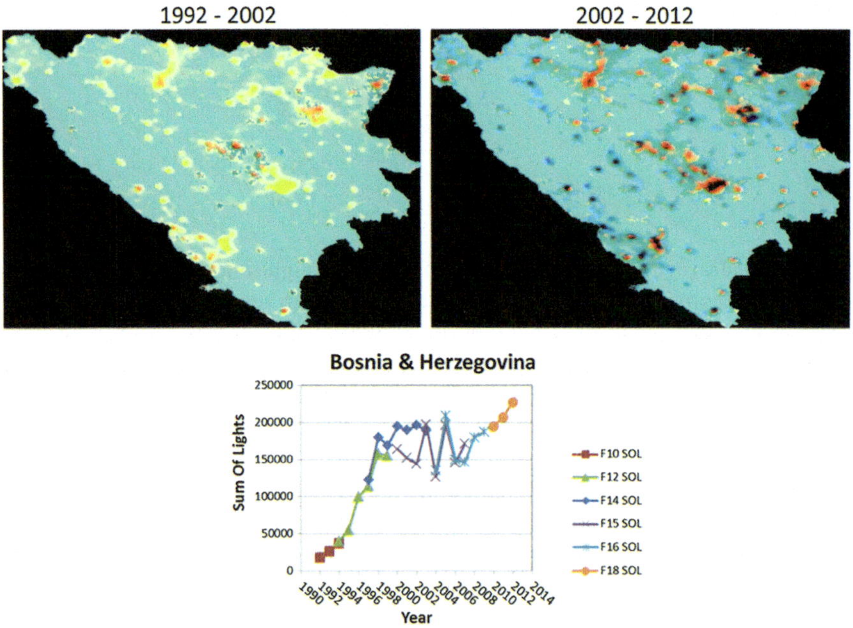

Fig. 4 Nighttime light change in Bosnia-Herzegovina from 1992–2002 to 2002–2012 in addition to the sum-of-lights plot, which is based on data from each satellite

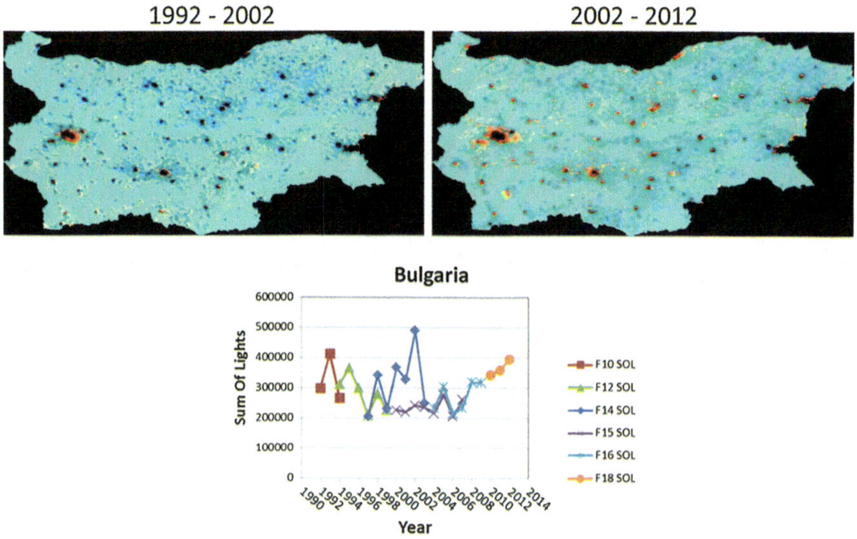

Fig. 5 Nighttime light change in Bulgaria from 1992–2002 to 2002–2012 in addition to the sum-of-lights plot, which is based on data from each satellite

Croatia: Figure 6 shows the nighttime light changes in Croatia. The temporal chart shows rapid lighting growth from 1992 to 1998 and steady growth thereafter. The image from the first decade shows urban lighting growth and the development of new lighting in rural areas to the south and east. During the second decade, the lighting growth was concentrated in the major urban areas, with no clear loss of rural lighting. In both decades, the lighting growth was most pronounced in and around the capital city of Zagreb.

Czech Republic: Figure 7 shows the nighttime light changes in the Czech Republic. The temporal chart shows an erratic pattern throughout the two decades. The lights image from the first decade illustrates urban lighting growth in the south as well as in the western half of the country. New rural lighting developed in the far south and southwest corner. Lighting losses occurred in the eastern corner. The lighting brightness increased around several of the cities during the second decade.

Estonia: Figure 8 shows the nighttime light changes in Estonia. The temporal chart shows growth in the early and late portions of the period. During the first decade, the lighting brightness increased in the cores of major cities. In addition, new lighting was developed near the urban cores and in many rural areas. The lighting growth was concentrated in the major urban areas during the second decade.

Germany: Figure 9 shows the nighttime light changes in Germany. East Germany is outlined with a yellow line. The temporal chart shows that lighting grew in the 1990s and was largely stable in the 2000s. The lights images show growth in both urban and new rural lighting during the first decade. The lighting growth was concentrated in major urban areas during the second decade.

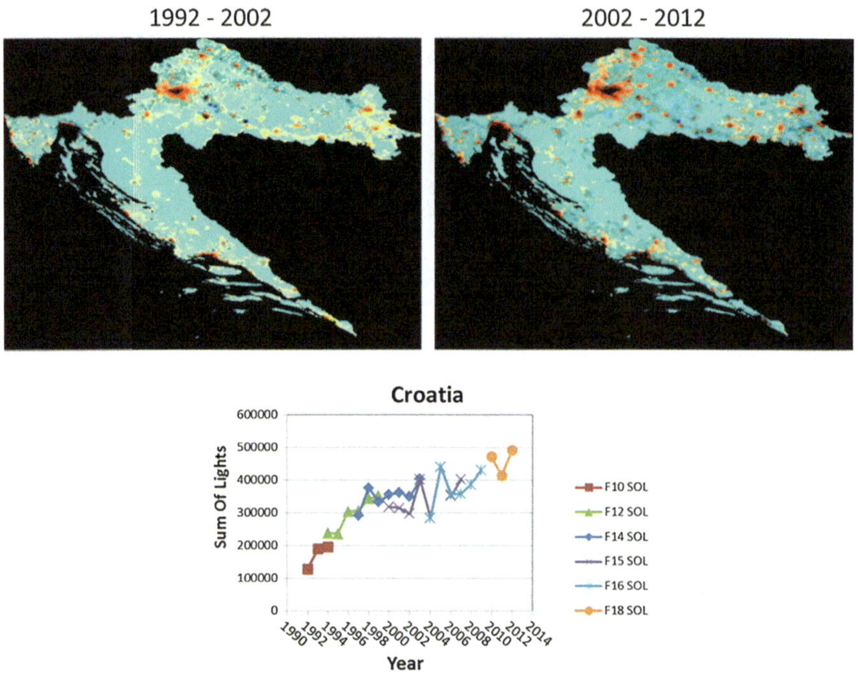

Fig. 6 Nighttime light change in Croatia from 1992–2002 to 2002–2012 in addition to the sum-of-lights plot, which is based on data from each satellite

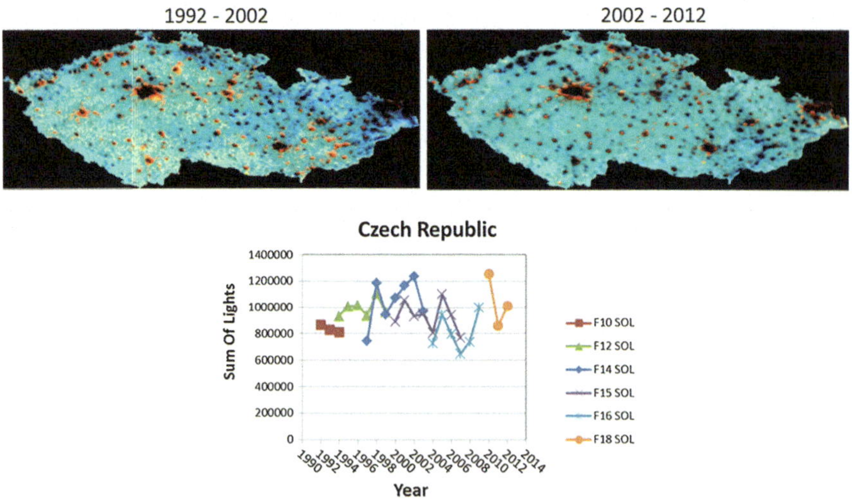

Fig. 7 Nighttime light changes in the Czech Republic from 1992–2002 to 2002–2012 in addition to the sum-of-lights plot, which is based on data from each satellite

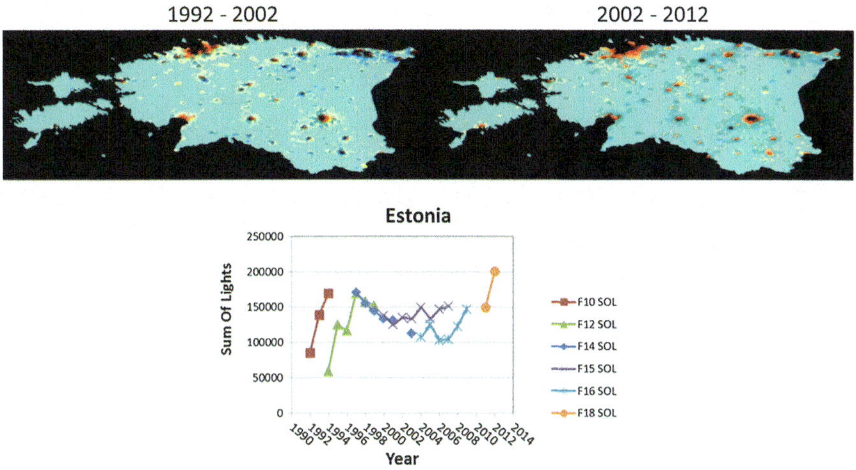

Fig. 8 Nighttime light changes in Estonia from 1992–2002 to 2002–2012 in addition to the sum-of-lights plot, which is based on data from each satellite

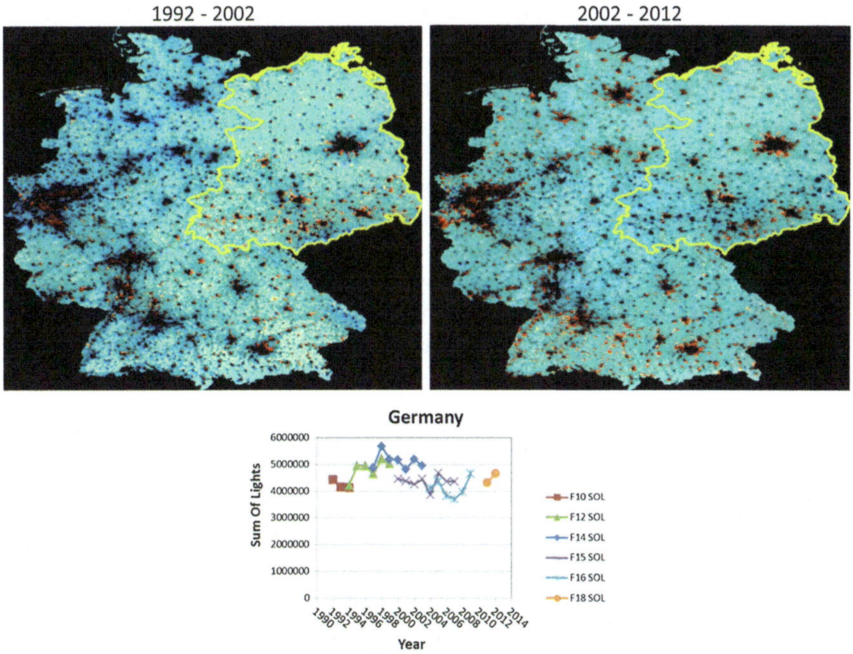

Fig. 9 Nighttime light changes in Germany from 1992–2002 to 2002–2012 in addition to the sum-of-lights plot, which is based on data from each satellite

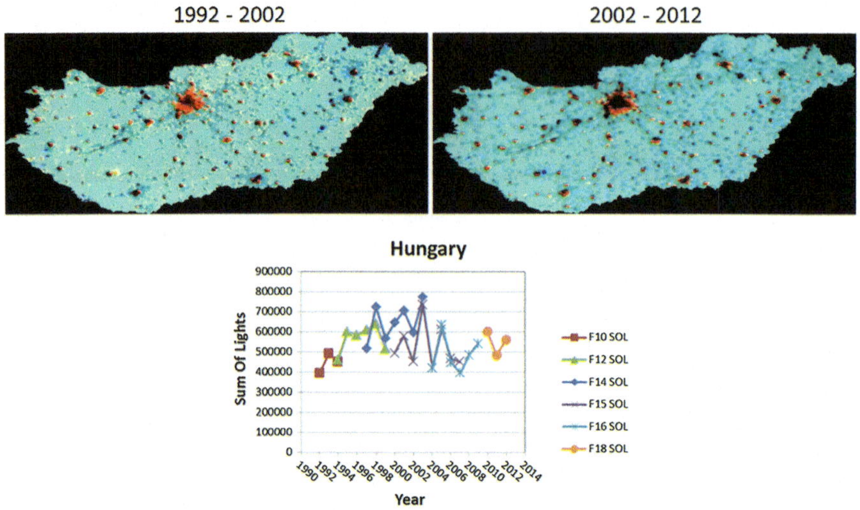

Fig. 10 Nighttime light changes in Hungary from 1992–2002 to 2002–2012 in addition to the sum-of-lights plot, which is based on data from each satellite

Hungary: Figure 10 shows the nighttime light changes in Hungary. The temporal chart shows an erratic pattern, with a general increasing trend during the first decade. The trend then stabilizes during the second decade. The lights image for the first decade is complicated. Some rural lighting loss occurred in the east, while approximately half of the urban areas experienced lighting growth. Additionally, several urban and rural areas developed new lighting. Lighting growth occurred in more than half of the cities and towns during the second decade.

Latvia: Figure 11 shows the nighttime light changes in Latvia. The time series is largely flat until 2010 followed by two years (2011 and 2012) of lighting growth. The lights image for the first decade indicates some local urban and rural lighting losses, some urban lighting growth and some areas of new urban and rural lighting. The lighting brightness increased in nearly all cities and towns during the second decade.

Lithuania: Figure 12 shows the nighttime light changes in Lithuania. The time series shows an erratic lighting trend during the first decade, a level trend from 2003 to 2009 and an increasing trend from 2011 to 2012. The lights image for the first decade shows some rural lighting losses, urban lighting growth, new rural lighting and new urban lighting in the extreme west. Some loss of urban and rural lighting occurred in the far west during the second decade. Urban lighting growth and new areas of rural lighting can also be observed in the second decade image.

Macedonia: Figure 13 shows nighttime light changes in Macedonia. The time series shows rapid lighting growth during the first decade, stable lighting from 2002 to 2010 and increased lighting from 2011 to 2012. The first decade lights image shows extensive urban lighting growth and new rural lighting. The second decade illustrates that the lighting growth rate decreased in urban areas.

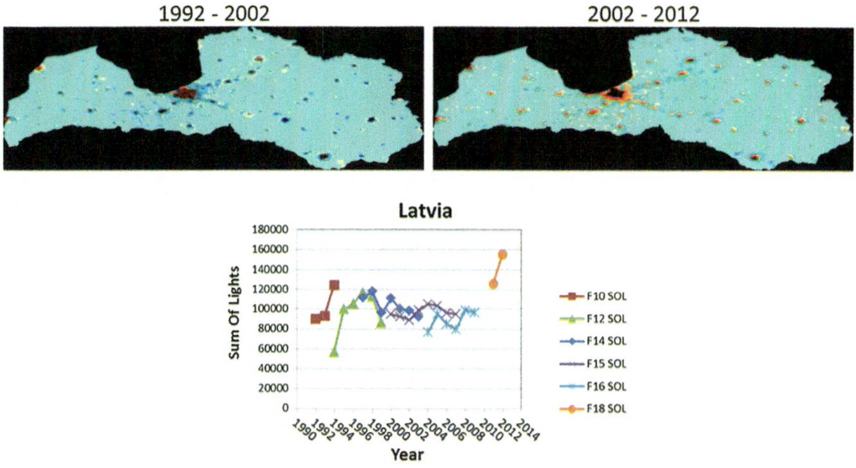

Fig. 11 Nighttime light changes in Latvia from 1992–2002 to 2002–2012 in addition to the sum-of-lights plot, which is based on data from each satellite

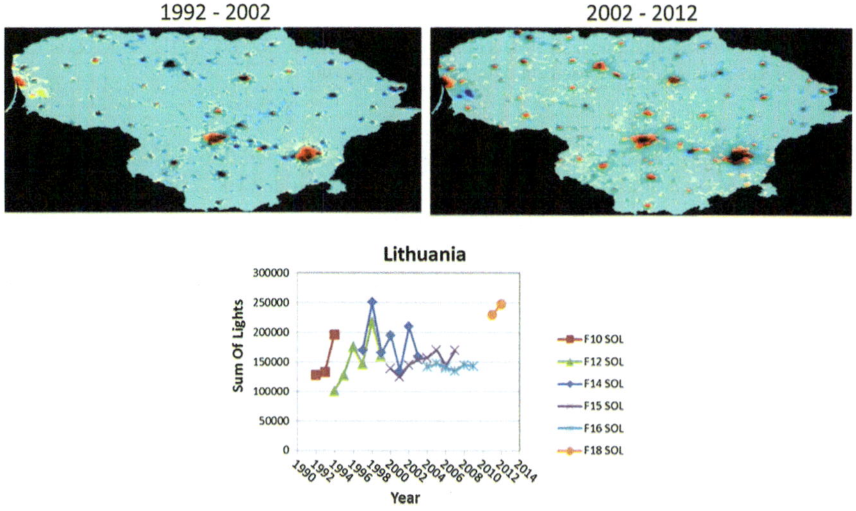

Fig. 12 Nighttime light changes in Lithuania from 1992–2002 to 2002–2012 in addition to the sum-of-lights plot, which is based on data from each satellite

Moldova: Figure 14 shows nighttime light changes in Moldova. The time series shows a rapid decline in lighting from 1992 to 1999. The lighting then generally stabilizes until 2010 when it increases for the remainder of the period. The lights image for the first decade shows a uniform loss of lighting in urban and rural areas. The second decade image shows growth in lighting in the capital and several of the other cities, with some rural areas developing new lighting.

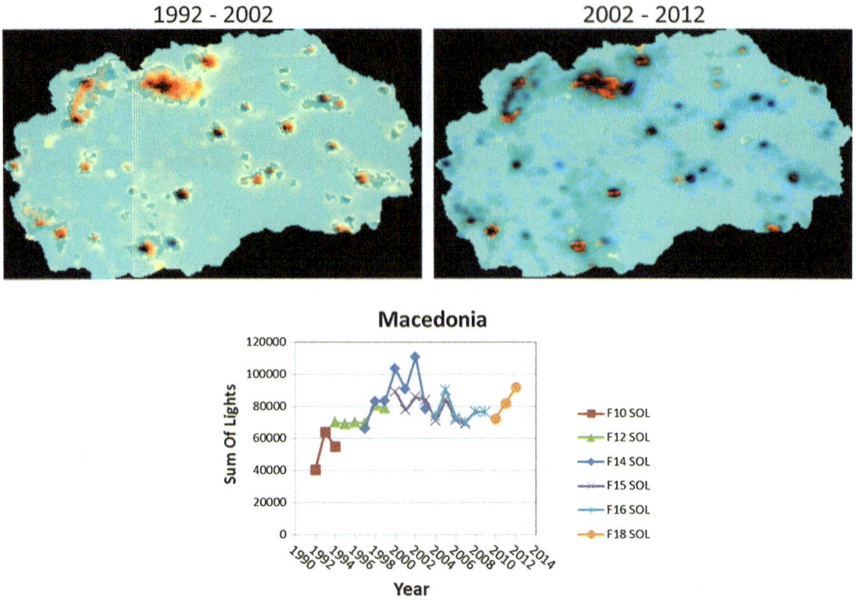

Fig. 13 Nighttime light changes in Macedonia from 1992–2002 to 2002–2012 in addition to the sum-of-lights plot, which is based on data from each satellite

Montenegro: Figure 15 shows the nighttime light changes in Montenegro. The time series shows steady lighting growth across two decades. The variation image for the first decade shows urban lighting growth and some new rural lighting. Urban lighting is also observed during the second decade.

Poland: Figure 16 shows the nighttime light changes in Poland. The time series shows an erratic pattern but clear upward trend over the two decades. The lights image for the first decade illustrates brighter lighting in the centers of major cities. Additionally, new lighting was developed around the urban cores and in many rural areas in the east and north. The lighting growth was concentrated in the major urban areas during the second decade, with new lighting developed in rural northeastern areas.

Romania: Figure 17 shows the nighttime light changes in Romania. The time series shows an erratic pattern from 1992 to 2007. Substantial light growth occurred from 2007 to 2012. The lights image from the first decade shows widespread development of new lighting in both urban and rural areas as well as lighting growth in most urban areas. The lights image from the second decade shows continued urban lighting growth and new lighting development in some rural areas.

Russia: Figure 18 shows the nighttime light changes in western Russia. The temporal chart shows a lighting decline from 1992 to 2006, which is followed by

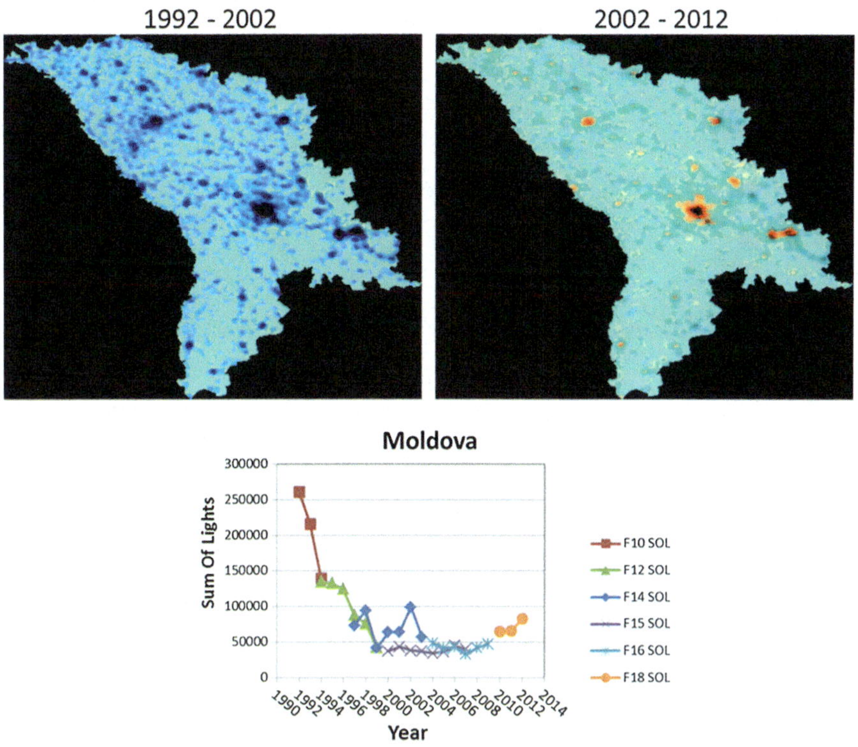

Fig. 14 Nighttime light changes in Moldova from 1992–2002 to 2002–2012 in addition to the sum-of-lights plot, which is based on data from each satellite

growth through 2011. The lights image from the first decade shows widespread lighting losses. The only exception is urban lighting growth in parts of Moscow and St. Petersburg. Urban lighting growth was widespread during the second decade. In addition, rural lighting was developed along the border with the Ukraine.

Serbia: Figure 19 shows the nighttime light changes in Serbia. The time series shows steady lighting growth over the two decades. Urban lighting growth occurred in approximately one third of the cities and towns during the first decade. New rural lighting development also occurred in the central part of the country. Nearly all of the cities and towns experienced urban lighting growth during the second decade.

Slovakia: Figure 20 shows the nighttime light changes in Slovakia. The time series shows an erratic pattern with a slight downward trend. The lights image from the first decade shows widespread lighting losses in the north and east, urban lighting growth in the west and limited new rural lighting development in the south. Urban lighting growth occurred in approximately 20 % of the cities and towns during the second decade.

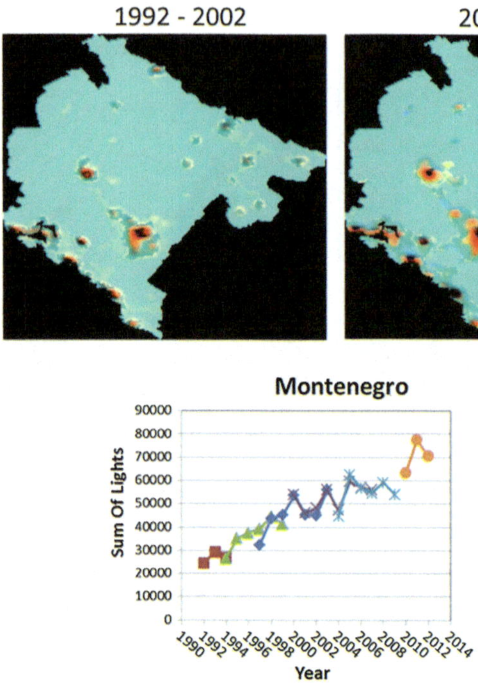

Fig. 15 Nighttime light changes in Montenegro from 1992–2002 to 2002–2012 in addition to the sum-of-lights plot, which is based on data from each satellite

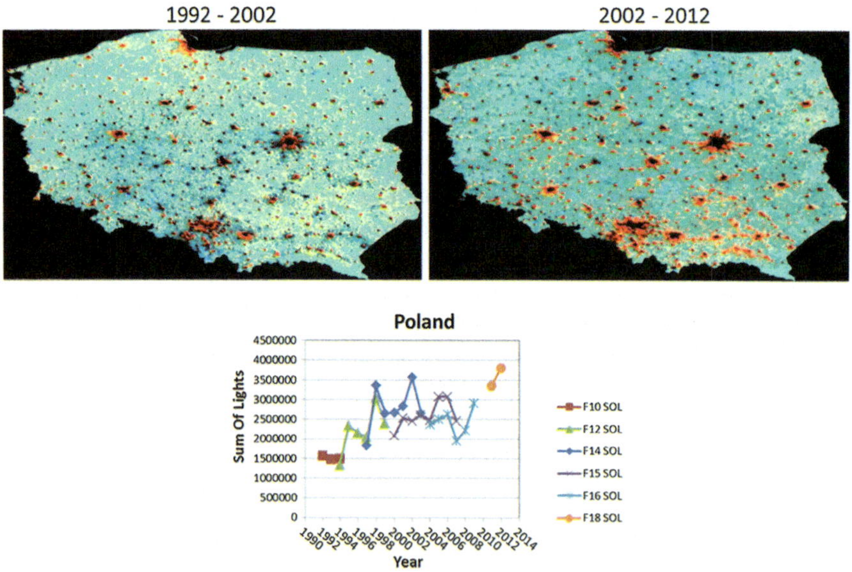

Fig. 16 Nighttime light changes in Poland from 1992–2002 to 2002–2012 in addition to the sum-of-lights plot, which is based on data from each satellite

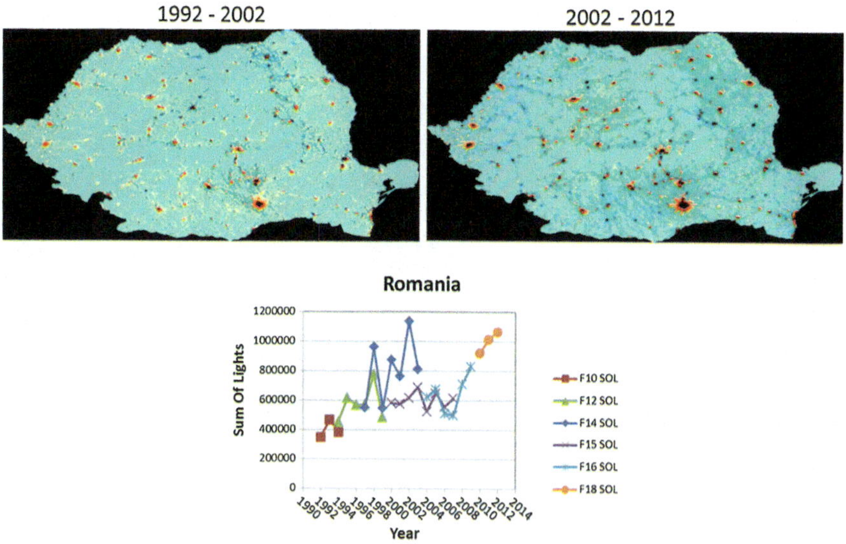

Fig. 17 Nighttime light changes in Romania from 1992–2002 to 2002–2012 in addition to the sum-of-lights plot, which is based on data from each satellite

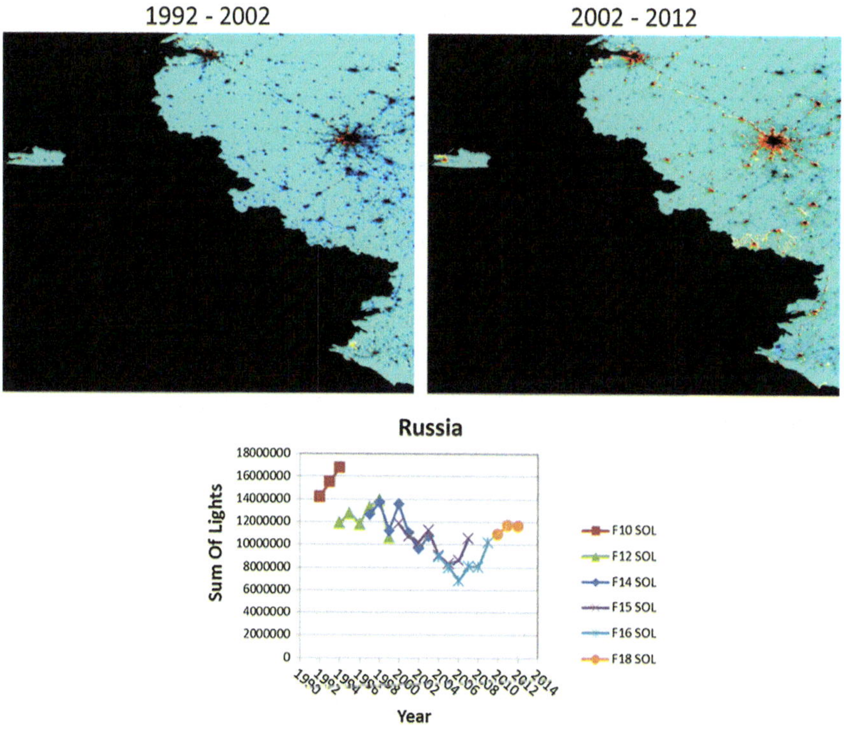

Fig. 18 Nighttime light changes in western Russia from 1992–2002 to 2002–2012 in addition to the sum-of-lights plot, which is based on data from each satellite

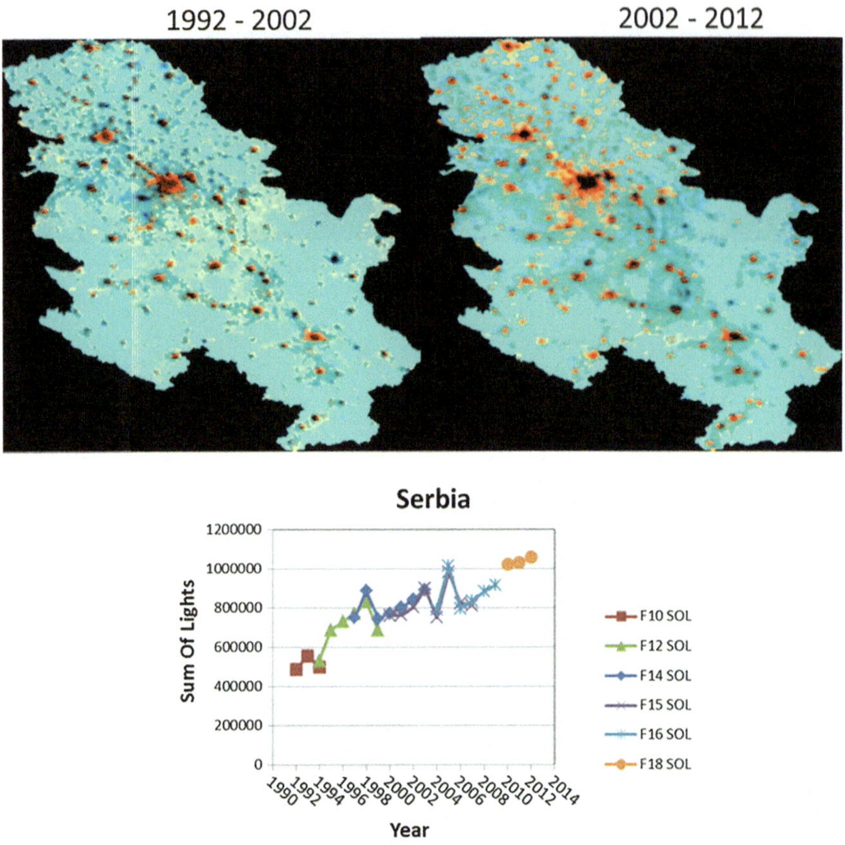

Fig. 19 Nighttime light changes in Serbia from 1992–2002 to 2002–2012 in addition to the sum-of-lights plot, which is based on data from each satellite

Slovenia: Figure 21 shows the nighttime light changes in Slovenia. The time series shows steady lighting growth from 1992 to 2000, which was followed by erratic but largely stable lighting through 2012. The lights image for the first decade shows urban lighting growth and many areas of new rural lighting. Urban lighting growth also occurred during the second decade.

Ukraine: Figure 22 shows the nighttime light changes in Ukraine. The time series shows a steady decline in lighting from 1992 to 2007, which is followed by lighting growth through 2012. The lights image from the first decade shows nationwide urban and rural lighting losses. The trend reverses during the second decade based on urban lighting growth and new rural lighting development.

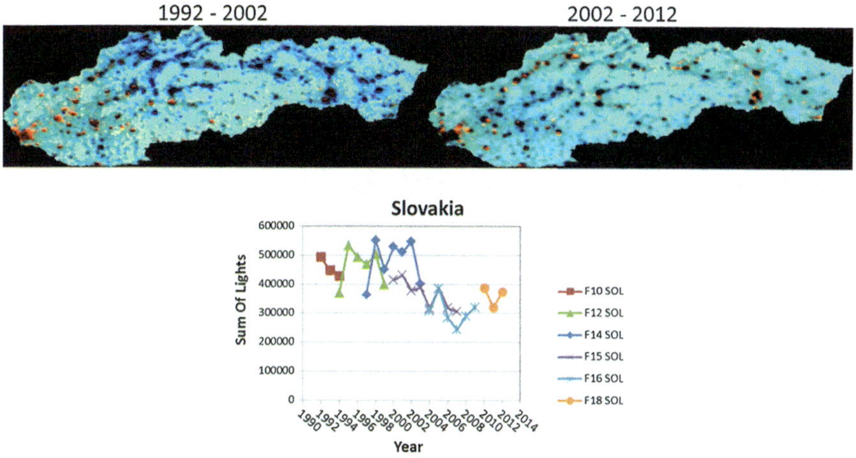

Fig. 20 Nighttime light changes in Slovakia from 1992–2002 to 2002–2012 in addition to the sum-of-lights plot, which is based on data from each satellite

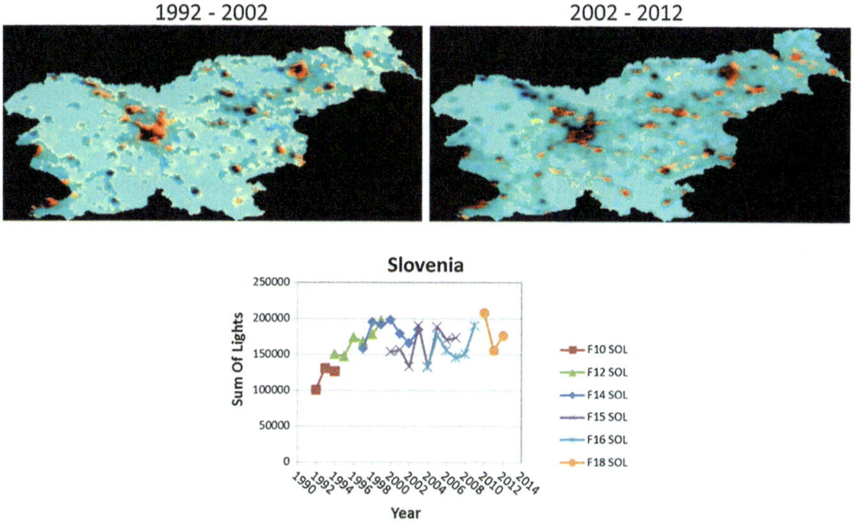

Fig. 21 Nighttime light changes in Slovenia from 1992–2002 to 2002–2012 in addition to the sum-of-lights plot, which is based on data from each satellite

4 Discussion

The countries can be broadly divided into three groups (Table 1). Approximately half of the countries experience rapid lighting growth during the first decade in addition to growth in the brightness of existing lighting and development of new

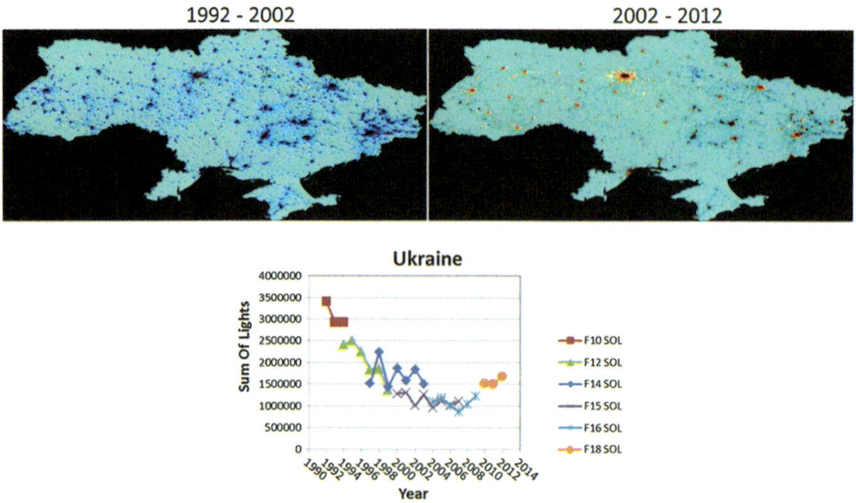

Fig. 22 Nighttime light changes in Ukraine from 1992–2002 to 2002–2012 in addition to the sum-of-lights plot, which is based on data from each satellite

lighting in rural areas and some urban areas. Lighting growth continues in urban areas during the second decade. The rapid lighting growth countries include Albania, Bosnia and Herzegovina, Croatia, Estonia, Macedonia, Montenegro, Poland, Romania, Serbia and Slovenia.

The second group includes countries that experienced slower lighting growth. This group is characterized by lights images with fewer red or yellow areas and stable or erratic temporal records. This group includes six countries: the Czech Republic, East Germany, Hungary, Latvia, Lithuania and Slovakia. In the case of Slovakia, there were both losses and gains in lighting in different parts of the country. Slovakia experienced relatively slow growth of its infrastructure from 1990 to 2006 (Feranec and Soukup 2012).

A third group of countries experienced widespread lighting losses during the first decade and urban lighting recovery during the second decade. This group includes Belarus, Bulgaria, Moldova, Russia and Ukraine. The collapse of the Soviet Union resulted in diminished electric power generation capacity for a solid decade, which likely affected these countries.

Figure 23 denotes the three groups in the original Fig. 1. The rapid growth countries include most of the GDP-centric countries as well as countries that plot in the upper portion of the diagonal. The slow growth countries plot along the lower half of the diagonal, excluding Latvia and Lithuania, which exhibit weak GDP-centric behavior. The loss and recovery group includes population-centric countries (Moldova, Ukraine and Russia) as well as Belarus and Bulgaria, which plot near the (0, 0) position on the chart.

Table 1 Summary of satellite-observed lighting patterns across two decades

	Country	1992–2002			2002–2012		
		Losses	DN increase	New lighting	Losses	DN increase	New lightin growth
Rapid growth	Albania		U	U & R		U	U
	Bosnia-Herzegovina		U	U & R		U	
	Croatia		U	U & R		U	U
	Estonia		U	U & R		U	
	Macedonia		U	R		U	
	Montenegro		U	R		U	U
	Poland		U	R		U	R
	Romania		U	R		U	R
	Serbia		U	R		U	
	Slovenia		U	R		U	
Slow grɔwth	Czech Republic	U & R	U	R	U & R	U	
	East Germany		U	R		U	
	Hungary	R	U	U & R		U	
	Latvia	U & R	U	U		U	
	Lithuania	R	U	U & R	U & R	U	R
	Slovakia	U & R	U	R		U	
Loss fol owed by growth	Belarus	U & R				U	U & R
	Bulgaria	U & R	Capital city	R		U	R
	Moldova	U & R				U	R
	Russia	U & R	Moscow & St. Petersburg			U	R
	Ukraine	U & R				U	R

U—urban, *R*—rural

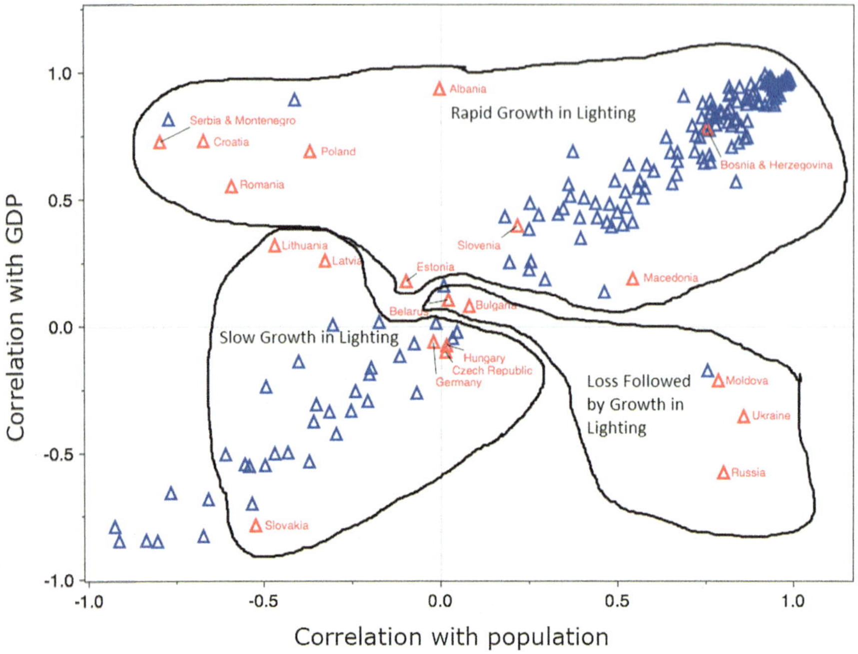

Fig. 23 The countries of Eastern Europe can be divided into three groups based on their lighting behaviors over time: rapid lighting growth, slow lighting growth and lighting loss followed by growth

5 Conclusion

The two-decade long record of nighttime lighting in Eastern Europe demonstrates rapid nighttime lighting changes that were observed using satellites. There is no other region of the world with stronger spatial and temporal lighting patterns between adjacent countries. The countries of Eastern Europe lived for decades under communist rule, and state ownership constrained the economies. The countries had different responses to the collapse of the Soviet Union. These differences are expressed by their nighttime light behaviors.

Nine countries experienced rapid lighting expansion associated with rapid economic growth. Lighting in these countries is positively correlated with GDP, thus falling in the upper half of Fig. 1. Lighting expansion outpaced population growth in five of these countries, resulting in those countries plotting above the diagonal in Fig. 1. In these cases, the correlation with population was negative. These countries may have embarked on rapid transitions from state ownership to free enterprise economic systems.

Six countries experienced slow lighting growth, plotting in the lower half of the diagonal. Four of these countries (Czech Republic, Germany, Hungary and

Slovakia) experienced regional lighting behavior differences. The other two, Latvia and Lithuania, weakly exhibited GDP-centric lighting behavior.

Five countries experienced widespread lighting losses during the first decade followed by lighting recovery in urban cores during the second decade. These countries include the population-centric countries (Moldova, Russia and Ukraine) as well as Belarus and Bulgaria. These countries appear to have experienced electric power generation capacity decreases following the collapse of the Soviet Union. Many years were needed to restore electric power in these areas.

Satellite-observed nighttime light data form part of the historical record associated with the development of these countries following the fall of the Soviet Union. The data suggest widespread lighting growth in most of the countries of the former Warsaw Pact and the Balkan states. However, data for most of the countries formed after the Soviet Union's breakup indicate that electric power services diminished during the 1990s and recovered during the 2000s.

These findings have implications for the use of satellite-observed nighttime light data sets as proxies for land-cover variables such as the density of constructed surfaces. We cannot determine whether, for example, the vast expansion of lighting in countries such as Poland was due to the expansion of human settlements, to lighting being installed at existing settlements or to a combination of both based on only the lighting data. Historical and recent aerial photos or Landsat data could be compared. These comparisons could then be potentially used to group countries when interpreting nighttime lights using satellite data.

Acknowledgments The DMSP orbital data used in this study were provided to the NGDC by the U.S. Air Force Weather Agency (AFWA).

References

Agnew J, Gillespie TW, Gonzalez J, Min B (2008) Baghdad nights: evaluating the US military surge using nightlight signatures. Environ Plann A40(10):2285–2295

Baugh K, Elvidge C, Ghosh T, Ziskin D (2010) Development of a 2009 stable lights product using DMSP-OLS data. Proc Asia Pac Adv Netw 30:114–130

Doll CNH (2008) CIESIN thematic guide to night-time light remote sensing and its applications. Center for International Earth Science Information Network of Columbia University, Palisades

Elvidge CD, Tuttle BT, Sutton PC, Baugh KE, Howard AT, Milesi C, Bhaduri B, Nemani R (2007) Global distribution and density of constructed impervious surfaces. Sensors 7:1962–1979

Elvidge CD, Sutton PC, Ghosh T, Tuttle BT, Baugh KE, Bhaduri B, Bright EA (2009a) Global poverty map derived from satellite data. Comput Geosci 35:1652–1660

Elvidge CD, Ziskin D, Baugh KE, Tuttle BT, Ghosh T, Pack DW, Erwin EH, Zhizhin M (2009b) A fifteen year record of global natural gas flaring derived from satellite data. Energies 2 (3):595–622

Feranec J, Soukup T (2012) Trend of changes in Czechia's and Slovakia's artificial surfaces 1990–2006 represented on map. In: Bičík I et al (eds) Land use/cover changes in selected regions in the world, vol VII. IGU and Charles University, Prague, pp 67–70

Ghosh T, Powell R, Elvidge CD, Baugh KE, Sutton PC, Anderson S (2010) Shedding light on the global distribution of economic activity. Open Geogr J 3:148–161

Hodler L, Raschky PA (2010) Foreign aid and enlightened leaders. Monash University, Department of Economics. Discussion paper 54/10

Hsu F-C, Elvidge CD, Matsuno Y (2013) Exploring and estimating in-use steel stocks in civil engineering and buildings from night-time lights. Int J Remote Sens 34(2):490–504. doi:10.1080/01431161.2012.712232

Matsumura K, Hijmans RJ, Chemin Y, Elvidge CD, Sugimoto K, Wu WB, Lee YW, Shibasaki R (2009) Mapping the global supply and demand structure of rice. Sustain Sci 4(2):301–313

Rayner PJ, Raupach MR, Paget M, Peylin P, Koffi E (2010) A new global gridded data set of CO2 emissions from fossil fuel combustion: methodology and evaluation. J Geophys Res 115: D19306. doi:10.1029/2009JD013439

Sutton P (1997) Modeling population density with night-time satellite imagery and GIS. Comput Environ Urban Syst 21:227–244

Witmer F, Loughlin J (2011) Detecting the effects of wars in the Caucasus regions of Russia and Georgia using radiometrically normalized DMSP-OLS nighttime lights imagery. Geogr Inf Sci Remote Sens 4:478–500

Zhao N, Ghosh T, Currit NA, Elvidge CD (2011) Relationships between satellite observed lit area and water footprints. Water Resour Manag 25(9):2241–2250

Land Change in the Carpathian Region Before and After Major Institutional Changes

Catalina Munteanu, Volker Radeloff, Patrick Griffiths, Lubos Halada, Dominik Kaim, Jan Knorn, Jacek Kozak, Tobias Kuemmerle, Juraj Lieskovsky, Daniel Müller, Katarzyna Ostapowicz, Oleksandra Shandra and Premysl Stych

Abstract The Carpathian region represents an ideal showcase of several land change theories and their implications for conservation because this region shares the long geo-political and socio-economic history of Eastern Europe while also being a bio-diversity hotspot. With a long history of abrupt socio-economic and institutional shifts, the Carpathians exemplify how ecosystems may or may not be pushed into an

---

The original version of this chapter was revised: The spelling of the fifth author name was corrected. The erratum to this chapter is available at DOI 10.1007/978-3-319-42638-9_11

C. Munteanu (✉) · V. Radeloff
SILVIS Lab, Department of Forest and Wildlife Ecology, University of Wisconsin-Madison, 1630 Linden Drive, Madison, WI 53706, USA
e-mail: cmunteanu@wisc.edu

V. Radeloff
e-mail: radeloff@wisc.edu

P. Griffiths · J. Knorn · T. Kuemmerle · D. Müller
Geography Department, Humboldt-University Berlin, Unter den Linden 6, 10099 Berlin, Germany
e-mail: patrick.griffiths@geo.hu-berlin.de

J. Knorn
e-mail: jan.knorn@geo.hu-berlin.de

T. Kuemmerle
e-mail: tobias.kuemmerle@geo.hu-berlin.de

D. Müller
e-mail: mueller@iamo.de

L. Halada · J. Lieskovsky
Institute of Landscape Ecology, Slovak Academy of Sciences Bratislava, Branch Nitra, Akademická 2, 949 01 Nitra, Slovakia
e-mail: lubos.halada@savba.sk

J. Lieskovsky
e-mail: juraj.lieskovsky@savba.sk

G. Gutman and V. Radeloff (eds.), *Land-Cover and Land-Use Changes in Eastern Europe after the Collapse of the Soviet Union in 1991*,
DOI 10.1007/978-3-319-42638-9_4

alternative stable state following shocks such as the collapse of empires, world wars or the collapse of socialism. Furthermore, ecosystem changes may or may not experience time-lags in response to shocks, and over long time periods, historic land-use practices may produce land-use legacies that persist on the landscapes for decades or centuries. Here, we analyze the long-term drivers of land change and their land-use outcomes in the Carpathian region, with a particular focus on forests, agriculture and grasslands, and provide examples of how ecosystems respond to shocks using examples of alternative stable states, time-lags and land-use legacies. Understanding how and why land change patterns vary over time and space is important for balancing land-use decisions, especially in biodiverse regions with a high conservation value.

1 Introduction

The Carpathian region (here defined as the Carpathian Mountains and the surrounding lowlands of the Pannonian plains) has experienced several episodes of drastic land-use change over the past two centuries and is therefore an interesting

D. Kaim · J. Kozak · K. Ostapowicz
Institute of Geography and Spatial Management, Jagiellonian University, Gronostajowa 7, 30-387 Kraków, Poland
e-mail: dkaim@gis.geo.uj.edu.pl

J. Kozak
e-mail: jkozak@gis.geo.uj.edu.pl

K. Ostapowicz
e-mail: kostapowicz@gis.geo.uj.edu.pl

T. Kuemmerle
Germany and Integrative Research Institute on Transformations in Human Environment Systems (IRI THESys), Humboldt-University Berlin, Unter den Linden 6, 10099 Berlin, Germany

D. Müller
Leibniz Institute of Agricultural Development in Transition Economies (IAMO), Halle (Saale), Germany

D. Müller
Integrative Research Institute on Transformations in Human Environment Systems (IRI THESys), Humboldt-University Berlin, Unter den Linden 6, 10099 Berlin, Germany

O. Shandra
Department of Physical Geography and Geoecology, Taras Shevchenko National University of Kyiv, Kiev, Ukraine
e-mail: alya.shandra@gmail.com

P. Stych
Faculty of Science, Charles University in Prague, Albertov 6, 128 43 Prague 2, Czech Republic
e-mail: premysl.stych@natur.cuni.cz

'natural experiment' (Gehlbach and Malesky 2014) for the study of overarching issues at the frontier of land change science: land change following socioeconomic and institutional shocks, the importance of drivers of land change, the time-lag and legacy effects of past changes and the uncertain futures of land cover. Moreover, as the "green backbone of Europe" and a biodiversity hotspot (UNEP 2007; Björnsen-Gurung et al. 2009; Hazeu et al. 2010), the Carpathian region harbors some of the largest old-growth forests on the European continent (Veen et al. 2010), as well as high nature-value grasslands (Akeroyd and Page 2011; Fischer et al. 2012), and provides habitat for a large range of species (Nabuurs et al. 2008; Schulp et al. 2008; Halada et al. 2010; Bálint et al. 2011).

Due to the geo-political context of Eastern Europe and the multiple abrupt shifts in institutions, politics and economics, the ecosystems across the Carpathians are at risk. The underlying driving forces and proximate causes (Lambin et al. 2001; Lambin and Meyfroidt 2010), of which institutional and socio-economic forces are the most important in the Carpathians (Kozak et al. 2013b; Griffiths et al. 2014; Munteanu et al. 2014), heavily affect natural ecosystems. Over the past two centuries, various shocks, such as the fall of empires, the collapse of socialism, and the accession of the EU, have caused several shifts in land management and have drastically affected the type and magnitude of land changes (Munteanu et al. 2015). Economic and institutional drivers have caused land-use intensification, while other socio-demographic and policy changes have resulted in land abandonment. As a consequence, a broad range of patterns of land change have occurred, allowing for a comparison of the effects of the underlying driving forces of change over a relatively short time period and across a small region. The effects of the interplay between drivers reverberate throughout the ecosystem, affecting biodiversity, carbon sequestration, ecosystem services, and rural livelihoods (DeFries et al. 2004; Foley et al. 2005; Ellis et al. 2013).

The drivers of land change and their interactions may cause immediate or delayed responses in ecosystems. Land change can occur immediately following a shift in policy or in economic conditions, or it may take as long as several decades for the effects to become quantifiable. For example, forest harvesting provided a source of immediate income in the period of economic depression following the collapse of the Soviet Union (Griffiths et al. 2012, 2014; Knorn et al. 2012a), and the effects of the loss of tree cover were observed only few years after 1989. However, delays in the legislative framework and the restitution process caused intensified logging to occur only after a time-lag in some regions of the Carpathians (Knorn et al. 2012a; Griffiths et al. 2014). In response to conservation practices and land cover changes in the post-Soviet era, large mammal populations rebounded only after a time lag of approximately 10 years following the collapse of the Soviet Union (Bragina et al. 2015; Rozylowicz et al. 2010; Moura et al. 2013).

Land-use futures are associated with high uncertainty because land changes may or may not persist over long time period: the land system may revert to previous land-use patterns following a shock event, or it may be pushed into an alternative stable state. The volatile history of Eastern Europe makes the region a valuable case study for examining alternative stable states. The regions attractiveness for development restricts future land-use possibilities because built up areas are unlikely to be reverted to a previous use in the future. Conversely, abandoned fields encroached upon by shrubs may or may not revert to agricultural use (Gerard et al. 2010; Griffiths et al. 2013). Due to their agricultural suitability and fertile soils, the lowlands of the Carpathian region have the potential for agricultural intensification and increased food production (Foley et al. 2011), but once an area has undergone forest succession, the cost of reverting it to agricultural use is high, and the system can stabilize in a state of forest cover. Integration into the EU's common market may entail both agricultural intensification and the abandonment of traditionally farmed areas in this region (Elbakidze and Angelstam 2007), and such changes may be permanent or temporary, depending on future economic or policy incentives, such as those provided by the EU Common Agricultural Policy.

The magnitude of recent land changes is modulated—alongside economic, institutional, and demographic drivers—by centuries of human impact on natural ecosystems. The legacies of past land-use patterns shape recent changes and affect ecosystem structure and function, as well as the type, magnitude and timing of more recent land change processes. Multiple, repeated shifts in land management over a relatively short time span (the collapse of the Habsburg Monarchy and the Austro-Hungarian Empire, two world wars, the effects of socialism, and the transition to market economies and EU accession) have caused repeated changes in land management that have affected land-use practices and have generated several land-use legacies (Munteanu et al. 2015). The rates and magnitude of change are generally higher in areas with shorter land-use histories (e.g., post-Soviet agricultural abandonment is more likely in areas cleared for agriculture during the Soviet era than in areas cleared for agriculture during the Habsburg Empire). The legacies of Soviet industrial pollution affect the recent health of forests (Main-Knorn et al. 2009), and natural disturbances are rooted in past forest management decisions (Falťan et al. 2009).

Overall, the land-use dynamics in the Carpathian Region represent an excellent case study for investigating aspects of land change related to drivers of change, shocks, land-use legacies, and time-lags. In the following we outline the environmental and socio-political background of the Carpathian region and review the past two centuries of land change across the eight countries in the region, with a particular focus on the time period around the collapse of socialist regimes in Eastern Europe. We show that institutional shocks can severely affect land change, but that the magnitude of change differs among regions and that the response to abrupt policy changes can sometimes occur with a substantial time-lag. We show that land-use legacies can persist for decades or even centuries, affecting the current landscape composition, pattern and potential for change. Understanding land change patterns and the potential land-use conflicts arising from these changes are essential for land management in the region.

2 The Carpathian Region

2.1 The Environmental Setting

The Carpathians, often referred to as the ecological backbone of Central and Eastern Europe (Hazeu et al. 2010), are the largest and the most geographically, culturally, and ecologically diverse mountain range in the region. They span 1400 km and cover an area of over 160,000 km^2 (Kozak et al. 2013a). Extending between latitudes 43 and 50°N, the Carpathians are a young mountain range, formed during the Alpine orogenesis, and are composed of five subunits, each with unique geological and environmental characteristics: the Northwestern Carpathians (including the Tatra Mountains); the Northeastern Carpathians (between the Tisza and Cheremosh rivers); the Eastern Romanian Carpathians; the crystalline Southern Carpathians (the Transylvanian Alps); and the lower elevation Southwestern Carpathians (UNEP 2007) (Fig. 1). The mean elevation is 850 m, but the highest peaks reach over 2500 m above sea level in the northwest and south (Kozak et al. 2013a). Elevations of approximately 400 m are common in valleys. The temperate climate is characterized by precipitation levels between 400 mm in the southeast

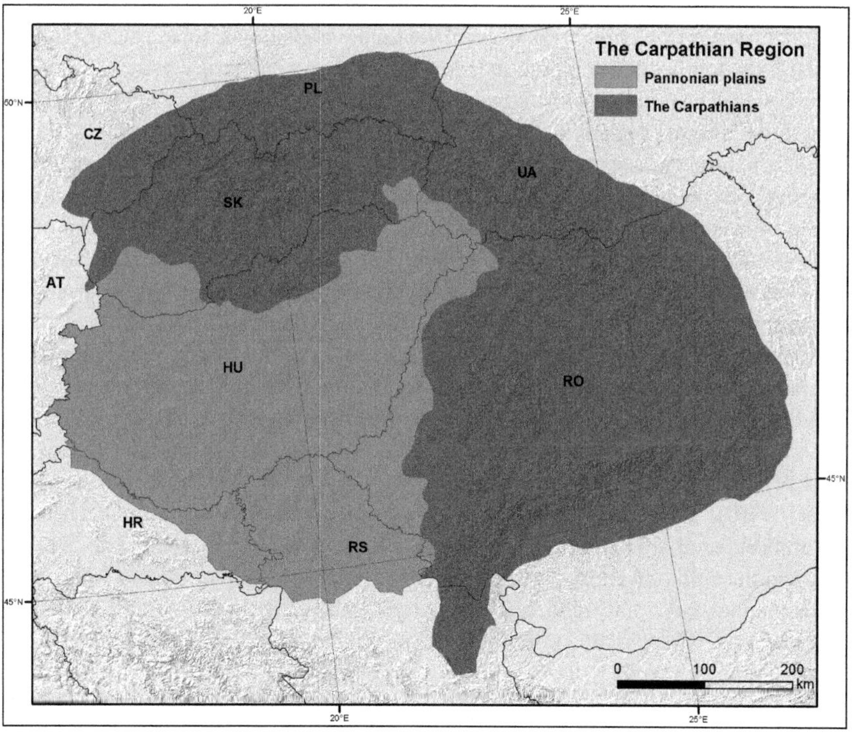

Fig. 1 The Carpathian region: *Graphic*: D. Kaim

and over 2000 mm in the high mountains. Mean annual temperatures range between −2 and 2 °C (UNEP 2007) in high mountains and between 4 and 6 °C at lower elevations. The geological and climatic diversity of the region results in a high level of environmental and biological diversity.

The Carpathians harbor some of the largest areas of continuous forests in Europe, with high conservation value and high biomass production (UNEP 2007; Keeton et al. 2013). Of these, approximately 200,000 ha are old-growth forests found only in Romania (Knorn et al. 2012b). Forests are important for carbon sequestration (Holeksa et al. 2009; Keeton et al. 2010), the provision of habitat and sustaining biodiversity (Knorn et al. 2012a), as well as for aesthetics and providing recreational amenities. Common landscape features consist of a highly diverse mosaic of forests and grasslands intermixed with wetlands along major river valleys. The Carpathian foothills are covered by mixed deciduous forests dominated by oak (Quercus sp.) and hornbeam (*Carpinus betulus*). At higher elevations, beech forests (*Fagus sylvatica*) mixed with silver fir (*Abies alba*) and Norway spruce (Picea abies) are common. Historically, some broadleaved forests have been replaced by Norway spruce (*Picea abies*) that has been introduced outside of its natural range for timber production since the early 20th century (Munteanu et al. 2008, 2016). However, at elevations over approximately 1100 m (with small regional differences), coniferous forests occur naturally, composed mostly of *Abies alba*, *Larix decidua* and *Pinus cembra*. Above the timber line (1400–1900 m), dwarf pine (*Pinus mugo*), and juniper (*Juniperus communis*) shrubs make up the transitional zone leading to alpine landscapes dominated by alpine meadows with high species diversity (Grodzińska et al. 2004). The relatively low elevation of the mountains does not allow glaciers to persist, even at the highest locations (Kozak et al. 2013b).

Semi-natural grasslands with high biodiversity (Bezák and Halada 2010; Akeroyd and Page 2011) and high mountain grasslands (in Ukraine, Poland and Slovakia, also called 'poloniny') are distinctive and typical landscapes of the Carpathians, and they harbor a high number of species that are threatened or endangered due to the overgrazing and shrub encroachment that has occurred over the past 60 years (Baur et al. 2007). Additional grassland communities of the Carpathians include dry and wet meadows, semi-natural mesophilic meadows, and extensively used pastures and fens, all of which have a high diversity and are rich in endemic plant and insect species (Bezák and Halada 2010; Kricsfalusy 2013). The regions forests and grasslands provide habitat for a significant portion of Europe's biodiversity. Large mammals, such as the brown bear (*Ursus arctos*) (Rozylowicz et al. 2010), European bison (*Bison bonasus*) (Perzanowski and Olech 2007; Kuemmerle et al. 2010), lynx (*Lynx lynx*), and wolf (*Canis lupus*), are abundant in this region, unlike in most parts of Western Europe (Kozak et al. 2013b).

Human activity has modified the natural landscapes of the region for over 2000 years. Characteristic cultural landscapes in the mountains consist of relatively small fields, scattered settlements, and large tracts of forests. In the lowlands, agriculture is practiced at a larger scale. Mining became an intensive activity in the medieval period and remains an important driver of land change, especially in central Slovakia (Bugár et al. 2010) and in the Western Romanian Carpathians

(Munteanu et al. 2014). Changes in forest structure and composition have resulted from intense natural resource use, especially for timber and pulpwood production. The intensive use of grasslands followed by their recent abandonment has altered their species diversity and composition (Halada et al. 2008, 2010; Kricsfalusy 2013). Urban cover has expanded, especially around larger historic cities. The past and recent land-use dynamics affect more recent aspects of landscape structure, composition, ecosystem functioning, and species diversity (Turnock 2002), posing great challenges to land management in this region.

2.2 *Geo-political and Socio-economic Context*

The current geo-political setting of Eastern Europe is relatively young. The study region includes a small part of Austria and the Czech Republic, most of Slovakia, southern Poland, Hungary, western Ukraine, most of Romania and a small part of northern Serbia. Administrative boundaries in the area have shifted multiple times during the past two centuries, causing repeated landscape changes. Most of the major land changes in the Carpathians were marked by changes in institutional systems and administrative boundaries (Bideleux and Jeffries 1998). Throughout the 19th century and up to the beginning of the 20th century, most of the Carpathian provinces were under the control of the Habsburg Empire and the Austro-Hungarian Monarchy (Munteanu et al. 2014), with the main mountain ridge in Romania constituting the border with the Ottoman Empire. After World War I, the multi-national Austro-Hungarian Empire broke up, and several national states emerged (Seton-Watson 1945). However, the boundaries of interwar Eastern Europe differed significantly from the present boundaries. Poland and Czechoslovakia included the current Ukrainian Carpathians, while the southern part of the Carpathians was part of the Kingdom of Yugoslavia. The regions of Bukovina, Bessarabia, Moravia, Galicia, and Transylvania were annexed to different states (Seton-Watson 1945; Bideleux and Jeffries 1998) (Fig. 2). With the redefinition of state boundaries, the core-periphery relationships changed, breaking old links between settlements and transportation systems and establishing new ones (Kozak et al. 2013b). The legacies of these repeated political changes still manifest themselves in the current land cover and patterns of land cover change in the region.

Following World War II, most of the Carpathian countries (except Austria and Yugoslavia) fell under the influence of the Soviet Union and established socialist regimes. During this socialist period, land use policies and management became relatively homogenous across the Eastern Block—with exception of Poland and Serbia, where the collectivization of agriculture was not broadly applied. The Carpathian countries became part of the Council for Mutual Economic Assistance (COMECON), and joint enterprises with Soviet partners targeted the exploitation of natural resources (Banu 2004; Kligman and Verdery 2011). Land nationalization and the collectivization of agriculture led to some intensification of land use and in a few instances its expansion, along with logging, in Romania, Hungary, Ukraine,

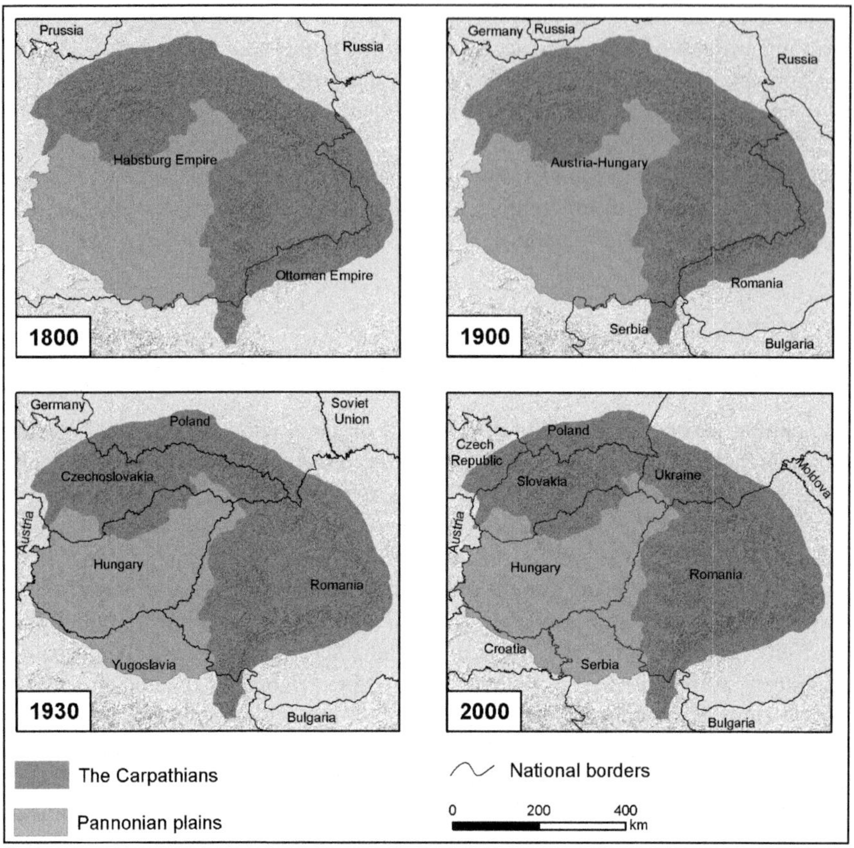

Fig. 2 Historic geo-political boundaries in the Carpathian Region during the Habsburg Empire (1800) Austro-Hungarian Monarchy (1900), interwar period (1930) and post-WWII (2000). *Graphic*: D. Kaim, *Data source*: euratlas.com (1800, 1900)

Slovakia, and the Czech Republic. In the mountainous regions of Eastern Poland, forest expansion was the dominant trend after World War II (Kozak 2010). Changes in Ukraine were particularly drastic because the country was part of the Soviet Union, and it became a main source of timber and agricultural products during Soviet times (Brain 2011).

The transition to market economies started between 1989 and 1991 in the Carpathian countries, when the Iron Curtain lifted and the Soviet Union collapsed. This political shift had major implications for the economies, demographics, and institutions of these countries and ultimately affected land change and conservation. Land reforms occurred in the form of restitution or redistribution of land (Lerman et al. 2004; Hartvigsen 2014), and markets opened up for trade with the Western World. The timing and speed of the transition to market economies varied among the countries, a process partly reflected in the timing of each country's accession into the European Union. The Czech Republic, Slovakia, Hungary and Poland joined the EU in 2004, while Romania joined three years later, and Serbia and Ukraine remain non-members. However, the Carpathian countries cooperate in regard to nature conservation and sustainable development under the umbrella of the Carpathian Convention, which was established in 2003.

The EU accession of most of the Carpathian countries represents the most recent shift in economics and institutions that have affected land change in this region. EU accession impacted both resource management and nature conservation. Regulations such as the Common Agricultural Policy (CAP) aim to build common markets for agricultural products, remove tariffs, and support environment-friendly agricultural land management. However, the implementation of the CAP and the levels of subsidies differ substantially among countries, affecting in turn the recent rates and extent of change (Young et al. 2007; Knowles 2011). Furthermore, several other EU policies, such as the Water Framework Directive and the Natura 2000 Network, affect the future of land change in the region by encouraging or restricting specific uses, such as wetland restoration and habitat conservation (Kallis 2001; Donald et al. 2002; Maes et al. 2012).

2.3 Demographic Changes

Along with institutional and economic forces, socio-demographic processes are important underlying drivers of land change in Eastern Europe (Haase et al. 2007). Across the eight countries, the region was home to approximately 17 million people at the turn of the 21st century (CERI 2001), who mostly lived in rural settlements. A few large cities at the fringe of the Carpathian Mountains experienced substantial population increases during the socialist period, and urban sprawl has increased since 1990. The highest population densities occur in the Czech Republic and the Polish Carpathians (>175 inhabitants/km^2), while the Romanian Carpathians are the least populated (<100 inhabitants/km^2). Supporting and encouraging population growth was one explicit goal during the communist regime, for example, in Romania (Schreiber 2003). The post-war trend of rural depopulation and the marginalization of rural areas led to increases in the populations of urban centers at the edges of the Carpathians. Population aging and a growing inequality between rural and urban areas has become a major concern in the post-Soviet Carpathians (UNEP 2007). The population growth rate since the collapse of the Soviet Union has oscillated between −1 and +0.5 %, with the highest rates since 2005 recorded for the Slovak Republic. In absolute numbers, the population of all Carpathian countries has declined slowly since 1990, except for those of Austria, Poland, and the Slovak Republic (Fig. 3). Due to the opening of borders after 1990, migration to Western Europe and to large cities outside the Carpathians increased substantially. Shifts in employment from the agricultural sector to the service sector continue to foster agricultural abandonment (Schreiber 2003; Kozak 2003). However, the region has recently become more attractive for recreation and tourism. The

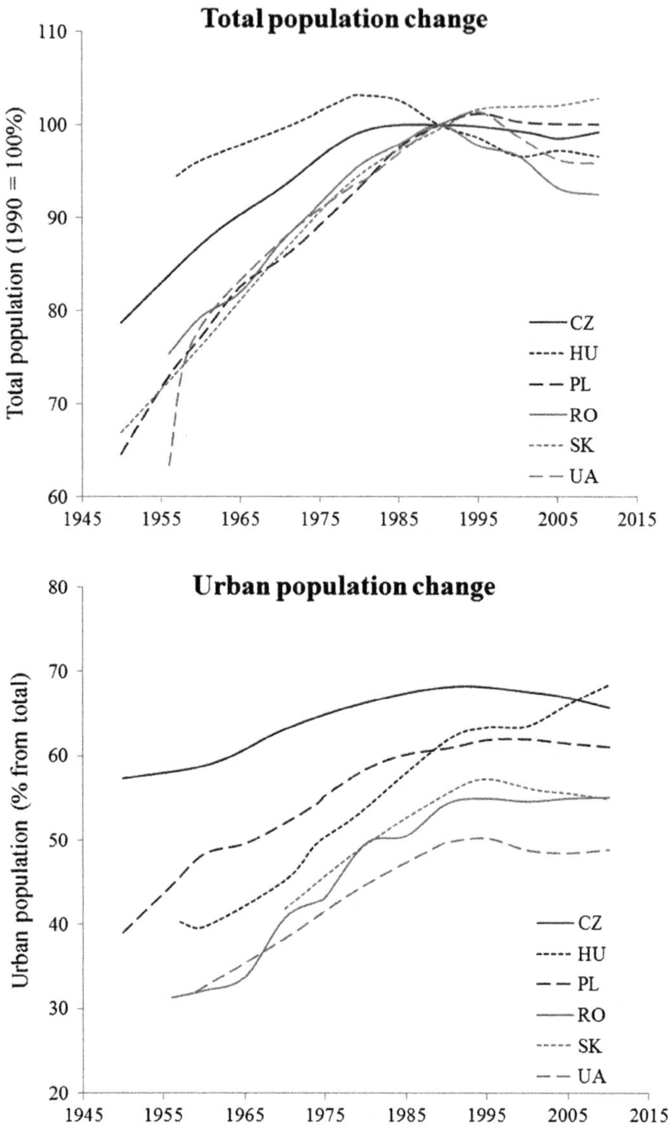

Fig. 3 Population dynamics and changes in the percent of the population in urban areas in the Carpathian countries since 1945. *Country codes CZ* Czech Republic, *HU* Hungary, *PL* Poland, *RO* Romania, *SK* Slovakia, *UA* Ukraine. *CZ and UA* Carpathian district level data (*CZ* Jihomoravský, Moravskoslezský, Olomoucký, Zlínský; *UA* Zakarpatska, Lvivs'ka, Ivano-Frankivska, Chernivets'ka oblasts). HU, PL, SK, RO country level data. *Data source* http://pop-stat.mashke.org/; http://ukrcensus.gov.ua; http://demoscope.ru/; UN demographic yearbooks. *Graphic*: O. Shandra

development of second homes in natural settings (Mika 2013) and employment in the service sector (Süli-Zakar 1998) are recent processes that lead to population increases in both rural and urban areas.

2.4 *Land Cover and Land-Use Changes*

Both the rates and extent of land change in the Carpathians have increased substantially over the past three decades. Most of these changes are due to socio-economic and political shocks related to the Soviet Union's collapse (Kuemmerle et al. 2008, 2011; Baumann et al. 2011). However, many changes are rooted in the regions long land-use history (Kozak 2010), and those land-use legacies may persist for centuries into the future (Bellemare et al. 2002; Foster et al. 2003; Munteanu et al. 2015). Over long time periods, several patterns have emerged for the Carpathians. Forest cover has increased over the past century, with a forest transition—the shift from decreasing to increasing forest area—occurring during the interwar period (Kozak et al. 2004; Kuemmerle et al. 2011; Munteanu et al. 2014). Agricultural land has experienced the opposite process, i.e., agriculture generally expanded up to the early 20th century, followed by the abandonment of marginal areas starting during the socialist period and accompanied by the intensification of existing cropland (Munteanu et al. 2014). Agricultural land abandonment and forest succession accelerated after the collapse of the Soviet Union in most of the Carpathians (Griffiths et al. 2013, 2014). During Soviet times, natural grasslands were converted to arable land (Feranec et al. 2000), especially in lowland and moderately hilly areas. Grazing pressure generally increased in the high mountain grasslands up to the 1990s, but declined thereafter (Sitko and Troll 2008; Shandra et al. 2013). After 1990, many grasslands were abandoned and reforested. However, following EU accession and the nature conservation efforts that accompanied it, extensively managed grasslands (pastures and hayfields) are re-appearing in Carpathian landscapes. Other prominent land changes in the Carpathians include a substantial loss of wetlands and an increase in urban sprawl (Ronnås 1982; Konkoly-Gyuró et al. 2011; Huzui et al. 2012). Wetland drainage peaked during the socialist period, but restoration efforts are being made thanks to EU regulations and incentives (Günther-Diringer 2000; Horváth et al. 2012). Increasing urban sprawl was common for many decades but accelerated substantially after 1990 (Ronnås 1982; Huzui et al. 2012). The regions attractiveness for tourism and second-home development led to the abandonment of fields in the proximity of urban settlements, which have been permanently taken out of agricultural production (Mika 2013).

While the overall trends in land change are strong, local and temporal variation occurs across the Carpathians (Munteanu et al. 2014). For example, in Habsburg times, forest cover decreased in the Romanian, Ukrainian, and Slovakian Carpathians, while it increased in the Polish Carpathians and was stable in the Czech Republic (Munteanu et al. 2014). Following the collapse of socialism, most countries experienced elevated rates of forest disturbance; however, while disturbances peaked right after 1990 in Ukraine, in Romania, the highest disturbance rates occurred only after 1995 (Griffiths et al. 2012, 2014). Low but positive annual rates of agricultural expansion occurred in Romania and southeast of Hungary during the socialist period, while in Poland, agricultural land decreased by up to 5 % per year (Woś 2005). Overall, the hotspots of land change (i.e., areas that experienced the most change in land cover) occur at the interface of land covers so that the same land change driver is mirrored in multiple processes (e.g., agricultural abandonment/grassland conversion is related to forest cover increase, wetland loss is related to agricultural expansion).

In the following section, we focus on the most important types of land cover in the Carpathians, highlighting their specific characteristics, and discuss land-use history, prominent land cover changes prior to and after the collapse of the Soviet Union, and the drivers of change. The same change processes often affect multiple land cover types, and therefore, they are discussed in the respective sections. While the main focus is on post-socialist land change, the broader historical context is important because past land changes modulate the extent and magnitude of post-Soviet processes. Using these examples, we demonstrate how land change processes exhibit legacies, time lags, alternative stable states, and we highlight some of their effects on biodiversity.

2.5 Forests

Forests represent the dominant land cover type of the Carpathians, making up approximately a quarter of the land cover of the region, with higher values for the mountainous areas (Kozak et al. 2013b; Griffiths et al. 2014), and less forest cover in adjacent lowlands, and they have a long history of human management. As a dynamic system over the past century, forests in this region have experienced multiple changes in terms of total area (Fig. 4), heavy logging related to shifts in composition, natural disturbances and succession over abandoned agricultural areas. The substantial changes in forest cover were driven by socio-economic and institutional shocks, of which the collapse of the Soviet Union was the most complex in terms of interactions between drivers and in terms of the magnitude of the effects on forest management and thus on forest structure and composition (Griffiths et al. 2014; Munteanu et al. 2015).

Over the last 200 years, the overall long-term trend has been an increase of forested areas. On average, forest cover has increased over the entire Carpathian region since the Soviet Union's collapse, with a net increase of 153,000–157,000 km^2 (Griffiths et al. 2014). Romania has experienced an increase of up to 7 % in forest cover since the mid-1980s, largely due to successional encroachment of deciduous species onto abandoned land, while mixed and coniferous forests substantially declined between 1985 and 2010, as was the case in most other countries in the Carpathians (Griffiths et al. 2014). Forest cover has become more continuous, but the habitat structure and the ecosystem services have also changed

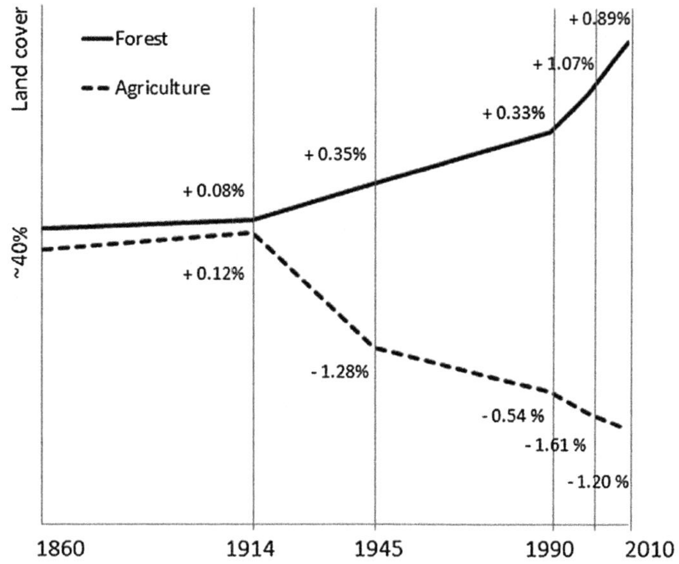

Fig. 4 Conceptual reconstruction of the long term forest and agricultural land cover dynamics in the Carpathians. Mean annual rates of land change based on case studies from Munteanu et al. (2014)

substantially. The smallest increases in forest cover have occurred in the Czech Republic and Austria, where despite similar trends in compositional shifts, forest cover has only increased by approximately 1 % over the past 25 years (Fig. 5). The increase in forest cover began much earlier however, with forest transitions occurring in the interwar period across the Carpathian region (Kozak et al. 2004; Kuemmerle et al. 2011; Munteanu et al. 2014).

A large proportion of the recent increase in forest cover in the Carpathians is due to agricultural abandonment, both in marginal lands (Müller et al. 2009) and on large agricultural fields that were previously owned by state farms (Kozak 2010; Baumann et al. 2011). Additionally, some reforestation of previously degraded forests has contributed to the overall increase in forest cover. During the 20th century, with the decline of transhumance, considerable forest cover increases occurred at the timberline, which was mostly composed of coniferous forest (Shandra et al. 2013). In Soviet times, despite heavy logging, forest cover losses were somewhat compensated for by the planting of trees outside of forest ranges, for example in Hungary (Munteanu et al. 2014). Afforestation also played an important role in marginal areas, such as the border area of Poland, Slovakia and Ukraine (Kozak 2010). Furthermore, local conditions, such as the special case of forced depopulation following WWII in the Biesczady Mountains of Eastern Poland, also generated a regional hotspot of forest increase in Poland, leading to extensive forest succession on abandoned land across an area including over 50 % of the total landscape (Wolski 2001; Warcholik 2005; Woś 2005).

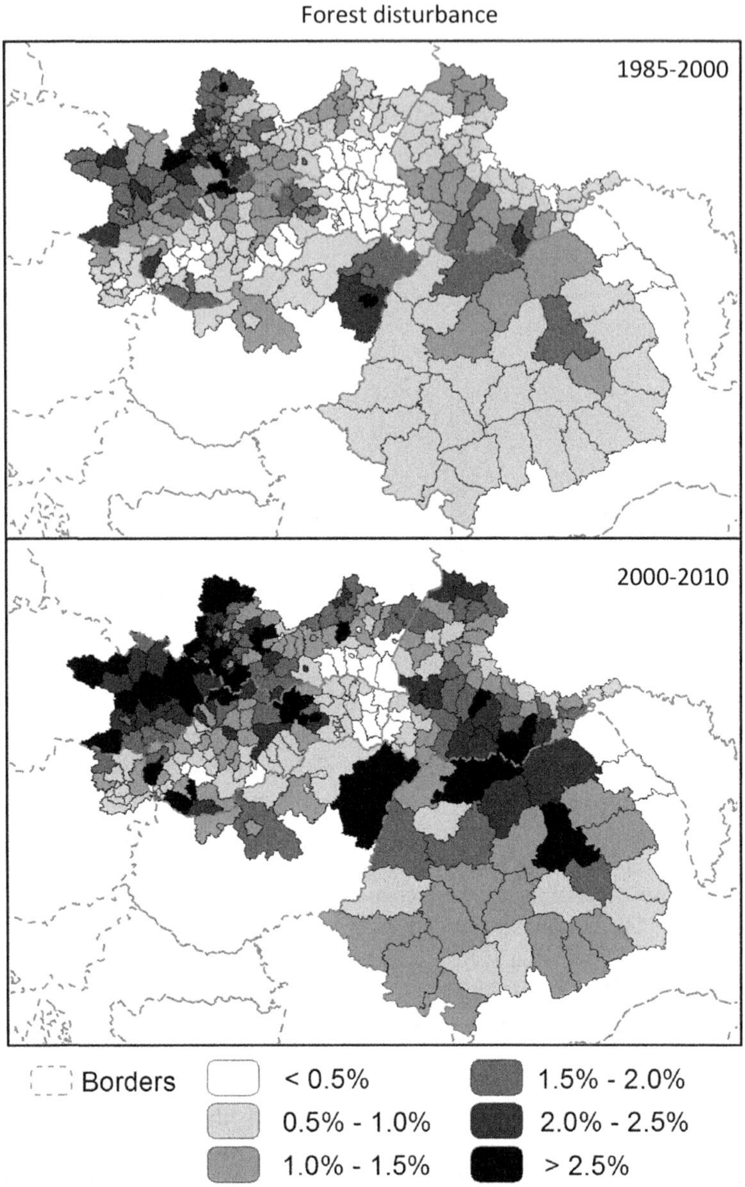

Fig. 5 Annual forest disturbance rates for the periods 1985–2000 and 2000–2010. Disturbance rates consider all stand-replacing disturbances at a 30 m pixel level. Country boundaries are shown in *dashed grey lines*. Disturbance rates are provided at the level of administrative units (district level for Czech Republic, Poland, Slovakia and Ukraine, and county level for Austria, Hungary and Romania). Scale of the maps is 1:8,750,000. Modified from Griffiths et al. (2013)

Despite an overall increase in forest cover, forest management and logging activities affect the structure and health of the Carpathian forests. Forestry as an economic activity developed mostly during the times of the Habsburg Empire and Austro-Hungarian Monarchy due to the high timber demands in the Empire. Mining, a prominent activity since the 13th century, intensified during this period, further raising timber demands and increasing the extent of deforestation (Turnock 2002). Coal and metal mining caused especially widespread deforestation in the Ostrava region of Moravia, the Romanian Banat, Apuseni, and in northern Hungary (Turnock 2002; UNEP 2007). Forest clearing for agriculture and for pastures, especially above the timberline, transformed the Carpathian landscapes in the Middle Ages and up to the 19th century.

Forest management for sawtimber and pulp production increased dramatically during Soviet times (Banu 2004). Large areas were clear-cut and restocked with monocultures. Such intense forest management caused forests to progressively become younger and less dense (Turnock 2002). Stand-replacement disturbances affected forest structure and composition, and were widespread across the Carpathians (Griffiths et al. 2014). In some areas (e.g., Ukraine) forest harvesting increased at surprising rates, especially after the collapse of the Soviet Union. In most cases, disturbances occurred in the decade between 1985 and 1995. Thereafter, disturbance rates dropped but then picked up again after 2005, mostly in Poland, the Czech Republic (due to a combination of natural processes (Main-Knorn et al. 2009)), Slovakia and Romania (mostly driven by changes in institutions and ownership (Griffiths et al. 2012)). However, in Romania and Ukraine, disturbances only peaked 10 years after the institutional shift. This time lag effect may be related to the speed of the restitution process and the speed at which countries transitioned to market economies (Bideleux and Jeffries 1998; Hartvigsen 2014). After the collapse of the Soviet Union and land restitution or redistribution, forest became an immediate source of revenue in many countries (Irimie and Essmann 2009; Mantescu and Vasile 2009), leading to both legal and illegal logging (Kuemmerle et al. 2009; Knorn et al. 2012a). The opening of timber markets, increased exports, shifts in forest ownership and poor regulatory frameworks became the main drivers of forest harvesting for timber production (Ioras and Abrudan 2006; Irimie and Essmann 2009). In Romania and Ukraine, the transition to market economies was slower, and drivers started having effects later than in the Western Carpathian countries. Overall, across the region, the intense use of forests caused forest loss within those areas and resulted in an overall shift to younger forests with shorter rotation cycles (Nijnik and van Kooten 2000; Griffiths et al. 2014).

Intensive and sometimes poorly regulated forest management was, over time, accompanied by substantial shifts in forest composition and patterns. During Habsburg and Soviet times, a shift in forest composition occurred towards a higher proportion of coniferous species (Kozak et al. 2007; Wiezik et al. 2007; Munteanu et al. 2016) due to timber demands and the need to restock with fast-growing species. Since 1985, the trend has reversed, with deciduous forests increasing by

9.0 % in the Carpathians, while mixed forest have decreased by 2.3 % and the coniferous forests by 7.0 % (Fig. 5).

The increase of monocultures for timber and pulp production from the 19th and 20th century had a cascading effect, increasing the occurrence natural disturbances, such as insect attacks or wind throws. Due to the vulnerability of spruce plantations to snow, wind throws, and pest outbreaks, the occurrence and extent of disturbances have increased over the past decades (Main-Knorn et al. 2009). In 2004 in the High Tatra Mountains, approximately 12,000 ha of forest were disturbed by a strong windfall (Falťan et al. 2009). In 1994, 9000 ha of mostly spruce monoculture was disturbed by wind in South Eastern Transylvania (Turnock 2002). Such events have led to increased forest management awareness. Forest management is currently shifting away from managing monocultures for pulpwood and timber production (Keeton and Crow 2009) and toward mixed forest management. Forest restoration and forest protection plans are being implemented in many areas (Turnock 2002; Keeton et al. 2013; Macicuca and Diaconescu 2013). As a result, forest recovery from disturbance has been more substantial in areas that were disturbed after 1990 than in those harvested prior to 1990. However, the effectiveness of protective and restorative endeavors remains questionable in many Carpathian countries (Knorn et al. 2012a).

Forest changes in the early 19th century were mostly driven by proximate causes, such as infrastructural development and timber harvesting. As the Soviet Union expanded its sphere of influence into Eastern Europe, centralized policies and economies became the most important underlying drivers of forest change (Munteanu et al. 2014). Infrastructure and tourism development were proximate causes of deforestation in Romania (Huzui et al. 2012) and the Tatra Mountains (Gerard et al. 2010). Population displacement, on the other hand, drove reforestation in the Polish-Slovakian-Ukrainian border region (Woś 2005; Kozak 2010). With the collapse of the Soviet Union, a multitude of proximate and underlying drivers interacted at various levels across the Carpathian countries. As countries developed their own policies, redefined property rights, and accessed new markets, a complex suite of underlying drivers become increasingly important for determining patterns of forest change. For example, loopholes in laws related to the use of forests and illegal harvesting caused severe disturbance patterns in Romania and Ukraine (Irland and Kremenetska 2009; Kuemmerle et al. 2009; Knorn et al. 2012a), affecting valuable ecosystems such as old-growth forest (Knorn et al. 2012b). Increasing migration to western Europe reduced the pressure on the land and allowed for forest succession to occur on abandoned grasslands and agricultural areas (Kozak 2003; Munteanu et al. 2008; Smaliychuk 2012). The extent and timing of disturbance patterns differed among countries depending on the time that drivers became active, causing time lags of up to 10 years following the institutional shift to occur in Romania (Griffiths et al. 2012, 2014). Overall, Carpathian forest cover was severely affected by the socio-economic shock caused by the collapse of the Soviet Union. The general trend was a slight overall increase in forest cover accompanied by substantial shifts in forest composition and structure (Shandra et al. 2013; Griffiths et al. 2014). Recent human-induced disturbances are currently affecting ecosystem health, mostly due to selective logging and shorter rotation cycles (Nijnik and van Kooten 2000). Natural disturbances can be exacerbated by past forest management practices for wood production (e.g., spruce monocultures), but their environmental effects sometimes only became clear decades later (Main-Knorn et al. 2009). All of the changes discussed here were driven by a complex web of proximate causes and underlying forces, whose intensity after the collapse of the Soviet Union increased substantially but at different rates in the various Carpathian countries (Fig. 5).

2.6 Agriculture

Following forestry, agriculture is the second important economic sector influencing the land cover of the Carpathian region and the one that experienced most change over the past three decades. Agricultural land covers approximately a third of the Carpathian Region (Ruffini 2008) with land use patterns ranging from large-scale, intensively used fields in lowlands to small-scale subsistence farming, predominantly in marginal mountain areas. Agricultural land uses include arable land, managed meadows and pastures but our focus here is mostly on arable use, although the reviewed change processes may include land dynamics at the interface of arable land and managed meadows and pastures.

Farm size varies with location and time-period: small-scale subsistence farms in the mountainous areas are typically 1–5 ha in size, while large-scale intensive farms established following collectivization reached over 1000 ha in size, mostly in the lowlands (Jepsen et al. 2013). The main crops in the region are cereals (wheat, corn, barley, rye, and oats) and legumes (potatoes, sugar beets, and peas). In recent years, bioenergy crops such as rapeseed have gained in importance (Griffiths et al. 2013). Fruit orchards, hop fields, and vineyards constitute the majority of the perennial crops, and they are particularly common in the Carpathian foothills and in Slovakia (Špulerová et al. 2011). The dynamics of field size, type of crops, expansion into other land cover types and abandonment have changed over the past two centuries. The overall trends in agricultural land change since the 18th century consist of expansion up to WWI, intensification during Soviet times, widespread abandonment after the collapse of the Soviet Union and recultivation in the past decade.

Traditional agricultural practices, prior to industrialization, included small-scale agriculture based on two-field rotation cropping (Irland and Kremenetska 2009) and livestock farming, including transhumant sheep farming, especially in Romania (Turnock 2002). During the Habsburg and Austro-Hungarian Empires, arable land was expanded into grasslands and wetlands at a rate of approximately 0.1 % annually (Munteanu et al. 2014). In the 19th century and the beginning of the 20th century, Ukrainian smallholders cleared forest patches for agricultural use (Vepryk 2001). Forest clearing for agriculture was also common in Hungary (Konkoly-Gyuró et al. 2011). Agricultural expansion in the Carpathians continued throughout the interwar period, but following the World War II and extensive land reforms, the process gave way to intensification and land abandonment.

The intensification of agriculture was a typical process from the 1950s to the 1970s for the whole of Europe (Young et al. 2007), partly due to efforts to bolster food security after the food shortages caused by WWII. The establishment of socialist regimes after WWII across most of the Carpathian countries represented a major shock for cropland dynamics in the Carpathians. Agricultural land was nationalized in Ukraine and collectivized in the countries outside of the Soviet Union. Agricultural production became highly subsidized, and policies supported the emergence of large collective, state-owned farms at the expense of small family farms (Turnock 1996). The change in field size was particularly striking: the former mosaic of small and narrow patches of land was replaced by large areas of arable land intermixed with intensive grasslands. Large, homogenous, uniformly utilized patches drastically changed the formerly diverse sub-montane landscape (Halada et al. 2008). Agricultural mechanization and the use of mineral fertilizers increased rapidly. Political goals included an increase in agricultural production and caused the expansion of arable land in the Czech Republic (Demek et al. 2008; Skokanová et al. 2009) and Romania (Schreiber 2003). Environmental constrains (e.g., topography and wetlands) were not regarded as a limitation to agricultural use during Soviet times, which often made the allocation of production factors in socialist agricultural systems inefficient (Müller et al. 2013).

Despite the high level of intensification of land use, agricultural abandonment occurred in the Carpathian Mountains soon after WWII (Munteanu et al. 2014). Rural populations increasingly found employment in the industrial sector (Schreiber 2003; Munteanu et al. 2014), abandoning small-scale mountain agriculture. Marginal areas were abandoned in Slovakia (Gerard et al. 2010), while in Poland, private farms persisted during communism, leading to a higher persistence of mosaic landscapes that included small agricultural fields, forest and some tourist-related development (Turnock 2002). The peak of agricultural land abandonment in the Carpathians was reached only after the collapse of the Soviet Union, a trend which is consistent with other post-socialist European countries (Müller and Munroe 2008; Prishchepov et al. 2013).

Most Eastern European countries enacted comprehensive land reforms after the collapse of the socialism, and agricultural markets were liberalized. However, the implementation of the land reforms and agricultural policies and the availability of subsidies varied substantially across countries, which contributed to different patterns of abandonment in the different post-socialist countries (Lerman et al. 2004; Rozelle and Swinnen 2004; Müller et al. 2013). In the Carpathians, approximately 40,000 km^2 of cropland had been abandoned by 2000, and an additional 7100 km^2 by 2010 (Griffiths et al. 2013)—mostly due to cropland-to-grassland conversion (Fig. 6). The highest abandonment rates occurred in Romania and Ukraine, predominantly in marginal areas (45.8 and 58.9 %, respectively, abandoned by 2000). Regional differences depended on the countries' post-Soviet political, institutional and economic situation, as well as on the individual restitution methods (Hartvigsen 2014), resulting in, for example, Slovakia experiencing less abandonment (13.1 %) than Ukraine (58.9 %) (Fig. 6). The highest abandonment rates were associated

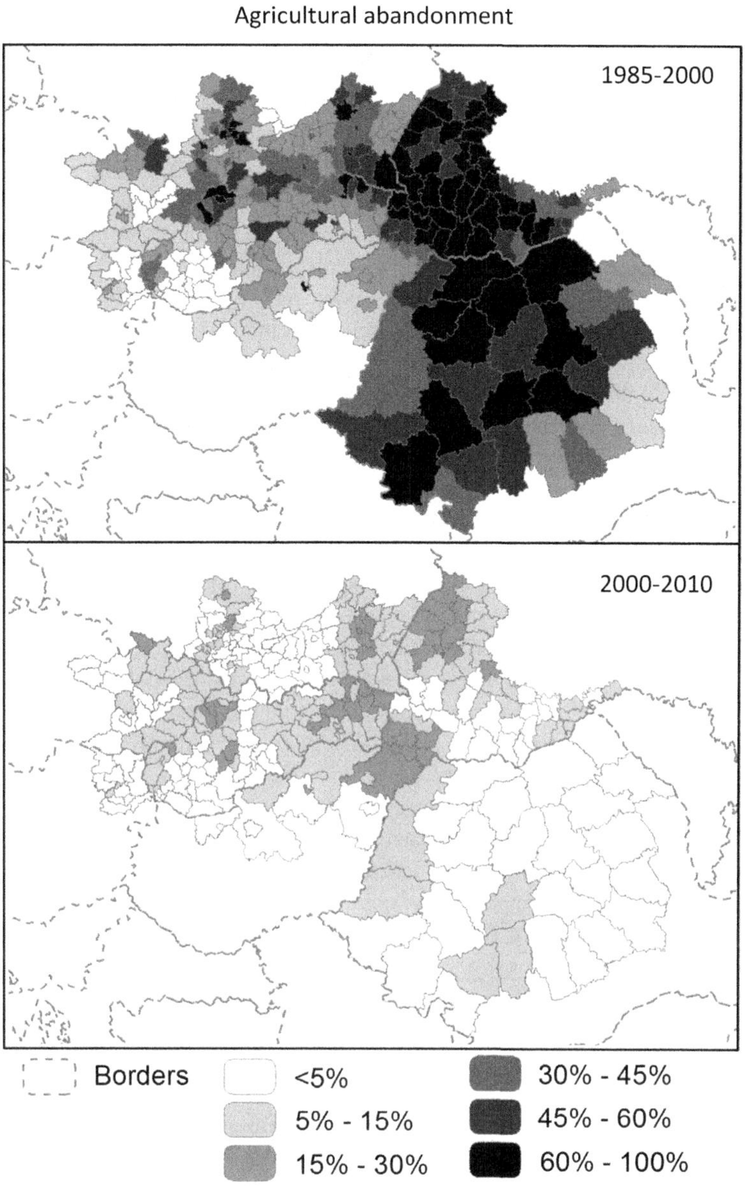

Fig. 6 Agricultural land abandonment for the periods 1985–2000 and 2000–2010. Abandonment is calculated relative to the 1985 and 2000 cropland area per unit. Country boundaries are shown in *dashed grey lines*. Abandonment is calculated on the level of administrative units (district level for Czech Republic, Poland, Slovakia and Ukraine, county level for Austria, Hungary and Romania). Scale of the maps is 1:8,750,000. Modified from Griffiths et al. (2013)

with areas of intermediate and low agricultural suitability (Müller et al. 2013; Griffiths et al. 2013).

Abandonment rates decreased around 2000, and the recultivation of formerly abandoned cropland occurred in agriculturally suitable areas. Of the land that had been abandoned by 2000, 18 % was brought back into production by 2010 (Griffiths et al. 2013). Most of the recultivation occurred in Romania and Hungary, while abandonment and forest succession persisted in Ukraine, parts of Poland and Hungary. With the opening of agricultural markets and access to land, abandoned agricultural land in post-socialist countries represents an attractive target for foreign investors (Deininger 2011; Visser and Spoor 2011). Additionally, the EU CAP includes income subsidies, supporting farmers to increase production efficiency in order to be competitive on the world market (Zanten et al. 2013) and to help manage land in an environmentally friendly way. Although access to subsidies for agricultural production may lead to increased cropping and the re-establishment of large-scale agricultural operations in the lowlands, the access of local, small landowners to CAP subsidies, especially in mountainous areas, remains limited (Bezák and Mitchley 2014). However, the effectiveness of such programs is questionable due to different rates at which farmers apply for subsidies, differences in farmers` attitudes, and a lack of landscape-scale coordination (Zanten et al. 2013) potentially causing issues for the conservation of biodiversity and for the maintenance of traditional agricultural practices (Špulerová 2013).

The reform of institutions, land ownership and agricultural policies following the collapse of the Soviet Union, combined with socio-demographic processes caused by the opening of borders, were major drivers for agricultural land change. The strength of the institutions and the timing of reforms significantly modulated the intensity of agricultural change (Turnock 1996; Hartvigsen 2014) and affected the rates of post-Soviet land abandonment and recultivation. During the Soviet era, institutional and economic factors such as the forced industrialization and intensified production of food supported by state subsidies caused an expansion of agricultural land in the lowlands (Turnock 1996). Following the collapse of the system, the lack of agricultural subsidies, decreased profitability, and the bankruptcy of large agricultural enterprises caused widespread abandonment (Turnock 1996; Lieskovský et al. 2013; Hartvigsen 2014). Providers of agricultural services, such as machinery owners, shifted their activities to the more productive lowlands, further contributing to the abandonment of the marginal mountainous areas (Müller et al. 2013). The underlying drivers of abandonment were related to the land restitution process, which did not take into account the landowners' agricultural activities (Kuemmerle et al. 2008). Migration to Western Europe caused a decrease in employment in agriculture the Carpathian region (Hartvigsen 2014). As the remaining rural population aged, less land was used for agricultural production. Following the bankruptcy of large industrial operations, urban populations moved back to the countryside, but few returned to farming. Many agricultural parcels were taken out of production to the benefit of urban sprawl.

Land abandonment was a short-term process in many areas, which did not necessarily push the land system into a new stable state. For example, areas in western Romania were abandoned following the collapse of the Soviet Union due to the convoluted restitution process, only to be brought back into production a decade later (Fig. 6). While the land-use future of some abandoned areas is still uncertain, recultivation in others is likely in the future due to the fertility of their soils and their suitability for intensified food production (Foley et al. 2011). However, marginal lands with low agricultural suitability may transition to an alternative state of continuous forest cover, such as the Bieszczady mountains in Eastern Poland (Wolski 2001; Woś 2005). The length of time that areas were used for agriculture before abandonment may influence the speed of abandonment and cause forests to be established faster. The implications of these changes are manifold in terms of habitat provision for key species. Many large mammals, such as European bison (Bison bonasus), benefit from increasing forest cover (Perzanowski and Olech 2007), while forest edge species or generalist species, such as brown bears (*Ursus arctos*) (Gula et al. 1995), generalist birds (Angelstam 1992), and insect species (Magura 2002), may prefer a more fragmented landscape pattern. The future of small-scale farms, extensive agricultural practices and land management for agro-biodiversity is still uncertain in this highly diverse area, and a uniform spatiotemporal land-use future (Aldwaik and Pontius 2012) for the entire Carpathian region seems unlikely.

2.7 Grasslands, Pastures and Hayfields

The grasslands of the Carpathians represent one of the most vulnerable components of the overall land system due to their biological diversity—and the one where changes in species richness and diversity have varied substantially over time. About one third of the Carpathian region is made up of semi-natural habitats, most of which are grasslands (Turnock 2002) whose dynamics are closely related to the other land cover types. In the lowlands, steppe-like grasslands occur predominantly in Hungary, Transylvania, and Western Romania, which are largely remnants of primary steppes or forest steppes (Biró et al. 2012). These grasslands have a high diversity of plants and invertebrates, many of which are endemic, and provide refuge for numerous threatened open-land species (Cremene et al. 2005). Such lowland grasslands were mostly altered due to the intensification of agriculture after the WWII, post-Soviet conversion to urban and agricultural landscapes (Biró et al. 2012), and decreased grazing (Cremene et al. 2005) leading to abandonment and shrub encroachment. In hilly and mountainous areas, meadows of woodland origin, also called 'poloniny' (from the old Slavic 'polonina'), were established in the Middle Ages through deforestation and development for pasture use (Pietrzak 1998; Turnock 2002). Mountain meadows have a remarkably high species richness and contain many endemic species and medicinal plants (Halada et al. 2010). However, the intensified use of these grasslands, with increased human pressure, overgrazing and abandonment, simplified the local agro-biodiversity of their flora and fauna (UNEP 2007). Mountain meadows and grasslands experienced extensive abandonment due to decreasing livestock numbers (Bezák and Petrovic 2006), but in some regions of the Carpathians, such as Poland, grasslands have been brought back under management following agro-environmental EU policies. Alpine grasslands occur above the timber line, and despite their high species diversity and richness and the amount of endemism, these grasslands experienced conversion to woodlands as the elevation of timberlines tends to increase following a decrease in grazing and mowing.

Grasslands in the Carpathians have a very dynamic history, with a succession of multiple uses over time. They went through periods of intensification, expansion and conversion, all impacting the species diversity and richness of the ecosystem. In mountainous areas, deforestation for pasture and hayfields was common starting in the 14th century and up to the first half of the 20th century when grasslands were used for hay-making and the grazing of domestic animals (Rabbinge and van Diepen 2000). Transhumant sheep grazing is still practiced today as a traditional use of mountain pastures, especially in Romania. Following WWII, not all land was nationalized, and subsistence livestock farming was still practiced in mountainous regions (Turnock 2002). The increased demand for meat and dairy products together with Soviet policies of land-use intensification in the fertile lowlands led to increasing pressures on natural mountain grasslands due to the shift in grazing of privately owned livestock to the higher elevations. Natural pastures were affected by intensive grazing in northern Romania (Munteanu et al. 2008) and in the Southern Romanian Carpathians even after 1990, degrading local biodiversity (Baur et al. 2007). During the socialist period, grasslands in the lowlands were either hayed several times in a year or grazed by large herds of cattle and sheep. As a result of nationalization and collectivization, intensive livestock farming units were developed in the Carpathian lowlands, increasing the pressure on the semi-natural grasslands (Turnock 2002). Overall, grasslands suffered the most degradation through overgrazing under the Communist regimes (Turnock 2002).

Since the collapse of the Soviet Union, grasslands have expanded (Fig. 7), but the nature and composition of these grasslands is different from that of historic grasslands, mostly due to the nature of the underlying land change processes. Post-Soviet grassland expansion is mostly the result of cropland abandonment. Approximately 24 % of the cropland of 1985 had been abandoned by 2010 and subsequently transitioned to grasslands, but the rate increase in grassland cover on abandoned agricultural fields dropped to 9 % by 2010 (Fig. 6). The change from cropland to grassland was most prominent in Romania and Ukraine. However, due to their agricultural history, several decades after abandonment, these fields in the process of converting to grasslands have a different species composition than the historic semi-natural grasslands (Ruprecht 2006). Depending on the length of agricultural use, such agricultural legacies can persist for centuries on the landscape (Foster et al. 2003).

Grassland expansion is only a recent process that is counter-acting the past extensive grassland loss to other land covers. In the lowlands, conversion to arable land and intense grazing affected grasslands before and during the socialist period. Large areas of grasslands were brought into agricultural production by means of

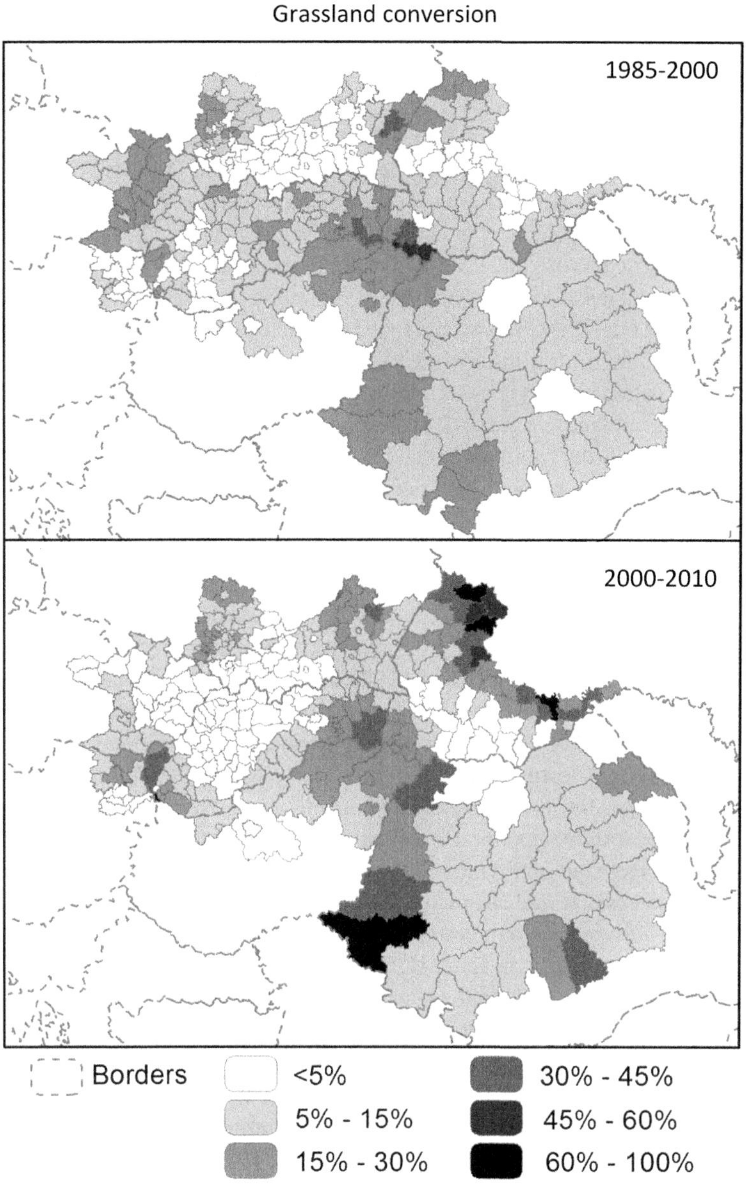

Fig. 7 Grassland conversion for the periods 1985–2000 and 2000–2010. Grassland conversion is calculated relative to the 1985 and 2000 grassland area per administrative unit. Country boundaries are shown in *dashed grey lines*. Grassland conversion is calculated on the level of administrative units (district level for Czech Republic, Poland, Slovakia and Ukraine, county level for Austria, Hungary and Romania). Scale of the maps is 1:8,750,000. Modified from Griffiths et al. (2013)

terrain adjustment, drainage, ploughing, sowing of more productive varieties of grasses and clovers, intensive fertilization, drainage, and the destruction of springs (Biró et al. 2012). Grasslands along river valleys were converted to arable land due to their high soil quality (Halada et al. 2008), leading to an overall substantial reduction of grassland cover: in Romania, steppe grasslands were reduced significantly to expand arable land by 39.2 % (Ioras 2003), and in Slovakia, more than 50 % of extensively used semi-natural grasslands were lost to arable land or to intensive meadows (Halada et al. 2010). In Hungary, wetlands were drained and converted to agricultural lands (Munteanu et al. 2014), although the grassland areas remained unchanged during Soviet times according to Hungarian statistics (Biró et al. 2012). After 1990, the afforestation of grasslands was a common process throughout the region (Griffiths et al. 2013) (Fig. 6).

Since the collapse of the Soviet Union, one of the largest threats to natural and semi-natural grasslands is related to land abandonment, followed by forest succession, which reduces the diversity and area of pastures and meadows (Kricsfalusy 2013). As livestock numbers declined in most Carpathian countries, so did the intensity of grazing and pasturing, allowing for forest succession on mountain pastures. The lack of management led to a loss of biodiversity (Cremene et al. 2005). In many cases, the abandonment of mountain meadows lead to a decline in species richness of up to 50 %, with unique grassland communities disappearing entirely (Bezák and Halada 2010). Lowland grasslands were also affected by abandonment: in Hungary, between 1988 and 1999, the annual grassland loss reached 1.3 %, in contrast to relatively static conditions during prior decades (Biró et al. 2012). Grasslands were mostly lost to urban development and forest cover in Hungary (Konkoly-Gyuró et al. 2011; Biró et al. 2012). In Western Ukraine and Western Romania, the rate of grassland loss to other covers has increased since 2000, largely due to agricultural recultivation (28 and 19 % of previously abandoned land in Romania and Ukraine, respectively).

With EU accession and an increased awareness of the importance of conservation, the environmental value of grasslands was recognized, and agro-environmental programs (part of the CAP) now support extensive uses and the conservation of biodiversity. In the Carpathian Mountains, traditional sheep herding is increasingly encouraged for tourism purposes, while maintaining the traditional extensive use of meadows and pastures, and supporting local food production (UNEP 2007). In Romania, subsidies for the management of high nature value grasslands aim to combat the effects of land abandonment and to support traditional agricultural practices (Akeroyd and Page 2011), despite the questionable suitability of the preservation approach of the currently available subsidy systems of the CAP (Fischer et al. 2012).

The underlying drivers of grassland change are closely interwoven with proximate environmental factors such as suitability for agriculture, elevation, and patch size (Biró et al. 2012). While the loss of grasslands in the socialist period was the result of centralized policies, recent grassland loss is mostly the result of the socio-demographic adaptations of individual farmers and land owners to the changing institutional and economic post-Soviet environment. In marginal mountainous areas, migration and the aging of the population, combined with a decrease in livestock farming, has caused a decrease in grassland management and resulted in conversion to forests (Munteanu et al. 2008, 2014). In Hungary, recultivation and afforestation both caused a decrease in grassland cover, and areas close to settlements and roads were more likely to be converted to other uses (Biró et al. 2012). Increasing foreign investment (Jordan 2013), EU CAP incentives for the recultivation of areas suitable for production, and increased commodity prices are expected to lead to further grassland loss in the Carpathian region (Fig. 6). Recent policy payments increased attention to conservation and support environmental grassland management in the Carpathians (Akeroyd and Page 2011). Management practices that are beneficial for supporting species richness and diversity, such as mowing and the removal of biomass (Galvánek and Lepš 2011) or extensive grazing (Cremene et al. 2005), are encouraged through agro-environmental schemes, a now commonly used instrument for landscape management. These schemes are based on contracts between land managers and public authorities, which provide payments for extensive management of land (especially grasslands and hedgerows). Recent grassland dynamics in the Carpathians indicate that the abandonment trend will not persist and current grassland cover may revert to agriculture or managed meadows and pastures (Griffiths et al. 2013).

2.8 *Other Land Cover Dynamics*

Two other land cover types in the Carpathians experienced drastic change over the course of history: wetlands and urban areas. Although entirely different, these two land cover types present a common characteristic: their trajectory of change over long periods of time has followed the same direction but with increasing intensity. Urban cover in the Carpathians has increased considerably over the past 250 years. Large urban centers developed in the lowlands and continued to rapidly sprawl after the collapse of the Soviet Union. In mountainous areas, tourism development and the building of second houses (Mika 2013) are the main causes of recent increase in built-up areas. In turn, wetland areas have experienced a continual decline since the middle of the 19th century. Many wetlands were drained and embanked for agricultural development and, in many cases, to support later urban development. The process was most prominent in the lowlands of Hungary.

Urban areas represent approximately 13 % of the Carpathians and are increasing in extent (UNEP 2007; Gerard et al. 2010). The conversion of land to built-up surfaces, including infrastructure development, industries and housing, is one of the six main land cover changes affecting the Carpathians (Gerard et al. 2010). Colonization during the 19th century increased the population of the Carpathians and subsequently caused an increase in built-up areas. The urbanization policies enacted during socialist rule led to the rapid conversion of farmland and grasslands to grow settlements, and increase industrial and infrastructure development. Industrial development pressures correlated with a heavy exploitation of natural resources that led to urban and industrial development at the edges of the Carpathians, such as Resita and Hunedoara in Romania, Upper Silesia and Krakow in Poland and Kosice in Slovakia. Since the collapse of socialism, many industrial operations have been abandoned, but the land did not revert to past land covers. Furthermore, urban sprawl and suburbanization have increased in magnitude (UNEP 2007). Loopholes in regulations led to unplanned development on recently privatized land (Munteanu et al. 2008). The development of tourism infrastructure, especially in the Tatra Mountains and the Southern Romanian Carpathians, led to additional increases in built-up areas. The Ukrainian Carpathians lag behind in tourism compared to the bordering Maramures (Romania), potentially due to their limited connectivity both by rail and road (Jordan 2013).

Wetland areas are the land cover type that have been changed to largest extent. The mosaic of wetlands and grasslands along the Danube and Tisza rivers have been altered continuously since the middle of the 19th century (UNEP 2007; Biró et al. 2012). River flow control regulations and embankments to extend agricultural production caused dramatic reductions in wetlands, down to less than half of their historic area along the Danube River (Ioras 2003). Most of the embanked areas were converted to agricultural production in Hungary and Romania (Ioras 2003; Konkoly-Gyuró et al. 2011). The straightening and shortening of streams, the construction of dikes and drainage canals, and wetland drainage became common socialist policies from the 1950s to the 1970s, aimed at supporting agricultural production and self-sufficiency (Kligman and Verdery 2011). In many cases in the Hungarian lowlands, the loss of wetlands was nearly total. For example, in Northern Hungary, wetlands made up approximately 17.7 % of the landscape in the 18th century but were reduced to 5.4 % by 1998 (Konkoly-Gyuró et al. 2011). With the adoption of EU regulations, such as the Water Framework Directive, and adherence to international agreements, such as the Ramsar Convention on Wetlands, wetland protection and restoration are becoming more important in conservation practices, yet the efforts required to restore such wetlands have kept the scale of wetland restoration low.

The changes in urban and wetland cover in the Carpathians provide good examples of permanent land cover changes. Once areas have transitioned to an urban landscape or away from wetlands, it is very unlikely that its future use will revert to its historical state. In this context, a sustainable, consistent management of these land covers over time is essential. However, in the Carpathians, most regions have experienced very little or no continuity in land management systems due to the multiple shifts that have occurred in political, economic and socio-demographic conditions. Poor regulations, missing cadastral planning and loopholes in development regulations, as well as shifts in property size and ownership, have all caused chaotic urban sprawl (Björnsen-Gurung et al. 2009; Suditu et al. 2010). Increasing food demands and forced industrialization historically led to massive wetland loss, and although this trend has slowed down in the past years, the effects of past wetland loss on ecosystem diversity are still reverberating. Because the loss of wetlands and urban expansion are mostly permanent changes, adequate planning that considers land as a limited resource is essential for managing these cover types.

3 Conclusions

The Carpathian Mountains, with their long land-use history, are a great showcase of how changes in institutions and land management regimes are reflected in the land cover. The region is suitable for the conceptual study of land-use legacies, alternative stable states and time lags related to the effects of shocks on land change. The truly diverse socio-economic and institutional history and conditions of the Carpathian countries provide a great natural experiment and exemplify how (1) following a shock, land systems may or may not be pushed into an alternative stable state (e.g., agricultural land abandoned following the collapse of the Soviet Union being reverted to agricultural use after EU accession versus agricultural abandonment following World War II resulting in conversion to stable forest cover); (2) effects of institutional changes may or may not experience time-lags (e.g., different rates of forest disturbance following the collapse of the Soviet Union); and (3) land-use legacies may persist for centuries (e.g., past land cover affecting the rate and magnitude of recent land changes).

The establishment and the collapse of the Soviet Union were major events, with long lasting effects on the different types of land cover. Some of the highest rates of land change in the Carpathians occurred after the collapse of the Soviet Union, but often, these changes were rooted in the prior land management history of the area. Forest transitions occurred in the interwar period, mirrored by agricultural abandonment and intensification since the establishment of the Soviet Union. However, many abandoned lands are being brought back into agricultural production following accession to the EU. In the past decade, grassland cover mostly decreased in the Carpathians, and wetland loss and urban expansion have been active processes since the 18th century. In summary, the Carpathians experienced extensive land changes and, thus, offer a wealth of land change lessons related to the effects of shocks, land change trajectories, time-lags and land-use legacies. Land-use trends and patterns in the Carpathians are broadly relevant to land change science as a whole and are applicable to multiple regions that have experienced abrupt transitions due to the collapse of the Soviet Union or to other political and socio-economic events. Land change science is tackling the implications that economic development, globalization, land-use policies, land grabbing, land-use displacement, and resource scarcity may have for future land change trajectories. We show that alongside these factors, socio-economic and political shocks, such as wars or abrupt changes in political regimes, severely affect the magnitude of change. Understanding the diversity of the spatiotemporal patterns of land change is important in balancing land-use decisions regarding intensified production versus extensive use for nature conservation—an issue that is relevant not only for the Carpathians but also for all of the worlds' biodiverse areas.

References

Akeroyd JR, Page JN (2011) Conservation of High Nature Value (HNV) grassland in a farmed landscape in Transylvania, Romania. Contribuiii Botanice XLVI:57–71

Aldwaik SZ, Pontius RG (2012) Intensity analysis to unify measurements of size and stationarity of land changes by interval, category, and transition. Landscape Urban Plan 106:103–114. doi:10.1016/j.landurbplan.2012.02.010

Angelstam P (1992) Conservation of communities—The importance of edges, surroundings and landscape mosaic structure. In: Hansson L (ed) Ecological principles of nature conservation. Applications in temperate and boreal environments. Elsevier, Uppsala, p 454

Bálint M, Ujvárosi L, Theissinger K et al (2011) The carpathians as a major diversity hotspot in Europe. In: Zachos FE, Habel JC (eds) Biodiversity hotspots. Springer, Berlin, pp 189–205

Banu F (2004) Asalt asupra economiei Romaniei de la Solagra la SOVROM (1936–1956) (in Romanian) 215

Baumann M, Kuemmerle T, Elbakidze M et al (2011) Patterns and drivers of post-socialist farmland abandonment in Western Ukraine. Land Use Policy 28:552–562. doi:10.1016/j.landusepol.2010.11.003

Baur B, Cremene C, Groza G et al (2007) Intensified grazing affects endemic plant and gastropod diversity in alpine grasslands of the Southern Carpathian mountains (Romania). Biologia 62:438–445. doi:10.2478/s11756-007-0086-4

Bellemare J, Motzkin G, Foster DR, Forest H (2002) Legacies of the agricultural past in the forested present: an assessment of historical land-use effects on rich mesic forests. J Biogeorgr 29:1401–1420

Bezák P, Halada L (2010) Sustainable management recommendations to reduce the loss of agricultural biodiversity in the mountain regions of NE Slovakia. Mt Res Dev 30:194–204

Bezák P, Mitchley J (2014) Drivers of change in mountain farming in Slovakia: from socialist collectivisation to the Common Agricultural Policy. Reg Environ Chang 14:1343–1356. doi:10.1007/s10113-013-0580-x

Bezák P, Petrovic F (2006) Agriculture, landscape, biodiversity: scenarios and stakeholder perceptions in the Poloniny National Park (NE Slovakia). Ekologia (Bratislava) 25:82–93

Bideleux R, Jeffries I (1998) A history of Eastern Europe: crisis and change, p 685

Biró M, Czúcz B, Horváth F et al (2012) Drivers of grassland loss in Hungary during the post-socialist transformation (1987–1999). Landscape Ecol 28:789–803. doi:10.1007/s10980-012-9818-0

Björnsen-Gurung A, Bokwa A, Chełmicki W et al (2009) Global change research in the Carpathian Mountain Region. Mt Res Dev 29:282–288. doi:10.1659/mrd.1105

Bragina EV, Ives AR, Pidgeon AM et al (2015) Rapid declines of large mammal populations after the collapse of the Soviet Union. Conservation Biology 29(3):844–853

Brain S (2011) Song of the forest: Russian Forestry and Stalinist Environmentalism, 1905–1953, p 231

Bugár G, Petrovič F, Mojses M (2010) Effects of underground coal mining on the rate of landscape changes. In: Barančoková M, Krajčí J, Kollár J, Belčáková I (eds) Landscape ecology—Methods, applications and interdisciplinary approach. Institute of Landscape Ecology Slovak Academy of Sciences, Bratislava, pp 425–432

CERI (2001) The status of the Carpathians, p 67

Cremene C, Groza G, Rakosy L et al (2005) Alterations of Steppe-Like Grasslands in Eastern Europe: a Threat to regional biodiversity hotspots. Conserv Biol 19:1606–1618. doi:10.1111/j.1523-1739.2005.00084.x

DeFries RS, Foley JA, Asner GP (2004) Land-use choices: balancing human needs and ecosystem function in a nutshell. Frot Ecol 2:249–257

Deininger K (2011) Challenges posed by the new wave of farmland investment. J Peasant Stud 38:217–247. doi:10.1080/03066150.2011.559007

Demek J, Havlíček M, Chrudina Z, Mackovčin P (2008) Changes in land-use and the river network of the Graben Dyjsko-svratecký úval (Czech Republic) in the last 242 years. J Landscape Ecol 1:22–51
Donald PF, Pisano G, Rayment MD, Pain DJ (2002) The Common Agricultural Policy, EU enlargement and the conservation of Europe's farmland birds. Agric Ecosyst Environ 89: 167–182. doi:10.1016/S0167-8809(01)00244-4
Elbakidze M, Angelstam P (2007) Implementing sustainable forest management in Ukraine's Carpathian Mountains: the role of traditional village systems. For Ecol Manage 249:28–38
Ellis EC, Kaplan JO, Fuller DQ et al (2013) Used planet: a global history. Proc Natl Acad Sci USA 110:7978–7985. doi:10.1073/pnas.1217241110
Falťan V, Katina S, Bánovský M, Pazúrová Z (2009) The influence of site conditions on the impact of windstorms on forests: the case of the High Tatras Foothills (Slovakia) in 2004. Moravian Geogr Rep 17:44–52
Feranec J, Šúri M, Oťaheľ J et al (2000) Inventory of major landscape changes in the Czech Republic, Hungary, Romania and Slovak Republic 1970s–1990s. Int J Appl Earth Obs Geoinf 2:129–139. doi:10.1016/S0303-2434(00)85006-0
Fischer J, Hartel T, Kuemmerle T (2012) Conservation policy in traditional farming landscapes. Conserv Lett 5:167–175. doi:10.1111/j.1755-263X.2012.00227.x
Foley JA, DeFries RS, Asner GP et al (2005) Global consequences of land use. Science 309: 570–574. doi:10.1126/science.1111772
Foley JA, Ramankutty N, Brauman KA et al (2011) Solutions for a cultivated planet. Nature 478:337–342. doi:10.1038/nature10452
Foster DR, Swanson FJ, Aber J et al (2003) The importance of land-use legacies to ecology and conservation. Bioscience 53:77–88
Galvánek D, Lepš J (2011) The effect of management on productivity, litter accumulation and seedling recruitment in a Carpathian mountain grassland. Plant Ecol 213:523–533. doi:10.1007/s11258-011-9999-7
Gehlbach S, Malesky EJ (2014) The grand experiment that wasn't? New Institutional Economics and the Postcommunist Experience. Institutions, Property Rights and Economic Growth: The Legacy of Douglass North Washington
Gerard F, Petit S, Smith G et al (2010) Land cover change in Europe between 1950 and 2000 determined employing aerial photography. Prog Phys Geogr 34:183–205. doi:10.1177/0309133309360141
Griffiths P, Kuemmerle T, Kennedy RE et al (2012) Using annual time-series of Landsat images to assess the effects of forest restitution in post-socialist Romania. Remote Sens Environ 118: 199–214
Griffiths P, Müller D, Kuemmerle T, Hostert P (2013) Agricultural land change in the Carpathian ecoregion after the breakdown of socialism and expansion of the European Union. Environ Res Lett 8:045024. doi:10.1088/1748-9326/8/4/045024
Griffiths P, Kuemmerle T, Baumann M et al (2014) Forest disturbances, forest recover, and changes in forest types across the Carpathian ecoregion from 1985 to 2010 based on Landsat image composites. Remote Sens Environ 151:72–88
Grodzińska K, Godzik B, Fraczek W et al (2004) Vegetation of the selected forest stands and land use in the Carpathian Mountains. Environ Pollut (Barking, Essex: 1987) 130:17–32. doi:10.1016/j.envpol.2003.10.031
Gula R, Frackowiak W, Perzanowski K (1995) Current status and conservation needs of brown bears in the Polish Carpathians. Ursus 10:81–86
Günther-Diringer D (2000) Evaluation of wetlands and floodplain areas in the Danube River basin. In: Nijland H, Cals MJ (eds) River restoration in Europe practical approaches. Wagenian, The Netherlands, p 344
Haase D, Seppelt R, Haase A (2007) Land use impacts of demographic change—Lessons from eastern German urban regions. In: Use of landscape sciences for the assessment of environmental security. Springer, Dordrecht, pp 329–344

Halada L, Ružičková H, David S, Halabuk A (2008) Semi-natural grasslands under impact of changing land use during last 30 years: Trollio-Cirsietum community in the Liptov region (N Slovakia). Community Ecol 9:115–123

Halada L, David S, Ružičková H (2010) Biodiversity of Carpathian grasslands under impact of socio-economic changes. In: Forum Carpaticum. Integrating nature and society towards sustainability, pp 35–36

Hartvigsen M (2014) Land reform and land fragmentation in Central and Eastern Europe. Land Use Policy 36:330–341

Hazeu GW, Roupioz LFS, Perez-Soba M (2010) Europe's ecological backbone: recognising the true value of our mountains

Holeksa J, Saniga M, Szwagrzyk J et al (2009) A giant tree stand in the West Carpathians—An exception or a relic of formerly widespread mountain European forests? For Ecol Manage 257:1577–1585. doi:10.1016/j.foreco.2009.01.008

Horváth Z, Ferenczi M, Móra A et al (2012) Invertebrate food sources for waterbirds provided by the reconstructed wetland of Nyirkai-Hany, northwestern Hungary. Hydrobiologia 697:59–72. doi:10.1007/s10750-012-1170-5

Huzui AE, Călin I, Pătru-Stupariu I (2012) Spatial pattern analyses of landscape using multi-temporal data sources. In: 2011 international conference of environment-landscape-european identity. Procedia Environ Sci 14:98–110

Ioras F (2003) Trends in Romanian biodiversity conservation policy. Biodivers Conserv 12:9–23. doi:10.1023/A:1021254615841

Ioras F, Abrudan IV (2006) The Romanian forestry sector: privatisation facts. Int For Rev 8: 361–367. doi:10.1505/ifor.8.3.361

Irimie DL, Essmann HF (2009) Forest property rights in the frame of public policies and societal change. For Policy Econ 11:95–101. doi:10.1016/j.forpol.2008.10.001

Irland LC, Kremenetska E (2009) Practical economics of forest ecosystem management: the case of the Ukrainian Carpathians. In: Soloviy IP, Keeton WS (eds) Ecological economics and sustainable forest management: developing a transdisciplinary approach for the Carpathian Mountains. Ukrainian National Forestry University Press, Lviv, p 432

Jepsen MR, Reenberg A, Kümmerle T et al (2013) Visions of land use transitions in Europe. In: Technological, institutional and economic drivers of land use change, p 226

Jordan P (2013) Transcarpathia—Ukrainian border region at the edge of the EU. Internal and external representations. ISPRS–Int Arch Photogrammetry Remote Sens Spatial Inf Sci XL-4/W3:83–87. doi:10.5194/isprsarchives-XL-4-W3-83-2013

Kallis G (2001) The EU water framework directive: measures and implications. Water Policy 3:125–142. doi:10.1016/S1366-7017(01)00007-1

Keeton WS, Crow SM (2009) Sustainable forest management alternatives for the Carpathian Mountain region: providing a broad array of ecosystem service. In: Soloviy I, Keeton WS (eds) Ecological Economics and Sustainable Forest Management: developing a Trans-disciplinary Approach for the Carpathian Mountains. Ukrainian National Forestry University Press, Lviv, Ukraine, pp 109–126

Keeton WS, Chernyavskyy M, Gratzer G et al (2010) Structural characteristics and aboveground biomass of old-growth spruce–fir stands in the eastern Carpathian mountains, Ukraine. Plant Biosyst Int J Dealing Aspects Plant Biol 144:148–159. doi:10.1080/11263500903560512

Keeton WS, Angelstam P, Baumflek M et al (2013) Sustainable forest management alternatives for the Carpathian Mountain region, with a focus on Ukraine. In: Kozak J, Ostapowicz K, Bytnerowicz A, Wyzga B (eds) The Carpathians: integrating nature and society towards sustainability. Springer, Berlin, pp 331–352

Kligman G, Verdery K (2011) Peasants under siege: the collectivization of Romanian agriculture, 1949–1962. Princeton University Press, Princeton

Knorn J, Kuemmerle T, Radeloff VC et al (2012a) Forest restitution and protected area effectiveness in post-socialist Romania. Biol Conserv 146:204–212. doi:10.1016/j.biocon.2011.12.020

Knorn J, Kuemmerle T, Radeloff VC et al (2012b) Continued loss of temperate old-growth forests in the Romanian Carpathians despite an increasing protected area network. Environ Conserv 40:182–193. doi:10.1017/S0376892912000355

Knowles B (2011) Mountain Hay Meadows: the Romanian context and the effects of policy on high nature value farming, p 8

Konkoly-Gyuró E, Nagy D, Balázs P, Király G (2011) Assessment of land cover change in western Hungarian landscapes. In: Balázs P, Konkoly-Gyuró É (eds) TransEcoNet workshop on landscape history proceedings. University of West Hungary Press, pp 75–89

Kozak J (2003) Forest cover change in the Western Carpathians in the past 180 years. Mt Res Dev 23:369–375. doi:10.1659/0276-4741(2003)023[0369:FCCITW]2.0.CO;2

Kozak J (2010) Forest cover changes and their drivers in the Polish Carpathian Mountains since 1800. In: Nagendra H, Southworth J (eds) Reforesting landscapes linking pattern and process. Landscape series 10. Springer, Berlin, pp 253–273

Kozak J, Ostapowicz K, Szablowska-Midor A, Widacki W (2004) Land abandonment in the Western Beskidy Mts and its environmental background. Ekológia Int J Ecol Probl Biosph 23:116–126

Kozak J, Estreguil C, Vogt P (2007) Forest cover and pattern changes in the Carpathians over the last decades. Eur J For Res 126:77–90

Kozak J, Ostapowicz K, Byterowicz A, Wyzga B (2013a) The Carpathian Mountains: challenges for the Central and Eastern European Landmark. In: The Carpathians: integrating nature and society towards sustainability. Springer, Berlin, pp 1–11

Kozak J, Ostapowicz K, Bytnerowicz A, Wyżga B (2013b) The Carpathians: integrating nature and society towards sustainability. doi:10.1007/978-3-642-12725-0

Kricsfalusy VV (2013) Mountain grasslands of high conservation value in the Eastern Carpathians: syntaxonomy, biodiversity, protection and management. Thaiszia J Bot 23: 67–112

Kuemmerle T, Hostert P, Radeloff VC et al (2008) Cross-border comparison of post-socialist Farmland Abandonment in the Carpathians. Ecosystems 11:614–628

Kuemmerle T, Chaskovskyy O, Knorn J et al (2009) Forest cover change and illegal logging in the Ukrainian Carpathians in the transition period from 1988 to 2007. Remote Sens Environ 113:1194–1207

Kuemmerle T, Perzanowski K, Chaskovskyy O et al (2010) European Bison habitat in the Carpathian Mountains. Biol Conserv 143:908–916. doi:10.1016/j.biocon.2009.12.038

Kuemmerle T, Oloffson P, Chaskovskyy O et al (2011) Post-Soviet farmland abandonment, forest recovery, and carbon sequestration in western Ukraine. Glob Change Biol 17:1335–1349. doi:10.1111/j.1365-2486.2010.02333.x

Lambin EF, Meyfroidt P (2010) Land use transitions: socio-ecological feedback versus socio-economic change. Land Use Policy 27:108–118. doi:10.1016/j.landusepol.2009.09.003

Lambin EF, Turner BL, Geist HJ et al (2001) The causes of land-use and land-cover change: moving beyond the myths. Glob Environ Change 11:261–269. doi:10.1016/S0959-3780(01)00007-3

Lerman Z, Csáki C, Feder G (2004) Agriculture in transition: land policies and evolving farm structures in post-soviet countries, p 254

Lieskovský J, Kanka R, Bezák P et al (2013) Driving forces behind vineyard abandonment in Slovakia following the move to a market-oriented economy. Land Use Policy 32:356–365. doi:10.1016/j.landusepol.2012.11.010

Macicuca A, Diaconescu C (2013) Forest certification effects on sustainable management of Romanian forest ecosystems. Present Environ Sustain Dev 7

Maes J, Paracchini ML, Zulian G et al (2012) Synergies and trade-offs between ecosystem service supply, biodiversity, and habitat conservation status in Europe. Biol Conserv 155:1–12. doi:10.1016/j.biocon.2012.06.016

Magura T (2002) Carabids and forest edge: spatial pattern and edge effect. For Ecol Manage 157:23–37. doi:10.1016/S0378-1127(00)00654-X

Main-Knorn M, Hostert P, Kozak J, Kuemmerle T (2009) How pollution legacies and land use histories shape post-communist forest cover trends in the Western Carpathians. For Ecol Manage 258:60–70

Mantescu L, Vasile M (2009) Property reforms in rural Romania and community-based forests. Rom Sociol 7:95–113

Mika M (2013) Spatial patterns of second homes development in the Polish Carpathians. In: The Carpathians: integrating nature and society towards sustainability. Springer, Berlin, pp 497–512

Moura AE, Tsingarska E, Dąbrowski MJ et al (2013) Unregulated hunting and genetic recovery from a severe population decline: the cautionary case of Bulgarian wolves. Conserv Genet. doi:10.1007/s10592-013-0547-y

Müller D, Munroe DK (2008) Changing rural landscapes in Albania: cropland abandonment and forest clearing in the postsocialist transition. Ann Assoc Am Geogr 98:855–876. doi:10.1080/00045600802262323

Müller D, Kuemmerle T, Rusu M, Griffiths P (2009) Lost in transition: determinants of post-socialist cropland abandonment in Romania. J Land Use Sci 4:109–129. doi:10.1080/17474230802645881

Müller D, Leitão PJ, Sikor T (2013) Comparing the determinants of cropland abandonment in Albania and Romania using boosted regression trees. Agric Syst 117:66–77. doi:10.1016/j.agsy.2012.12.010

Munteanu C, Geitner C, Scharr K (2008) Consequences of historical and modern landuse on cultural landscapes and biodiversity of the Maramureş Mountains. In: Gaceu L (ed) Journal of EcoAgriTourism, proceeding of the international conference BIOATLAS 2008. Transilvania University of Brasov, Romania, Brasov, pp 91–98

Munteanu C, Kuemmerle T, Boltiziar M et al (2014) Forest and agricultural land change in the Carpathian region—A meta-analysis of long-term patterns and drivers of change. Land Use Policy 38:685–697. doi:10.1016/j.landusepol.2014.01.012

Munteanu C, Kuemmerle T, Keuler NS et al (2015) Legacies of 19th century land use shape contemporary forest cover Article. Glob Environ Change 34:83–94

Munteanu C, Nita M-D, Abrudan IV, Radeloff VC (2016) Historical forest management in Romania is imposing strong legacies on contemporary forests and their management. For Ecol Manage 361:179–193

Nabuurs GJ, Thürig E, Heidema N et al (2008) Hotspots of the European forests carbon cycle. For Ecol Manage 256:194–200

Nijnik M, van Kooten GC (2000) Forestry in the Ukraine: the road ahead? For Policy Econ 1:139–151

Perzanowski K, Olech W (2007) A future for European Bison *Bison bonasus* in the Carpathian ecoregion? Wildlife Biol 13:108–112

Pietrzak M (1998) Development of settlement and farming from the Neolithic period to date in the marginal zone of the Carpathian foothills between the Raba and Uszwica Rivers. Prace Geograficzne 103:15–53

Prishchepov AV, Müller D, Dubinin M et al (2013) Determinants of agricultural land abandonment in post-Soviet European Russia. Land Use Policy 30:873–884. doi:10.1016/j.landusepol.2012.06.011

Rabbinge R, van Diepen CA (2000) Changes in agriculture and land use in Europe. Eur J Agron 13:85–99

Ronnås P (1982) Centrally planned urbanization: the case of Romania. Geografiska Annaler Seri B, Human Geogr 64:143–151

Rozelle S, Swinnen JFM (2004) Success and failure of reform: insights from the transition of agriculture. J Econ Lit 42:404–456. doi:10.1257/0022051041409048

Rozylowicz L, Popescu VD, Pătroescu M, Chişamera G (2010) The potential of large carnivores as conservation surrogates in the Romanian Carpathians. Biodivers Conserv 20:561–579. doi:10.1007/s10531-010-9967-x

Ruffini V, Hoffmann C, Streifeneder T, Renner K (2008) SARD-M Report for the 941 Carpathian Convention Member States. Assessment of Policies, Institutions and Processes, 942 Regional Synthesis for Czech Republic, Hungary, Poland, Romania, Republic of Serbia, 943 Slovak Republic and Ukraine. In: EURAC (ed) EURAC 944, Bozen, p 63

Ruprecht E (2006) Successfully recovered grassland: a promising example from Romanian old-fields. Restor Ecol 14:473–480. doi:10.1111/j.1526-100X.2006.00155.x

Schreiber W (2003) Analiza peisajelor geografice din partea de vest a Câmpiei Transilvaniei (in Romanian). 135

Schulp CJE, Nabuurs G-J, Verburg PH (2008) Future carbon sequestration in Europe—Effects of land use change. Agric Ecosyst Environ 127:251–264

Seton-Watson H (1945) Eastern Europe between the wars, 1918–1941, p 442

Shandra O, Weisberg P, Martazinova V (2013) Influences of climate and land use history on forest and timberline dynamics in the Carpathian mountains during the twentieth century. In: Ostapowicz K, Bytnerowicz A, Wyżga B, Kozak J (eds) The Carpathians: integrating nature and society towards sustainability. Springer, Berlin, pp 209–223

Sitko I, Troll M (2008) Timberline changes in relation to summer farming in the Western Chornohora (Ukrainian Carpathians). Mt Res Dev 28:263–271. doi:10.1659/mrd.0963

Skokanová H, Stránská T, Havlíček M et al (2009) Land use dynamics of the South Moravan region during last 170 years. Geoscape 4:58–65

Smaliychuk A (2012) Land cover change in low mountain landscapes of Lviv Oblast (Ukrainian Carpathians) during last three decades. In: Conference abstracts of the 2nd forum Carpaticum: From data to knowledge—From knowledge to action, pp 155–156

Špulerová J (2013) Policy instruments and methods for the protection and maintenance of historical agricultural landscapes in Slovakia. The Carpathians: integrating nature and society towards sustainability. Springer, Berlin, pp 429–439

Špulerová J, Dobrovodská M, Lieskovský J et al (2011) Inventory and classification of historical structures of the agricultural landscape in Slovakia. Ekologia 30:157–170. doi:10.4149/ekol_2011_02_157

Suditu B, Ginavar A, Muic A, Iordă C (2010) Urban sprawl characteristics and typologies in Romania. Human Geogr J Stud Res Human Geogr 4.2:79–87

Süli-Zakar I (1998) Socio-geographical transition in the rural areas of the Carpathian Euroregion. GeoJournal 46:193–197. doi:10.1023/A:1006995814365

Turnock D (1996) Agriculture in Eastern Europe: communism, the transition and the future. GeoJournal. doi:10.1007/BF00186661

Turnock D (2002) Ecoregion-based conservation in the Carpathians and the land-use implications. Land Use Policy 19:47–63

UNEP (2007) Carpathians environment outlook. United Nations Publications

Veen P, Fanta J, Raev I et al (2010) Virgin forests in Romania and Bulgaria: results of two national inventory projects and their implications for protection. Biodivers Conserv 19:1805–1819. doi:10.1007/s10531-010-9804-2

Vepryk N (2001) Istoryko-landshaftoznavchiy analiz lisogospodarskogo pryrodokorystuvannya (na prykladi Pivnichnoi Bukoviny u kinci 18 – na pochatku 20 stolittya) (in Ukrainian). Naukoviy Visnyk Chernivetskogo Universitetu, Seria Geografia 120:156–171

Visser O, Spoor M (2011) Land grabbing in post-Soviet Eurasia: the world's largest agricultural land reserves at stake. J Peasant Stud 38:299–323. doi:10.1080/03066150.2011.559010

Warcholik W (2005) Rejestracja różnic w przebiegu granicy rolno-leśnej w Beskidzie Niskim na obszarze Polski i Słowacji (1933-1975) z wykorzystaniem GIS (in Polish). Problemy Zagospododarowania Ziem Górskich 51:59–69

Wiezik M, Svitok M, Dovčiak M (2007) Conifer introductions decrease richness and alter composition of litter-dwelling beetles (Coleoptera) in Carpathian oak forests. For Ecol Manage 247:61–71. doi:10.1016/j.foreco.2007.04.013

Wolski J (2001) Kierunki zmian krajobrazu okolic bieszczadzkiej wsi Caryńskie (in Polish). Prace Geograficzne IGiPZ PAN 179:149–167

Woś B (2005) Zmiany pokrycia terenu w wybranych gminach Beskidów w drugiej połowie XX w. na podstawie analizy zdjęć lotniczych (in Polish). Teledetekcja Środowiska 35:1–114

Young J, Richards C, Fischer A et al (2007) Conflicts between biodiversity conservation and human activities in the Central and Eastern European countries. AMBIO: J Human Environ 36:545–550. doi:10.1579/0044-7447(2007)36[545:CBBCAH]2.0.CO;2

Zanten BT, Verburg PH, Espinosa M et al (2013) European agricultural landscapes, common agricultural policy and ecosystem services: a review. Agron Sustain Dev. doi:10.1007/s13593-013-0183-4

Underlying Drivers and Spatial Determinants of post-Soviet Agricultural Land Abandonment in Temperate Eastern Europe

Alexander V. Prishchepov, Daniel Müller, Matthias Baumann, Tobias Kuemmerle, Camilo Alcantara and Volker C. Radeloff

Abstract Our goal was to understand the underlying drivers and spatial determinants of agricultural land abandonment following the collapse of the Soviet Union and the subsequent transition from state-command to market-driven economies from 1990 to 2000. We brought an example of agricultural land-use change in one agro-climatic zone stretching across Lithuania, Belarus, and Russia. Here, we provide an overview of the agricultural changes for the studied countries. We estimated the rates and patterns of agricultural land abandonment based on Landsat TM/ETM+ satellite images and linked these data with institutional changes regarding land use. Using spatially explicit logistic regressions, we assessed spatial determinants of agricultural land abandonment. The highest rates of land abandonment from 1990 to 2000 were observed in Russia (31 %), followed by Lithuania (19 %), and Belarus (13 %), and the differences in land abandonment rates reflected the contrasting strategies for transitioning toward a market economy.

A.V. Prishchepov (✉)
Department of Geosciences and Natural Resource Management,
University of Copenhagen, Øster Voldgade 10, 1350 København K, Denmark
e-mail: alpr@ign.ku.dk

A.V. Prishchepov · D. Müller
Leibniz Institute of Agricultural Development in Transition Economies (IAMO),
Theodor-Lieser str. 2, Halle (Saale) 06120, Germany
e-mail: d.mueller@geo.hu-berlin.de

D. Müller · M. Baumann · T. Kuemmerle
Geography Department, Humboldt-University Berlin, Unter den Linden 6,
10099 Berlin, Germany

C. Alcantara
Departamento de Ecología y Recursos Naturales, Centro Universitario de la Costa Sur,
Universidad de Guadalajara, Avenida Independencia Nacional 151, 48900 Autlan de
Navarro, Jalisco, Mexico
e-mail: camiloalcantara@gmail.com

V.C. Radeloff
Department of Forest and Wildlife Ecology, University of Wisconsin-Madison,
Madison, WI, USA
e-mail: radeloff@wisc.edu

G. Gutman and V. Radeloff (eds.), *Land-Cover and Land-Use Changes in Eastern Europe after the Collapse of the Soviet Union in 1991*,
DOI 10.1007/978-3-319-42638-9_5

The spatial patterns of agricultural land abandonment across Lithuania and Russia corresponded to the land rent theory of von Thünen, as sites with low crop yields that were distant from markets had higher rates of abandonment. However, this was not the case for Belarus, where the institutional environment regarding agricultural land use differed from neighboring Lithuania and Russia.

1 Introduction

People and the way they use land are the leading drivers of global land-cover change, which in turn is a major cause of biodiversity decline and the loss of ecosystem services (Foley et al. 2005; Lambin and Meyfroidt 2010). Ultimately, all land-use decisions are made by local actors (e.g., land owners), but their actions are constrained by macro-economic (underlying) drivers such as national policies (Fox et al. 2003). Increasingly, evidence suggests that these underlying drivers are at the heart of many trends in changing land-use and land-cover. Furthermore, globalization is changing the way in which countries interact and impact one another (Lambin et al. 2001). For example, in a teleconnected world, the drastic decline in Russian domestic beef production after the removal of production and consumer subsidies in 1990 resulted in a steep increase in beef imports (Novozhenina et al. 2009), which indirectly led to increased deforestation in the Brazilian Amazon (Kaimowitz et al. 2004). However, the effects of underlying drivers on changes in land-use and the interactions of these drivers with other spatial determinants of land-use change are not well understood. The role of drastic socio-economic and political changes is particularly unclear. Most previous conceptualizations of land-use transition theory predicted unidirectional land-use intensification over time (Foley et al. 2005). This may not be the case after drastic shocks, such as socio-economic and political transformations (Alcantara et al. 2013; Prishchepov et al. 2012a, b; Kuemmerle et al. 2006; Alix-Garcia et al. 2012), natural disasters, conflicts (Baumann et al. 2014) or major pollution events (Hostert et al. 2011) that profoundly alter the conditions underpinning land use decision making. At the same time, fundamental and abrupt changes may also provide opportunities to better understand the drivers and processes of rapid, nonlinear land-use change (Baumann et al. 2011; Hostert et al. 2011; Kuemmerle et al. 2008; Nikodemus et al. 2005). The collapse of communism in Eastern Europe provided a 'natural experiment' (Diamond 2001) to study the effects of massive macro-economic and political changes on the economic performance of various countries and on changes in land-use.

One important effect of the dissolution of the Soviet Bloc and transition from state-command to market-driven economics has been the change in agricultural land use and massive agricultural land abandonment in Eastern Europe (Fig. 1). However, there is generally a lack of detailed agricultural statistics on both agricultural and land-use change in the pre- and post-Soviet eras for Eastern European countries, and the quality of such statistics, where they exist, is dubious (Prishchepov et al. 2012c). The most successful solution to overcome the limitations on determining land-use

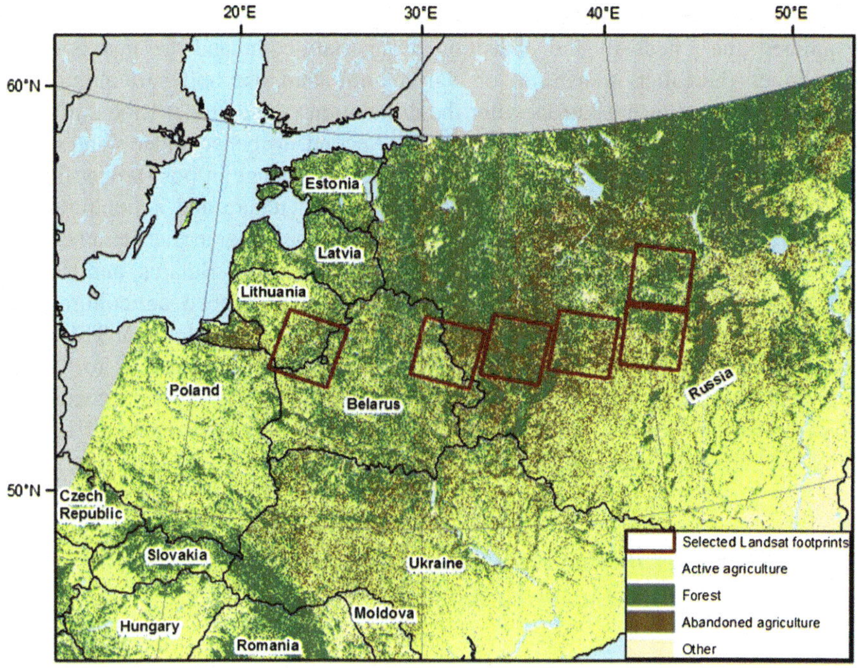

Fig. 1 Distribution of abandoned agricultural lands circa 2005–2007 for Central and Eastern Europe. *Dark red boxes* indicate selected Landsat footprints for detailed study of patterns and determinants of agricultural land abandonment from 1990 to 2000 . Source for land-cover change map: Alcantara et al. (2013)

change statistics is the use of satellite remote sensing (Alcantara et al. 2013; Prishchepov et al. 2012a, b; deBeurs and Ioffe 2013; de Beurs et al. 2017; Sieber et al. 2013). The 30-m resolution Landsat Thematic Mapper (TM)/Enhanced Thematic Mapper Plus (ETM+) images are well suited for applications in land-use analyses, particularly for retrospective mapping of agricultural land use, because image archives go back to the 1980s and all images are freely accessible. Furthermore, the spatial, spectral and radiometric resolution of Landsat TM/ETM+ images matches well with the field sizes in most Eastern European countries, thus providing an additional advantage for the accurate mapping of agricultural land use change in post-Soviet countries (Peterson and Aunap 1998; Václavík and Rogan 2009; Griffiths et al. 2013). For this reason, we employed Landsat satellite imagery to reconstruct agricultural land-use change for our study area.

We capitalized upon the 'natural experiment' that the collapse of Soviet Bloc provided to examine how national policies, land reform strategies, and change in land tenure affected abandonment rates and how underlying drivers and other spatial determinants (e.g., travel costs) shaped the pattern of agricultural land use in Eastern Europe. For our analysis, we focused on the first decade after the systemic change, i.e., from 1990 to 2000, because land-use responses were most dramatic during this period and macro-economic changes were most pronounced. First, we examined the effect of the transition from state-command to market-driven economies (hereafter, transition) on agricultural land use and agricultural land abandonment in Eastern Europe after the dissolution of the Soviet Bloc during the first decade of transition (circa 1990–2000). Second, to better under stand underlying and proximate drivers of agricultural land-use change in post-Soviet Eastern Europe, such as institutional changes regarding land tenure and agricultural production, we looked in detail at one uniform agro-climatic region that stretches from west to east across three former Soviet republics (Lithuania, Belarus, and Russia). To assess the dynamics of agricultural land use, we mapped agricultural land abandonment from 1990 to 2000 across selected countries using Landsat TM/ETM+ imagery and related agricultural land abandonment rates and patterns to the different institutional changes that affected land use. Finally, we used econometric spatially explicit logistic regressions to assess the spatial determinants of agricultural land abandonment across the study region.

1.1 The Effect of the Transition on Agricultural Land Use in Eastern Europe

The collapse of the Soviet Bloc in Eastern Europe and the adoption of the principles of an open-market economy brought about a number of policy changes that had important direct and indirect effects on the agricultural sector. Such policy transformations included the removal of state subsidies used to equilibrate output and input prices, which resulted in starkly deteriorating conditions for trade and therefore negatively affected agricultural profitability (Rozelle and Swinnen 2004). Moreover, post-socialist land reforms set the stage for a rapid shift from the dominance of collective and state farming to individualized land use (Lerman et al. 2004). A number of key developments outside of the agricultural sector contributed to the decrease in agricultural profitability. First, the overall contraction of the economy and the collapse of social security systems and social services resulted in a sense of high insecurity in rural populations, which led to massive international emigration rates and declining fertility (Kontorovich 2001). Second, the composition of the economy shifted towards the service sector, away from industry and agriculture (Fig. 2a, b). However, in Belarus, by 2000, the value added by agriculture to the total GDP remained high compared to other countries in our study area. Agriculture was an important economic sector for Belarus during the Soviet era and this legacy may have mediated the impact of the economic transition on agricultural land use in Belarus (Fig. 2a, b). Declining living standards and economic opportunities in rural areas also contributed to substantial domestic migration from rural to urban areas, which changed the face of post-socialist rural societies and contributed to rural labor scarcity (Ioffe et al. 2004; Müller and Sikor 2006; Müller et al. 2009). In sum, the deteriorating prospects for agricultural production across the region, especially in labor and finance intensive agricultural sectors such as livestock and dairy

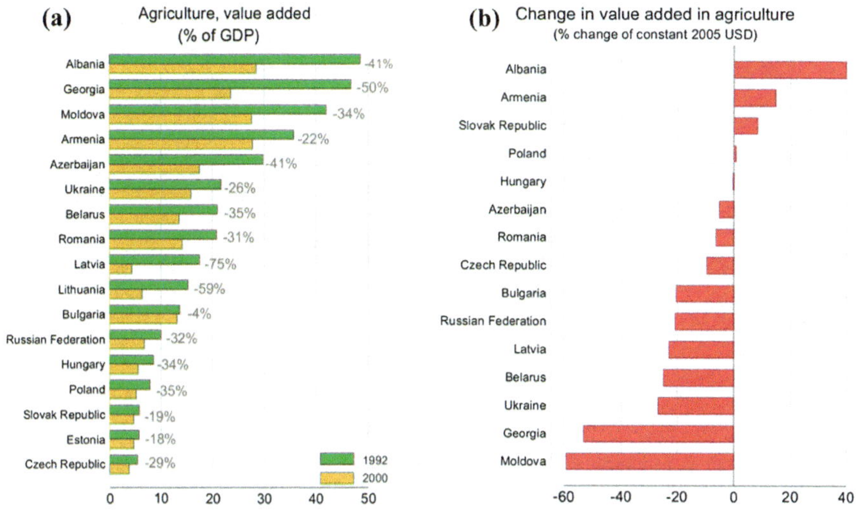

Fig. 2 **a** Agricultural production relative to the total GDP for both 1992 and 2000. **b** Change in value added by agriculture from 1992 to 2000, as a percent, relative to the base year 2005. *Source* World Bank (2013)

production, resulted in plummeting livestock numbers, abandonment of grazing lands and cessation of fodder production, though with some potential mediation of abandonment rates at the country level due to differences in the importance of agriculture (Nefedova 2011; Prishchepov et al. 2012b; Schierhorn et al. 2013).

1.2 *Study Area*

To better understand the effects of underlying drivers, such as institutional changes, and their impacts on agricultural land use and agricultural production, and to better understand the spatial determinants of agricultural land abandonment in Eastern Europe, a uniform agro-climatic region was set up. This region was determined based on the stratification of Europe by a number of agro-ecological values/products, including average annual mean temperature for January and July, the number of days with a mean temperatures over 5 °C, and average annual evapotranspiration (Prishchepov et al. 2012c). In such way, we controlled for agro-climatic variation and emphasized the socio-economic and institutional differences in agriculture and land use (e.g., governance and land tenure) after the collapse of the USSR.

The study area represents the temperate zone of Eastern Europe (Fig. 3; Table 1), which is well suited for agriculture, especially after melioration, liming, and fertilization of soils (Folch 2000). The primary summer crops in this region are barley, rye, oats, sugar beets, fodder maize and potatoes, and the primary winter

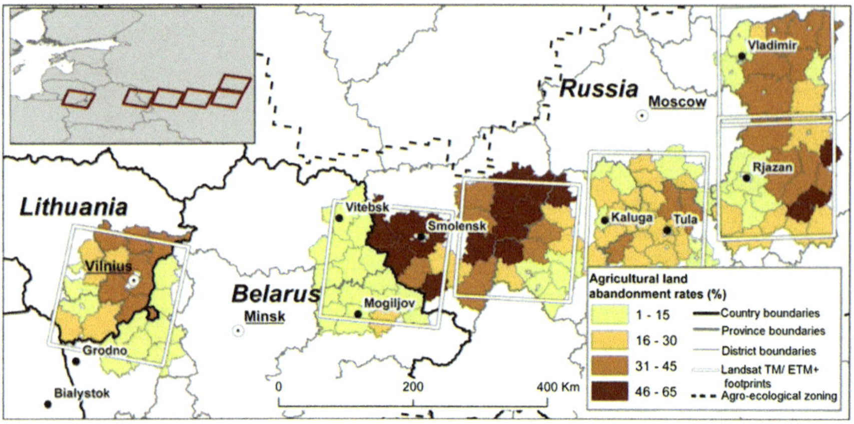

Fig. 3 Study area. Agricultural land abandonment rates from 1990 to 2000 summarized by districts. *Source* Prishchepov et al. (2012b)

crops are winter wheat, winter barley and winter rapeseed. Livestock production was extremely important in Lithuania, Belarus and Russia during the Soviet era; however, livestock numbers precipitously declined, particularly in parts of Lithuania and Russia, after the abolition of the Soviet state-command economy (Fig. 4). During the last decades of the Soviet era, the study region became one of the core agricultural areas of the Soviet Union, especially following the failure of the Soviet government to increase grain production during Khrushchev's Virgin Lands Campaign (Ioffe et al. 2004; Kraemer et al. 2015). After the collapse of the Soviet Bloc in Eastern Europe, Lithuania restituted collectivized farmland to its former owners and their heirs (Stuikys and Ladyga 1995). The Belarusian government privatized agricultural land early in the transition period, but in 1994 reversed its course and limited land ownership to small parcels, while also restricting land leases (Sakovich 2008). In Russia, agricultural land and the assets of former state and collective farms were privatized, and shares were distributed among former farm employees (Lerman et al. 2004). However, a moratorium on private agricultural land purchases and sales was introduced, which lasted until 2003 (Lerman and Shagaida 2007).

We generally assume that Lithuania, Belarus and Russia had a common starting point and that any differences regarding the rates of agricultural land abandonment in post-Soviet area are due to the varying transition approaches to a market economy. In reality, the differences in socio-economic and agricultural production already existed during the Soviet era. For instance, a clear west-east gradient in socio-economic development existed in the study area, with higher road density in Lithuania compared with Belarus and many Russian provinces (Prishchepov et al. 2012b) (Table 1). Moreover, Lithuania experienced a shorter period of collectivized agriculture comparable to Belarus and Russia and thus had better chances to adapt agriculture to market-economy conditions in an expedite mode.

Table 1 Climatic, socioeconomic conditions and agricultural production for study region in 1990

Provinces	Country after 1990	Temperature growing periods >5 °C[a]	Average annual precipitation[a]	Percentage of podzolic soils[b]	Rural population density (people/km^2)[c]	Road density (km/100 km^2)[c]	Tractors (hectares of arable land/tractor)[c]	Share of sown areas in total agricultural land for 1990 (%)[c]	Cattle density per total agricultural lands in 1990, heads/100 ha[c]	Milk production (kg/cow)[c]	Grain yield (centners/ha)[c]
–	Lithuania	2517	627	60	27	51	47	65	64	3733	28.5
Grodno	Belarus	2536	621	89	19.5	24.8	51	69	26.7	3223	30.7
Vitebsk	Belarus	2359	606	72	12.4	20.3	*	68	23.7	2914	20.7
Mogiljev	Belarus	2504	620	78	14.7	21.6	*	67	22.9	3058	25.8
Smolensk	Russia	2250	649	89	7.4	11	63	83	145	2478	11.3
Kaluga	Russia	2336	680	86	11.0	14	54	80	47	2527	13.8
Tula	Russia	2359	638	39	13.5	18	73	87	40	2645	19.2
Rjazan	Russia	2452	566	36	11.8	12	70	74	39	2881	16.8
Vladimir	Russia	2300	605	79	11.8	15	74	76	54	2880	16.2

[a]Climatic data from IIASA (2000)

[b]Soil data are taken from Batijes (2001)

[c]Statistical data from Belstat (2002), Goskomstat (1991), Goskomstat LitSSR (1989) and Sakovich (2008)

*-not available

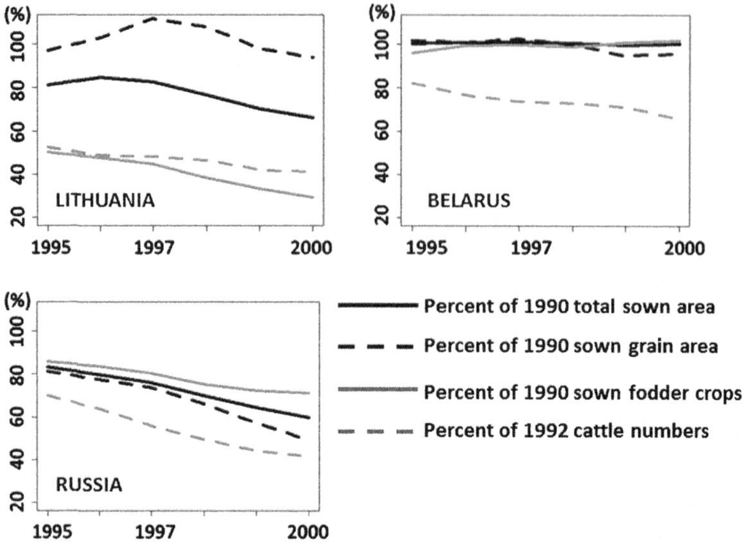

Fig. 4 Agricultural change in the study region. Data for Russia represent calculations based on numbers summed across the provinces of Smolensk, Kaluga, Tula, Rjazan and Vladimir. *Source* Belstat (2002), Goskomstat (1991), Goskomstat LitSSR (1989), Lithstat (2014) and Rosstat (2002)

If countries were equally affected by changing socio-economic conditions, we would assume that higher rates of abandonment would have occurred in Russia, with medium rates in Belarus and lower rates in Lithuania. This prediction relies solely on differences in infrastructure conditions prior the collapse of the Soviet Union and in the initial better entrepreneurial conditions in private farming (i.e., a shorter period of collectivization). However, official statistics suggested that there was a widespread contraction and abandonment of sown crops, particularly of fodder crops in Lithuania and Russia, and even a slight cropland expansion in Belarus (Fig. 4). At the same time, Belarusian official reports also indicated a decline in livestock (Fig. 4), which suggests that there may be some ongoing agricultural abandonment that is not reflected in the official statistics.

1.3 Mapping Agricultural Land Abandonment

Agricultural land use statistics that allow us to capture agricultural land abandonment are of varying quality and difficult to compare over time and across Eastern European countries (Prishchepov et al. 2012b). Thus, to produce reliable and consistent agricultural land abandonment maps across our study area and to reconstruct agricultural land use prior to the collapse of the Soviet Union (circa 1990), we relied on classification of 30-m resolution multi-temporal Landsat TM/ETM+ images for several 185 × 185 km footprints (Fig. 3). The decision to utilize Landsat TM/ETM+ images to map land-use changes was based on the good fit of the spectral, radiometric and spatial resolution to map agricultural land abandonment in our study area. Moreover, Landsat TM/ETM+ images for the Soviet period were available for our study area from the European Space Agency (ESA) and at the United State Geological Survey (USGS) archives, making Landsat TM/ETM+ imagery an excellent dataset for our study.

Previously, we classified stable, managed agricultural lands in circa 1990 and in circa 2000 (i.e., managed grasslands and arable lands) and agricultural land abandonment (i.e., managed agricultural land in circa 1990, but abandoned by 2000) using a nonparametric Support Vector Machines (SVM) classifier (Prishchepov et al. 2012a, b). The SVM classifier can be used to distinguish multimodal reflectance classes, such as managed agricultural areas and agricultural land abandonment, the classes, which are spectrally difficult to separate with parametric classifiers, such as Maximum Likelihood classifier. Because spectral profiles of different crops are complex and managed and non-managed grasses may have similar reflectance in certain parts of the year, it was important for us to select several images across the year to capture seasonal variation in reflectance for different crops and land-cover types. We mapped agricultural abandonment circa 1990 (images ranged between 1985 and 1992) and circa 2000 (images ranged between 1999 and 2002) and combined all satellite images simultaneously into one image composite or so-called "layerstack" prior the classification with SVM. We also conducted basic image pre-preprocessing steps, including co-registration, screening and masking out the clouds and shadows. The classification catalog consisted of five classes: 'forest and wetland', 'permanent shrubs, tree lines and riparian vegetation', 'stable agricultural land', 'abandoned agricultural land' and 'other' (Fig. 5). 'Stable agricultural land' consisted of both tilled agricultural land and grasslands that were intensively used for grazing and hay cutting during both the pre-transition era circa 1990 and after the first decade of transition circa 2000.

Agricultural land abandonment is a process that starts immediately when farmers terminate agriculture. However, it may take up to several years after abandonment before it is really possible to distinguish with satellite imagery if an agricultural plot is truly abandoned rather than left fallow as a part of the crop rotation cycle (Fig. 5). Thus, we defined 'abandoned agricultural land' as agricultural land that was used before 1990 for crops, hay cutting and livestock grazing but was no longer in use by 1999–2002 for any of the described land uses. Abandoned areas in temperate Europe are characterized by non-managed grasslands that often contain early successional shrubs (Fig. 5). Shrub encroachment in the study area usually takes place within three to five years after abandonment, with faster shrub advancement on well-drained and formerly plowed fields (Lyuri et al. 2010). Thus, assessment of abandonment was fairly conservative because some of the temporally fallow fields could have been classified as managed agricultural land. We assessed the accuracy of our land-use change maps using data collected during the field campaigns and

Fig. 5 Depiction of agricultural land abandonment in temperate Eastern Europe. **a** Depicts the idle agricultural field after 1–3 years after the abandonment. Field is covered by regrown grasses. **b** Depicts the idle agricultural field after 5–7 years after the abandonment. Field is covered by grasses and encroached shrubs and young trees intercepted within field. **c** Depicts the idle agricultural field after 15–20 years after the abandonment. Former agricultural field almost completely is encroached by young forest. Due to data gaps in Landsat archives and limited cloud-free imagery availability, often available Landsat imageries in Eastern Europe are suitable to detect accurately the encroachment of natural vegetation represented at the pictures **b** and **c**

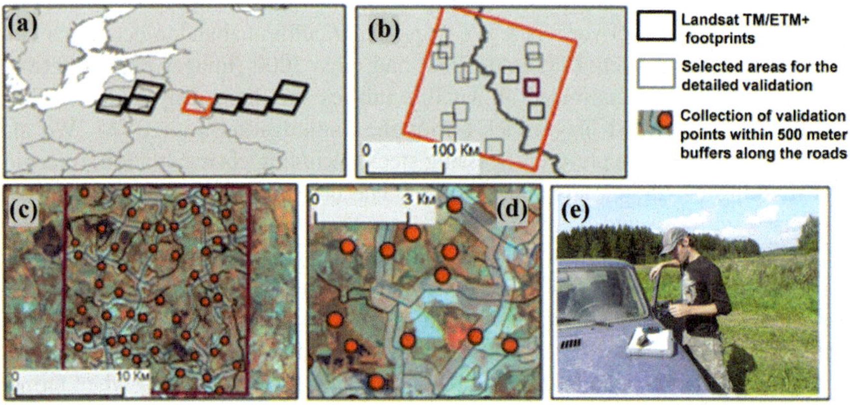

Fig. 6 Depiction of the three-step approach for the collection of validation data. **a** depicts an example of selection of Landsat TM/ETM + WRS-2 footprint path 182 row 22 for the detailed collection of training and validation data during the field campaign. **b** depicts stratified random generation of 20 * 20 km blocks within WRS-2 footprint path 182 row 22 for further collection of training and validation data. **c** depicts randomly generated points for validation of classification within 500 m buffer along the roads and with at least 500 m distance between the generated points. **d** depicts zoomed in area for the collection of validation points. **e** Allocation of validation points with non-differential GPS, recording land-cover/ and land-cover change types

high-resolution Quickbird™ and IKONOS™ satellite images (Fig. 6). Our resulting maps were accurate and suitable for detailed spatially explicit econometric modeling of determinants of agricultural land abandonment (Prishchepov et al. 2012c). For instance, the accuracy of 'abandoned agricultural land' varied for the selected footprints between 80.6 and 92.7 % (Prishchepov et al. 2012b).

1.4 Hypothesized Determinants of Agricultural Abandonment

According to the utility maximization theory, agricultural land abandonment can be driven by economic decisions based on profit maximization, such as utilization of agricultural lands with better socio-economic and agro-environmental endowment, or by abandonment of marginal areas (Irwin and Bockstael 2002; Prishchepov et al. 2013). Under this theory, an agricultural agent (e.g., a farmer) would terminate agricultural production when the production costs rise above the returns from production. For instance, we assumed areas with low yields and low population density that were located far from markets would exhibit higher production costs and, therefore, a higher likelihood of abandonment. Isolated agricultural areas within forest patches and in close proximity to forests may have a higher likelihood of abandonment due to a combination of factors (such as remoteness, and lower suitability for farming), which may trigger abandonment.

We acquired data from Belarus and Lithuania for the same explanatory variables as those used in our previous study of temperate European Russia (Prishchepov et al. 2013). These data captured the proximity to market centers, demographic changes that affected the labor supply, infrastructure, proxies for agricultural productivity and natural suitability of agricultural plots (Tables 2 and 3). Some explanatory variables may be endogenous to land-use change phenomenon (i.e., rural population change from 1990 to 2000 could have caused agricultural

Table 2 Explanatory variables

Variables (units)	Source	Spatial resolution
Biophysical		
Soil pH (units)	SOVEUR/SOTER 1:2,500,000 digital maps	Rasterized vector dataset
Elevation (m), slope (°)	Shuttle Radar Terrain Mission (SRTM)	Resampled raster 90 m dataset
Average annual evapotranspiration (mm)	AgroAtlas, 2010	Resampled raster 10 km dataset
Distance from nearest forest edge (100 m)	30 m Landsat TM/ETM + classifications	Pixel level calculations
Isolated agricultural areas within forest matrix in 1990 (dummy)	30 m Landsat TM/ETM + classifications	Pixel level calculations
Agricultural productivity		
Average grain yields in the late 1980s (centners/ha)	Belstat (2002), Goskomstat (1991), Goskomstat LitSSR (1989) and Rosstat (2002)	Rasterized district level statistics

(continued)

Table 2 (continued)

Variables (units)	Source	Spatial resolution
Population		
Interpolated population counts from settlements in the late 1980s (the proxy for population density) (number of people)	1:100,000 declassified Soviet topographic maps	Pixel level calculations
Distance variables		
Distance from provincial capital (km)	1:100,000 declassified Soviet topographic maps	Pixel level calculations
Distance from nearest district center (km)	1:100,000 declassified Soviet topographic maps	Pixel level calculations
Distance from nearest municipality center (km)	1:100,000 declassified Soviet topographic maps	Pixel level calculations
Distance from nearest settlement with more than 500 people (km)	1:100,000 declassified Soviet topographic maps	Pixel level calculations
Distance from nearest settlement (km)	1:100,000 declassified Soviet topographic maps	Pixel level calculations
Distance from nearest hard-surfaced road (100 m)	1:500,000 declassified Soviet topographic maps	Pixel level calculations
Infrastructure		
Road density in the late 1980s (km/100 km^2)	1:500,000 digital dataset	Rasterized district level statistics

abandonment, but abandonment of agricultural lands and marginalization of rural areas could also have caused the exodus of a rural population). To avoid such endogeneity, we included only time-invariant variables in our models, such as biophysical parameters and parameters that represent socio-economic conditions before the transition (i.e., 'average grain yields in the late 1980s'). Based on previous analyses for European Russia (Prishchepov et al. 2013), we selected 14 variables to analyze for Belarus and Lithuania, two of which were district-level variables ('average grain yields in the late 1980s' and 'road density in the late 1980s') and the remainder were pixel-level variables (Table 2).

Binary logistic regressions are suitable for modelling land-use change between times I and II, with land-use change phenomenon coded as "1" and non-changing land use coded as "0". We used binary logistic regression models and defined "1" as representing 'abandoned agricultural land' and "0" as 'stable agricultural land' between 1990 and 2000. Our spatial datasets presented an opportunity to improve the understanding of the spatial determinants of agricultural land abandonment because commonly used econometric models rely on aggregated data, such as district level statistics on dynamics of sown areas (Ioffe et al. 2004). The full dataset contained over 76 million 30-m resolution pixels; each pixel was a potential observation for the logistic regression model. Many such potential observations are redundant for statistical models and can exhibit strong spatial autocorrelation. Thus,

Table 3 Descriptive statistics for explanatory variables

Variables	Level	Unit	Mean	Median	Standard deviation	Minimum	Maximum
Lithuania							
Abandoned agricultural land	Pixel	Dummy (1/0)	0.500	0	0.500	0	1
Soil pH	Pixel	Units	6.08	6.00	0.88	2.5	8.0
Slope	Pixel	Degrees	2.10	1.51	2.13	0.00	21.8
Average annual evapotranspiration	Pixel	mm/100	6.04	6.03	0.189	5.3	6.5
Distance from nearest forest edge	Pixel	100 m	1.79	1.20	1.84	0.00	16.79
Isolated agricultural areas within forest matrix in 1990	Pixel	Dummy (1/0)	0.336	0.0	0.472	0	1
Average grain yields in the late 1980s	District	Centners/ha	18.98	17.30	4.24	14.0	26.7
Interpolated population from settlements in the late 1980s	Pixel	Number of people	511.5	68.3	6919.62	10.0	332,375.3
Distance from provincial capital	Pixel	km	61.1	57.9	30.47	1.05	138.8
Distance from nearest district center	Pixel	km	17.02	16.55	7.908	0.32	43.7
Distance from nearest municipality center	Pixel	km	4.78	4.55	2.42	0.03	17.49
Distance from nearest settlement with more than 500 people	Pixel	km	6.38	5.92	3.67	0.00	8.46
Distance from nearest settlement	Pixel	km	0.84	0.76	0.496	0.00	1.1
Road density in the late 1980s	District	km/100 km^2	48.2	45.0	10.68	33.0	52.0
Distance from nearest hard-surfaced road	Pixel	100 m	6.6	5.16	5.89	0.00	47.4
Number of sampled '0'	8500						
Number of sampled '1'	8529						
Belarus							
Abandoned agricultural land	Pixel	Dummy (1/0)	0.4903	0	0.455	0	1
Soil pH	Pixel	Units	5.4	5.3	0.8	3.82	7.7
Slope	Pixel	Degrees	1.7	1.068	1.77	0.00	19.49
Average annual evapotranspiration	Pixel	mm/100	6.7	6.2	0.86	5.21	8.23

(continued)

Table 3 (continued)

Variables	Level	Unit	Mean	Median	Standard deviation	Minimum	Maximum
Distance from nearest forest edge	Pixel	100 m	2.03	1.34	1.94	0.00	16.36
Isolated agricultural areas within forest matrix in 1990	Pixel	Dummy (1/0)	0.162	0.0	0.368	0	1
Average grain yields in the late 1980s	District	Centners/ha	23.98	24.00	4.568	12.9	39.00
Interpolated population from settlements in the late 1980s	Pixel	Number of people	241.25	111.42	8.041	10.3	42,480.27
Distance from provincial capital	Pixel	km	95.71	81.65	3.797	0.067	237.5
Distance from nearest district center	Pixel	km	16.49	13.67	7.89	0.00	97.09
Distance from nearest municipality center	Pixel	km	4.36	4.1	2.34	0.00	23.65
Distance from nearest settlement with more than 500 people	Pixel	km	6.2	5.6	5.9	0.00	38.36
Distance from nearest settlement	Pixel	km	0.81	0.75	0.52	0.00	5.72
Road density in the late 1980s	District	km/100 km^2	28.9	24.0	40.8	14.0	44.0
Distance from nearest hard-surfaced road	Pixel	100 m	12.25	9.42	7.327	0.00	73.57
Number of sampled '0'	4500						
Number of sampled '1'	4328						
Russia							
Abandoned agricultural land	Pixel	Dummy (1/0)	0.293	0	0.455	0	1
Soil pH	Pixel	Units	6.62	6.9	6.432	4.22	7.38
Slope	Pixel	Degrees	1.253	1	1.653	0.00	29.00
Average annual evapotranspiration	Pixel	mm/100	7.00	6.89	0.58	5.54	8.82
Distance from nearest forest edge	Pixel	100 m	7.169	4.37	7.864	0.00	70.59
Isolated agricultural areas within forest matrix in 1990	Pixel	Dummy (1/0)	0.13	0.0	0.34	0.0	1
Average grain yields in the late 1980s	District	Centners/ha	15.9	16	4.568	8.00	27.00

(continued)

Table 3 (continued)

Variables	Level	Unit	Mean	Median	Standard deviation	Minimum	Maximum
Interpolated population counts from settlements in the late 1980s (the proxy for population density)	Pixel	Number of people	267.81	115.77	8.041	0.16	85,337.80
Distance from provincial capital	Pixel	km	71.6	68.05	3.797	0.40	210.61
Distance from nearest district center	Pixel	km	15.62	14.7	7.898	0.11	52.30
Distance from nearest municipality center	Pixel	km	4.105	3.74	2.34	0.00	23.55
Distance from nearest settlement with more than 500 people	Pixel	km	6.784	5.7	5.9	0.00	39.47
Density of settlements in the late 1980s	District	Number of settlements/100 km^2	10.6	10.0	3.30	4.00	18.00
Road density in the late 1980s	District	km/100 km^2	34.4	35.1	4.8	25.7	44.9
Distance from nearest hard-surfaced road	Pixel	100 m	8.79	7	7.327	0.30	79.59
Number of sampled '0'	93,671						
Number of sampled '1'	39,018						

we sampled observations with at least a 500-m distance between points to avoid spatial autocorrelation (after Prishchepov et al. 2012b). This resulted in approximately 158,000 observations for our statistical models over the entire study area (Table 3).

Samples within the same administrative unit (i.e., the same district) may not be truly independent (Gellrich et al. 2007; Müller and Munroe 2008). For instance, farmers can be equally affected by specific socio-economic and environmental conditions within the same district and by the decisions of the district administration regarding subsidies and rural development programs. To account for such clustering effects, we used the Huber–White sandwich estimator, which controls for potentially correlated error terms, in our models without affecting the coefficients (Huber 1967; Müller et al. 2009; White 1982). We also explored the contribution of each independent variable in our models at the country and province level. To do so, we used hierarchical partitioning to assess the contribution of each independent variable at the country and province level by calculating the percentage of the total variance explained by each statistically significant variable ($p < 0.05$) (Baumann et al. 2011; Chevan and Sutherland 1991; Mac Nally 1996; Millington et al. 2007). All statistical analysis was performed with the use of R statistical software (R Development Core Team 2011).

2 Results

2.1 *Rates of Agricultural Land Abandonment Among Study Countries*

Our results indicated that widespread agricultural land abandonment had occurred across our study area. Within the classified Landsat footprints, statistically adjusted estimates of agricultural land abandonment showed that 6.9 million hectares were in agricultural use in 1990, of which 24 % were abandoned by 2000. The highest rate of agricultural land abandonment at the national level was observed for the Russian part of study area, where 31 %, or 1.7 million ha, of the agricultural land managed in 1990 was abandoned by 2000. In the Lithuanian and Belarussian parts of study area, 19 and 13 % of agricultural land was abandoned by 2000, respectively. Agricultural land abandonment rates were consistently high across the Russian provinces: 30 % in Kaluga, 26 % in Tula, 28 % in Rjazan and 27 % in Vladimir provinces. The highest abandonment rate was observed in Smolensk province (46 %), which borders Belarus.

The abandonment rates at the district (rayon) level varied substantially in the Russian study area, reaching as high as 65 % (Fig. 3). Belarus had the lowest rates of agricultural land abandonment compared to the other study countries; abandonment rates were consistently low for both classified footprints that covered the Belarusian study area. In Lithuania, we observed variation in abandonment rates at the district level; districts with high abandonment rates were concentrated in north-eastern part of the classified Landsat footprint that covered part of Lithuania. Allocation of satellite footprints in cross-border areas allowed us to depict striking differences in agricultural land abandonment rates, e.g., across the national boundaries of Lithuania-Belarus and Belarus-Russia (Fig. 3).

2.2 *Regression Results*

The explanatory power of the models for each individual country were relatively low (adjusted R^2 = 0.170 for Lithuania, adjusted R^2 = 0.072 for Belarus and adjusted R^2 = 0.144 for Russia) (Table 4). However, it is common to have a low adjusted R^2 for spatially explicit pixel-based logistic regression models. Because the models explain the variability across a large number of samples (e.g., ~157,000 samples from Russia were used), this statistical measure has to be interpreted with caution (Gellrich et al. 2007; Müller and Munroe 2008). The model goodness-of-fit (area under the curve, AUC) for our logistic regression models was good (AUC = 0.708 for Lithuania, AUC = 0.635 for Belarus and AUC = 0.703 for Russia) (Pontius and Schneider 2001). This result is substantially better than the probability of separating stable agriculture and abandonment solely by chance (AUC = 0.5) (DeLeo 1993; Gellrich et al. 2007).

Our logistic regression models showed that agricultural land abandonment in Lithuania was significantly associated ($p < 0.05$) with low average grain yields in the late 1980s, isolated agricultural areas, and with increased distance from capital, district and municipality centers (Table 4). Hierarchical partitioning analysis showed that the 'low average grain yields in the late 1980s' variable, followed by 'isolated agricultural areas within forest matrix in 1990' and 'distances from provincial capital' (in the case of Lithuania-distance from Vilnius), had the greatest contribution for explaining agricultural land abandonment patterns (Fig. 7). Interpreting the odds ratios for Lithuania revealed that a decrease in crop yields of 0.1 ton/ha among districts (rayons) increased the probability of observing abandoned lands by 8 %, and each additional kilometer away from settlements increased the probability of abandonment of an agricultural plot by 38 % (Table 4). Isolated agricultural areas within the forest matrix in 1990 were 32 % more likely to be abandoned than non-isolated agricultural areas between 1990 and 2000.

In Belarus, in contrast to Lithuania, 'low average grain yields in the late 1980s' at the country level did not contribute to explaining the pattern of agricultural land abandonment. Variables depicting low environmental suitability for agriculture ('distance from nearest forest edge' and 'isolated agricultural areas within forest matrix in 1990'), proximities to settlements ('distance from nearest settlement') and infrastructure ('road density in late 1980s') were statistically significant ($p < 0.05$) and had high explanatory power, with a minor contribution of the variable 'slope' (Fig. 7). For each kilometer away from settlements, the likelihood of encountering an abandoned agricultural plot increased by 66 % (Table 4). In Belarus, isolated agricultural areas in 1990 were 34 % more likely to be abandoned by 2000.

Table 4 Regression results

Variable	Lithuania	Belarus				Russia					
	All	Belarus-all	Grodno	Vitebsk	Mogiljev	Russia-all	Smolensk	Kaluga	Tula	Rjazan	Vladimir
Soil pH	0.985	1.032	1.1543*	0.738*	1.119	0.96	0.98	1.167	1.448***	0.759*	1.26*
Slope	1.030**	0.968*	0.9612*	0.982	1.024	0.992	0.957**	0.966**	0.99	0.999	0.985
Average annual evapotranspiration	1.399	1.0465	1.458	0.551	3.11***	0.788	1.85	2.059*	0.741	0.411*	0.777
Distance from nearest forest edge	0.861	0.889***	0.887***	0.954	0.851***	0.961***	0.905***	0.887***	0.952***	0.962***	0.892***
Isolated agricultural areas within forest matrix in 1990	1.319*	1.346***	1.273**	2.473***	1.513**	1.484**	1.202	2.339***	0.982	0.891	2.48***
Average grain yields in the late 1980s	0.916***	1.028	1.0379	0.882***	0.945*	0.89***	0.933	0.898***	0.875***	0.851***	0.943*
Interpolated population counts from settlements in the late 1980s (the proxy for the population density)	1.0***	0.999	0.999	0.999	1.00	0.965*	0.949*	0.931	0.996	0.952	0.973
Distance from provincial capital	0.992***	1.002	1.001	1.028***	1.0176***	0.998	1.001	0.997	0.985***	1.006*	1.003
Distance from nearest district center	1.014*	1.004	1.006*	0.989	1.033***	1.006	1.007	0.996	1.025	0.998	1.019*

(continued)

Table 4 (continued)

Variable	Lithuania	Belarus				Russia					
	All	Belarus-all	Grodno	Vitebsk	Mogiljev	Russia-all	Smolensk	Kaluga	Tula	Rjazan	Vladimir
Distance from nearest municipality center	1.029**	0.978	1.008	0.989	0.936**	1.063***	1.105***	1.043	1.028	1.093***	1.019
Distance from nearest settlement with more than 500 people	1.014	1.023	0.969	1.0476**	1.0189	1.032***	1.017**	1.014	1.068***	1.059***	1.041*
Distance from nearest settlement	1.386***	1.664***	1.536***	1.596***	2.0928***	1.086*	1.256*	1.39***	1.074	0.971	0.964
Road density in the late 1980s	1.007	0.993	1.004	0.9849	0.996	1.001	1.082**	1.01	1.02	1.01	0.997
Distance from nearest hard-surfaced road	1.04***	1.0132***	1.023***	0.999	1.09*	1.004	1.015*	1.017**	1.02***	0.989	1.009
Number of sampled '0'	8500	4500	2500	805	1195	93671	14081	16285	15 314	33919	14072
Number of sampled '1'	8529	4328	2495	621	1212	39018	11083	5413	5 817	12109	4596
AUC	0.708	0.635	0.64	0.69	0.686	0.7	0.68	0.752	0.653	0.745	0.748
Adjusted R^2	0.17	0.072	0.08	0.146	0.138	0.144	0.131	0.213	0.085	0.203	0.199

Odds ratios in bold indicate significance at $p < 0.05$
*significance at $p < 0.05$
**significance at $p < 0.01$
***significance at $p < 0.001$

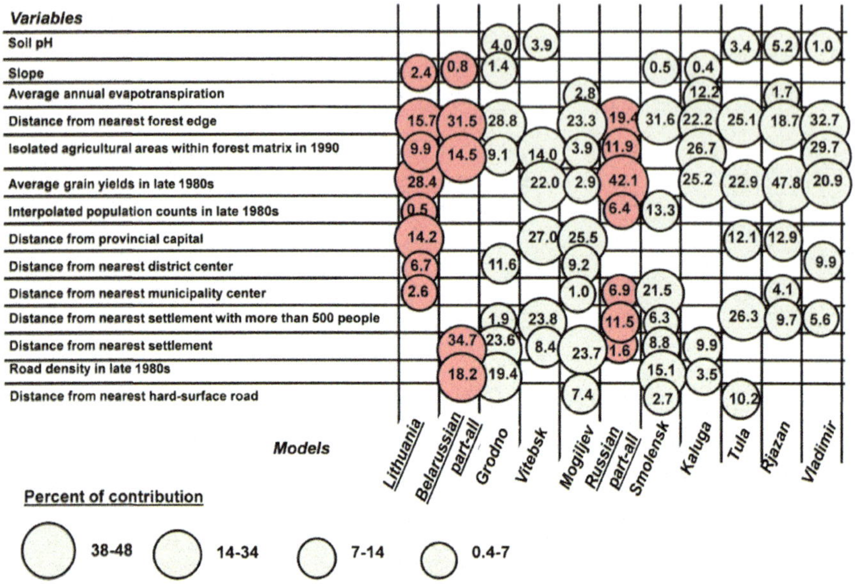

Fig. 7 Results of the hierarchical partitioning analysis for statistically significant variables. *Pinkish circles* represent the summaries for the studied parts of selected countries. *Light grey circles* represent summaries at province for the studied provinces of selected countries

Similar to Lithuania, the pattern of agricultural land abandonment in European Russia was significantly associated ($p < 0.05$) with low average grain yields in the late 1980s, indicators of low environmental suitability for agriculture, low rural population density, and increased distance from settlements (Table 4; Fig. 7). For instance, a decrease in crop yields of 0.1 ton/ha among districts (rayons) increased the probability of agricultural land abandonment between 1990 and 2000 by 11 %. For every kilometer away from settlements, the probability of agricultural land abandonment from 1990 to 2000 increased by 8 % (Table 4). As was the case in Lithuania, hierarchical partitioning analysis showed that in temperate European Russia the 'average grain yields in the late 1980s' variable had the highest explanatory power, followed by indicators of low environmental suitability for agriculture and proximities to settlements (Fig. 7).

Interestingly, in all three countries, the variables that were the least important for explaining abandonment often represented biophysical conditions ('soil pH', 'slope', 'average annual evapotranspiration'). Overall, the province-level results in Belarus and Russia were similar to the models for the country level, but there were differences in the strength of the explanatory determinants and their contribution to the overall model fit (Table 4; Fig. 7).

3 Discussion

Our analysis showed that the political, institutional, and socioeconomic changes in Eastern Europe after the collapse of the Soviet Union resulted in widespread agricultural land abandonment during the first decade of transition from state-command to market-driven economies. Because our study design minimized agro-climatic differences across Lithuania, Belarus and Russia, the differences of agricultural land abandonment rates most likely reflected the effects of underlying drivers, such as institutional changes regarding land tenure and land use and the importance of the agricultural sector in the economy at large. The disappearance of guaranteed markets for agricultural production within the Soviet Union and of subsidized fuel, machinery and fertilizers, as well as a lack of competitiveness compared to imported agricultural products, also contributed to the decline in agricultural production.

Despite our expectation of mid-level abandonment rates in Belarus if this country would experience "pure" liberalized economic conditions, we observed the lowest abandonment rates in Belarus. This was most likely due to ongoing strong governmental regulation of the agricultural sector in Belarus during the transition. Despite the initial plans to privatize agricultural lands, in 1994, the Belarusian government reversed the privatization of agricultural land and capital assets of state and collective farms. Similar to what happened during the Soviet era, subsidies and a complex system of offsets among Belarusian state enterprises ensured that state and collective farms continued to receive cheap fertilizers, fuel, and equipment and that farms could sell agricultural products at fixed prices. State and collective farms retained their key social role in the countryside, thus providing jobs, housing, and social infrastructure. Furthermore, among the studied countries, the contribution of the agricultural sector to the total GDP was highest in Belarus, both on the eve of the transition period and after one decade of transition. However, by 2000 the value added by agriculture had substantially declined (Fig. 2). Indeed, in the cross-border region of Belarus (Mogilev province) and Russia (Smolensk province), we observed that ~70 % of agricultural enterprises were unprofitable in Mogilev (Belarus) in 2000 but were kept running with the support of subsidies and other offsets (Belstat 2002). At the same time, approximately 80 % of the agricultural enterprises were unprofitable in the neighboring Smolensk province of Russia, but almost all agricultural activity ceased due to the absence of governmental support and well-functioning markets (Rosstat 2002). This might explain the differences in the rates of agricultural land abandonment among the neighboring countries (Fig. 3).

In Russia, however, we observed consistently high rates of agricultural land across all studied provinces, which may indicate that the underlying driving forces of abandonment (e.g., institutional changes and policies) operated at the national scale and affected the studied provinces equally.

Analysis of the spatial determinants of agricultural land abandonment in Lithuania, European Russia and Belarus suggested that agricultural land abandonment was highest in the areas that already had low productivity during the

Soviet era, as indicated by low grain yields in the late 1980s, and where the environmental conditions for agricultural production were less favorable for agriculture (proximity to forest edges and in isolated agricultural areas within the forest matrix). Areas far from populated places were also often prone to abandonment.

The models at the province level for European Russia generally reflected the same determinants as for the entire study area of European Russia. However, in the case of Belarus, there were some interesting differences between models at different scales. Vitebsk province in Belarus had lower agro-environmental suitability for crop production (i.e., for wheat production) than Grodno and Mogiljev provinces and was the only Belarusian province where 'average grain yields in late 1980s' variable contributed to explaining agricultural land abandonment patterns among selected Belarusian provinces. However, we do note that Vitebsk province had low abandonment rates. At first glance, this might be surprising, especially if the differences in abandonment rates at the province level are taken into account. The rates differed between Vitebsk and the neighboring Russian province of Smolensk, which had the most similar agro-environmental conditions of province pairs within our study area.

Interestingly, 'settlements with more than 500 people' was statistically significant variable explaining land abandonment in Belarusian and Russian provinces, but not in Lithuania. This variable is a proxy for important infrastructural networks, the support of which was crucial for agricultural production and social infrastructure during the Soviet era. However, in Lithuania, there is a complex settlement pattern compared Belarus and Russia. This reflects the historical legacies of co-existing individual farmsteads and large settlements in Lithuania, where state-and collective farms had their headquarters during the Soviet era. The travel distances between populated areas were much smaller in Lithuania than in Belarus and Russia. We hypothesize that such infrastructural differences were the reason why the variable 'distance to nearest settlement' contributed to explaining agricultural land abandonment patterns in Russia (Smolensk and Kaluga provinces) and in Belarus, but not in Lithuania.

Finally, the variable 'interpolated population counts from settlements in late 1980s' was statistically significant for the Russian province of Smolensk but contributed surprisingly little to explaining abandonment patterns in Lithuania. Such a departure from our initial hypothesis about the relationship of agricultural land abandonment and low population densities could be due to the disaggregation (interpolation) approach of population counts we chose. Another explanation is that the effects of demographic changes on land use were not yet visible by 2000. In fact, Smolensk was the only Russian province, where the variable 'interpolated population counts from settlements in late 1980s' had a modest contribution in explaining abandonment patterns. This may be due to the high percentage of sown areas out of the total available land used for agriculture in 1990 in Smolensk province and to the density of cattle, which was the densest of all the studied Russian provinces. In addition, rural population density, road density, and crop yields during the Soviet era were the lowest in Smolensk province among the studied Russian provinces (Table 1). Once subsidies were removed, this province experienced massive agricultural land abandonment, with the highest abandonment rates from 1990 to 2000 of all the studied Russian provinces.

To summarize the results for our case study, one of the main lessons from the regression results was that market principles increasingly shaped agricultural land use, particularly in Lithuania and Russia, during the first decade of transition from state-command to a market-driven economy. Agricultural land use patterns moved away from the subsidized Soviet-style agricultural pattern, particularly in Lithuania and Russia where the government fostered agricultural land expansion into marginal areas, and towards landscapes that were predominantly shaped by economic forces with much less governmental intervention. Interestingly, selected Russian provinces had a greater amount of sown areas of the total land used for agriculture in 1990s compared with Lithuania and Belarus, whereas infrastructure, population density and productivity were low (Table 2). It is likely after the collapse of the Soviet Union socio-economically and agro-environmentally marginal areas subsidized during the Soviet ere were abandoned first. Rapid decline in labor availability and removal of the subsidies, particularly from capital intensive livestock production (Schierhorn et al. 2013), caused unprecedent rates of agricultural land abandonment in our study area after the collapse of the Soviet Union.

4 Conclusion

Underlying drivers, e.g., changing institutional practices regarding land use, can play a key role in shaping land cover and land use. The countries with drastic institutional changes regarding land use, land tenure and agricultural production had the highest abandonment rates in our case study. However, pre-socialist land tenure legacies (i.e., in Lithuania) and variation in socio-economic development and agricultural performance during and after the Soviet era shaped agricultural land abandonment patterns in our study area. In general, socio-economically and agro-environmentally marginal areas were abandoned first once the subsidies were removed. Knowledge of the impact reforms have on land use is important because land reforms and institutional changes are common and many post-soviet countries are still in transition (i.e., Belarus and Russia). Thus, it is essential to be able to predict the potential impact of possible transition approaches for creation of effective land use policies. This study is also an excellent example of the possible integration of satellite remote sensing data on dynamics of land use and land-use modeling to better understand the rates and patterns of agricultural land abandonment and their underlying and proximate drivers. Moreover, the occurrence of idle agricultural and biofuel production potential (Schierhorn et al. 2014; Meyfroidt et al. 2016), the possibility for carbon sequestration (Kurganova et al. 2013; Schierhorn et al. 2013), and the effects on biodiversity when forests regrow in abandoned areas suggests important avenues for future research in analyzing the trade-offs on abandoned agricultural lands in post-Soviet Eastern Europe (Meyfroidt et al. 2016; Smailichuk et al. 2016). This opportunity for research is particularly relevant in the face of ongoing agricultural land abandonment and abandoned lands recultivation across Lithuania, Belarus and Russia, but also in other post-Soviet states.

Acknowledgments We gratefully acknowledge support for this research by the NASA Land Cover and Land Use Change program (NASA NNX13AC66G), the German Science Foundation (DFG, project LUCC-BIO-1), the European Commission (projects VOLANTE and HERCULES), the GERUKA project funded by the German Ministry of Food and Agriculture (BMLE) and EPIKUR (Leibniz Foundation).

References

Alcantara C, Kuemmerle T, Baumann M, Bragina EV, Griffiths P, Hostert P, Knorn J et al (2013) Mapping the extent of abandoned farmland in Central and Eastern Europe using MODIS time series satellite data. Environ Res Lett 8(3) (September 1):035035

Alix-Garcia J, Kuemmerle T, Radeloff VC (2012) Prices, land tenure institutions, and geography: a matching analysis of farmland abandonment in post-socialist Eastern Europe. Land Econ 88(3):425–443

Batijes NH (2001) Soil data for land suitability assessment and environmental protection in Central Eastern Europe—the 1:250 0000 scale SOVEUR project. Land 5(1):51–68

Baumann M, Kuemmerle T, Elbakidze M, Ozdogan M, Radeloff VC, Keuler NS, Prishchepov AV, Kruhlov I, Hostert P (2011) Patterns and drivers of post-socialist farmland abandonment in Western Ukraine. Land Use Policy 28(3):552–562

Baumann M, Radeloff VC, Avedian V, Kuemmerle T (2014) Land-use change in the Caucasus during and after the Nagorno-Karabakh conflict. Reg Environ Change

Belstat (2002) Regions of Belarus republic. Statistical compendium (Regioni Respubliki Belarus. Statisticheskii Sbornik), vol V1. BELSTAT, Minsk

Chevan A, Sutherland M (1991) Hierarchical partitioning. Am Stat 45:90–96

deBeurs KM, Ioffe G (2013) Use of landsat and MODIS data to remotely estimate Russia's Sown Area. J Land Use Sci 1–25, 22 May 2013

de Beurs K, Ioffe G, Nefedova T, Henebry G (2017) Land Change in European Russia: 1982–2011. In: Gutman G, Radeloff V (eds) Land-Cover and Land-Use Changes in Eastern Europe after the Collapse of the Soviet Union in 1991. doi:10.1007/978-3-319-42638-9_10

DeLeo J (1993) Receiver operating characteristic laboratory ROCLAB: software for developing decision strategies that account for uncertainty. In: Proceedings of the second international symposium on uncertainty, modeling and analysis. IEEE Computer Society Press, College Park, pp 318–325

Diamond J (2001) Ecology—Dammed experiments! Science 294(5548) (November 30):1847–1848

Folch R (2000) Encyclopedia of the biosphere: deciduous forests, vol 7. Gale Group, Detroit

Foley JA, DeFries R, Asner GP, Barford C, Bonan G, Carpenter SR, Chapin FS et al (2005) Global consequences of land use. Science 309(5734) (July 22):570–574

Fox J, Rindfuss RR, Walsh SJ, Mishra V (2003) People and the environment: approaches for linking household and community surveys to remote sensing and GIS. Springer, Berlin

Gellrich M, Baur P, Koch B, Zimmermann NE (2007) Agricultural land abandonment and natural forest re-growth in the Swiss Mountains: a spatially explicit economic analysis. Agric Ecosyst Environ 118(1–4):93–108

Goskomstat (1991) National economy of USSR in 1990 (Narodnoe Hozyaistvo v 1990). Finansy i Statistika, Moscow

Goskomstat LitSSR (1989) Statistical yearbook of Lithuanian SSR in 1988 (Statisticheskii Ezhegodnik Litovskoi SSR v 1988). Information-Publishing Center, Vilnius
Griffiths P, Müller D, Kuemmerle T, Hostert P (2013) Agricultural land change in the Carpathian ecoregion after the breakdown of socialism and expansion of the European Union. Environ Res Lett 8(4):045024
Hostert P, Kuemmerle T, Sieber A, Prishchepov AV, Lambin EF, Radeloff VC (2011) Rapid land use changes after socioeconomic disturbances: chernobyl and the collapse of the Soviet Union. Environ Res Lett
Huber PJ (1967) The behavior of maximum likelihood estimates under nonstandard conditions. In: Proceedings of the fifth Berkeley symposium on mathematical statistics and probability, vol 1. University of California Press, Berkeley, pp 221–233
IIASA (2000) Global agro-ecological zones (Global-AEZ) CD-ROM FAO/IIASA. Available: www.iiasa.ac.at/Research/LUC/GAEZ/index.htm. Accessed 2014. IIASA
Ioffe G, Nefedova T, Zaslavsky I (2004) From spatial continuity to fragmentation: the case of Russian farming. Ann Assoc Am Geogr 94(4) (December):913–943
Irwin EG, Bockstael NE (2002) Interacting agents, spatial externalities and the evolution of residential land use patterns. J Econ Geogr 2(1) (January 1):31–54
Kaimowitz D, Mertens B, Wunder S, Pacheco P (2004) Hamburger connection fuels Amazon destruction: cattle ranching and deforestation in Brazil's Amazon, 23 Jan 2009
Kontorovich V (2001) The Russian health crisis and the economy. Communist Post-Communist Stud 34(2):221–240
Kuemmerle T, Radeloff VC, Perzanowski K, Hostert P (2006) Cross-border comparison of land cover and landscape pattern in Eastern Europe using a hybrid classification technique. Remote Sens Environ 103(4) (August 30):449–464
Kuemmerle T, Hostert P, Radeloff V, van der Linden S, Perzanowski K, Kruhlov I (2008) Cross-border comparison of post-socialist farmland abandonment in the Carpathians. Ecosystems 11:614–628
Kurganova I, Lopes de Gerenyu V, Six J, Kuzyakov Y (2013) Carbon cost of collective farming collapse in Russia. Global Change Biol (20):938–947
Kraemer R, Prishchepov AV, Müller D, Kuemmerle T, Radeloff VC, Dara Andrey, Terekhov A, Frühauf M (2015) Long-term agricultural land-cover change and potential for cropland expansion in the former Virgin Lands area of Kazakhstan. Environ Res Lett 10(5):054012
Lambin EF, Turner BL II, Geist HJ, Agbola SB, Angelsen A, Bruce JW, Coomes OT, Dirzo R, Fischer G, Folke C, George PS, Homewood K, Imbernon J, Leemans R, Li X, Moran EF, Mortimore M, Ramakrishnan PS, Richards JF, Skanes H, Steffen W, Stone GD, Svedin U, Veldkamp TA, Vogel C, Xu J (2001) The causes of land-use and land-cover change: moving beyond the myths. Glob Environ Change 11(4):261–269
Lambin EF, Meyfroidt P (2010) Land use transitions: socio-ecological feedback versus socio-economic change. Land Use Policy 27(2):108–118
Lerman Z, Shagaida N (2007) Land policies and agricultural land markets in Russia. Land Use Policy 24(1):14–23
Lerman Z, Csaki C, Feder G (2004) Agriculture in transition: land policies and evolving farm structures in post-Soviet Countries. Lexington Books, Lanham
Lithstat (2014) Lithuanian statistics online database. Available: http://www.stat.gov.lt/en/. Accessed 2014
Lyuri DI, Goryachkin SV, Karavaeva NA, Denisenko EA, Nefedova TG (2010) Dynamics of agricultural lands of Russia in the XXth century and postagrogenic restoration of vegetation and soils (Dinamika selskohozjaistvennih zemel Rossii v XX veke I postagrogennoje vosstanovlenije rastitelnosti i pochv). GEOS, Moscow
Mac Nally R (1996) Hierarchical partitioning as an interpretative tool in multivariate inference. Aust J Ecol 21:224–228
Meyfroidt P, Schierhorn F, Prishchepov AV, Müller D, Kuemmerle T (2016) Drivers, constraints and trade-offs associated with recultivating abandoned cropland in Russia, Ukraine and Kazakhstan. Glob Environ Change 37:1–15

Millington J, Perry G, Romero-Calcerrada R (2007) Regression techniques for examining land use/cover change: a case study of a Mediterranean Landscape. Ecosystems 10(4):562–578
Müller D, Munroe DK (2008) Changing rural landscapes in Albania: cropland abandonment and forest clearing in the postsocialist transition. Ann Assoc Am Geogr 98(4):855–876
Müller D, Kuemmerle T, Rusu M, Griffiths P (2009) Lost in transition: determinants of post-socialist cropland abandonment in Romania. J Land Use Sci 4:109–129
Müller D, Sikor T (2006) Effects of postsocialist reforms on land cover and land use in South-Eastern Albania. Appl Geogr 26(3–4):175–191
Nefedova T (2011) Agricultural land in Russia and its dynamics. Reg Res Russia 1(3):292–295
Nikodemus O, Bell S, Grine I, Liepins I (2005) The impact of economic, social and political factors on the landscape structure of the Vidzeme uplands in Latvia. Landscape Urban Plann 70 (1–2) (January 15):57–67
Novozhenina O, Baharev I, Mollicone D (2009) 'Hard-Okorok'('Hard-Hock'). Gazeta.ru (January 23): 23 Jan 2009
Peterson U, Aunap R (1998) Changes in agricultural land use in Estonia in the 1990s detected with multitemporal landsat MSS imagery. Landscape Urban Plann 41(3–4) (July 6):193–201
Pontius RG, Schneider LC (2001) Land-cover change model validation by an ROC method for the Ipswich Watershed, Massachusetts, USA. Agric Ecosyst Environ 85(1–3):239–248
Prishchepov AV, Radeloff VC, Dubinin M, Alcantara C (2012a) The effect of landsat ETM/ETM+ image acquisition dates on the detection of agricultural land abandonment in Eastern Europe. Remote Sens Environ 126(November):195–209
Prishchepov AV, Radeloff VC, Baumann M, Kuemmerle T, Müller D (2012b) Effects of institutional changes on land use: agricultural land abandonment during the transition from state-command to market-driven economies in post-Soviet Eastern Europe. Environ Res Lett 7 (2) (June 1):024021
Prishchepov AV, Butsic V, Radeloff VC (2012c) Sensitivity of spatially explicit land-use logistic regression models to the errors land-use change maps. In: International environmental modelling and software society (iEMSs) 2012 international congress on environmental modelling and software managing resources of a limited planet, sixth Biennial meeting, vol 8, Leipzig, Germany
Prishchepov AV, Müller D, Dubinin M, Baumann M, Radeloff VC (2013) Determinants of agricultural land abandonment in post-Soviet European Russia. Land Use Policy 30(1) (January):873–884
R Development Core Team (2011) R: a language and environment for statistical computing, Vienna, Austria. Available: http://www.R-project.org. Accessed 2014
Rosstat (2002) Regions of Russia. Socio-economic indicators (Regiony Rossii. Sotsial'no-ekonomicheskie pokazateli). Russian federal service of state statistics, Moscow, Russia. Available: http://www.gks.ru. Accessed 2014
Rozelle S, Swinnen JFM (2004) Success and failure of reform: insights from the transition of agriculture. J Econ Lit 42(2):404–456
Sakovich VS (2008) Agriculture in Belarus between 1980–2007: tendency of the development (Selskoe Hozjaistvo v Respublike Belarus v 1980–2007g.: tendencii razvitija). Belorusskaja nauka, Minsk
Sieber A, Kuemmerle T, Prishchepov AV, Wendland KJ, Baumann M, Radeloff VC, Baskin LM, Hostert P (2013) Landsat-based mapping of post-Soviet land-use change to assess the effectiveness of the Oksky and Mordovsky protected areas in European Russia. Remote Sens Environ 133:38–51
Schierhorn F, Müller D, Beringer T, Prishchepov AV, Kuemmerle T, Balmann A (2013) Post-Soviet cropland abandonment and carbon sequestration in European Russia, Ukraine and Belarus. Global Biogeochem Cycles (November) (27):1–11
Schierhorn F, Müller D, Prishchepov AV, Faramarzi M, Balmann A (2014) The potential of Russia to increase its wheat production through cropland expansion and intensification. Glob Food Secur 3(3–4):133–141

Smailichuk A, Müller D, Prishchepov AV, Levers C, Kruhlov I, Kuemmerle T (2016) Recultivation of abandoned agricultural lands in Ukraine: patterns and drivers. Glob Environ Change

Stuikys V, Ladyga A (1995) Agriculture of Lithuania. Valystybinis Leidybos Centras, Vilnius

Václavík T, Rogan J (2009) Identifying trends in land use/land cover changes in the context of post-socialist transformation in Central Europe: a case study of the greater Olomouc Region, Czech Republic. GISci Remote Sens 46(1):54–76

White H (1982) Maximum likelihood estimation of misspecified models. Econometrica 50(1):1–25

World Bank (2013) World development indicators. Available: http://data.worldbank.org/data-catalog/world-development-indicators/

The Effects of Institutional Changes on Landscapes in Ukraine

V. Lyalko, S. Ivanov, V. Starodubtsev and J. Palamarchuk

Abstract Ukraine has a great variety of natural landscapes because the country contains parts of four landscape zones and two mountain regions. However, most of Ukraine's landscapes have been severely altered by human activities, especially agriculture, mining, and industrial activities. These landscape changes had profound effects on ecosystem processes. For example, according to our data for the western part of Ukraine, changes in vegetated areas decreased the amount of absorbed carbon in 2000 by one thousand tons compared to the 1990 value. Thus, a larger fraction of unabsorbed carbon remained in the atmosphere, contributing to a stronger "greenhouse effect". Similarly, the excessive use of water resources since the 19th century decreased the water and sediment inflow into the deltas of the Black Sea basin and changed their landscapes. In summary, socio-economic transformations after the breakup of the Soviet Union together with climate change effects have been widespread in Ukraine.

1 Dynamics of Landscape-Forming Processes

The landscape changes in Ukraine are, as in other parts of the world, determined by natural and anthropogenic factors. Ultimately, these changes are the result of complex interaction of exogenous, endogenous and anthropogenic landscape-forming processes, the first two groups of processes being caused by different climatic conditions and the dynamics of endogenous processes.

V. Lyalko
Scientific Centre for Aerospace Research of the Earth, Institute of Geological Sciences, National Academy of Sciences of Ukraine, Kiev, Ukraine

S. Ivanov (✉) · J. Palamarchuk
Odessa State Environmental University, Odessa, Ukraine
e-mail: svvivo@te.net.ua

V. Starodubtsev
National University of Life and Environmental Sciences, Kiev, Ukraine

G. Gutman and V. Radeloff (eds.), *Land-Cover and Land-Use Changes in Eastern Europe after the Collapse of the Soviet Union in 1991*,
DOI 10.1007/978-3-319-42638-9_6

The main climatic factors that play a role in the spatial organization and intensity of exogenous landscape-forming processes include solar radiation and atmosphere circulation, which are modulated by the surface heterogeneity, such as topographic variability, water areas, woods, swamps, agricultural lands, and so on. These main climate and landscape factors must be considered when examining landscape changes.

In this chapter, we focus on the effects of institutional changes after the collapse of the Soviet Union on landscape changes. In general, anthropogenic landscape-forming processes include construction activities (e.g., hydrotechnical, road, industrial, civil, and pipeline construction), the extraction of natural resources (e.g., coal, ores, construction materials, petroleum, gas, and water), and landscape changes from agricultural land use (e.g., farming and melioration), as shown in Fig. 1. The most drastic landscape changes are from mining activities, which are particularly intensive in the Donbas region of Ukraine. Since 2000, technogenic deposits have increased by 10 cm/year, and mine tailings and other disposals occupy an area over 220 km^2. Indeed, the total area of anthropogenic accumulative landscape forms in Donbas, i.e., areas where deposits cover the prior surface, increased by a factor of 11.5 during the 20th century. In 1998, destructive forms and cauldrons above mines (with an average depth of 2–5 m), which are even more widespread, covered an area over 10,000 km^2 (*Information Bulletin on the state of the geological environment of Ukraine in 1998, 2000*).

Mining is only one part of the overall anthropogenic landscape modifications and is not the most widespread, although it is very intensive where it occurs. For example, landscape changes from agricultural development before the 1990s occurred over 41.89 million ha (out of which 5.89 million ha were meliorated lands), transportation infrastructure occurred over 10.69 million ha, mining occurred over approximately 2.0 million ha, industrial and military construction occurred over 2.27 million ha, and settlements covered almost 8.42 million ha. Furthermore, the rates of recent anthropogenic processes often exceed those of natural processes by an order of magnitude or more.

However, mining activities represent a particularly drastic type of landscape change. For example, major transformations in the natural landscape occurred in the Krivoy Rog Iron Ore Basin, where the industrial mining of iron ores started in 1881. At the end of the 1990s, the depths of the open pit mines reached 160–390 m, and individual pits were 80–780 ha in size. Open pit mines of such scale result in major tailings, where rocks that are found above the ores are deposited. The height of these tailings is between 30 and 84 m, their area ranges from 12 to 900 ha, and their length ranges from 0.5 to 4 km. During the last 50 years alone, the volume of all the tailings in the basin increased by a factor of 25. The open pit mines in the Krivoy Rog Iron Ore Basin at the beginning of the 21st century covered 33.34 km^2, the tailings covered 60.0 and 52.4 km^2, the water reservoirs covered 50 km^2, and the surface subsidence zone above the mine fields covered approximately 34.71 km^2. Furthermore, industrial enterprises covered 159.0 km^2. On the whole, the natural landscape in the basin was changed over a total area of 389.75 km^2, which represents 55.7 % of the general ore claim area; sub-flooding zones,

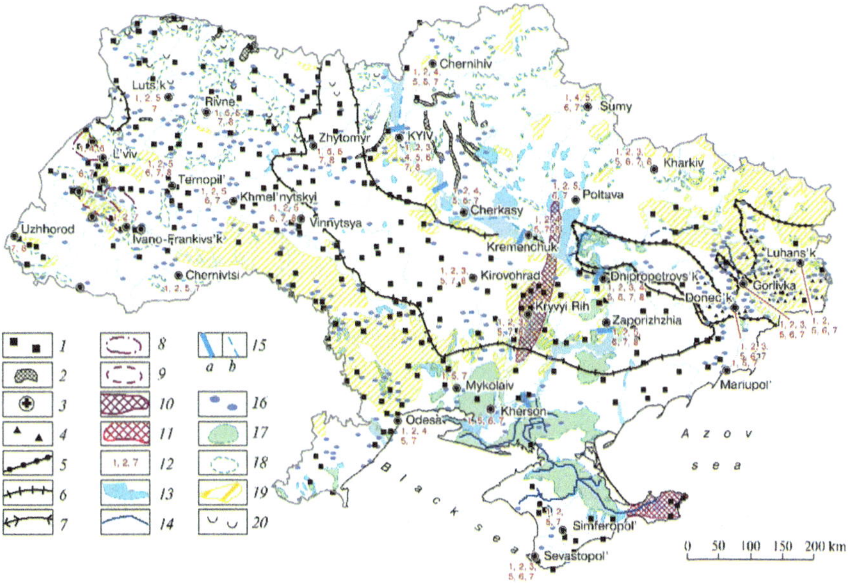

Fig. 1 Recent anthropogenic processes in Ukraine. Types of anthropogenic landscape transformation from *mining activity: 1*—quarries; *2*—peat removal; *3*—technogenic karst; *4*—mining tailings. *Mining regions with current and potential future anthropogenic landscape transformation: 5*—Donetsk mineral coal beds; *6*—Dnieper brown coal beds; *7*—Lviv-Volyn mineral coal beds; *8*—Yavoriv and Razdollia sulfur deposits; *9*—Stebnitsia-Kalush salt deposits; *10*—Krivoy Rog-Kremenchug iron ore; *11*—Kerch iron ore. *Construction activities within urbanized territories: 12*—towns with significant technogenic landscape transformations (*red digits* show the type of influence upon the landscape: (*1*) road construction; (*2*) rock and gravel extraction; (*3*) subsurface buildings; (*4*) washing up of terraces; (*5*) road embankments; (*6*) filling of ravines; (*7*) accumulation of technogenic (cultural) layers; and (*8*) dams and pools). *Water management activities: 13*—sub-flooding; *14*—channels; *15*—construction of **a** pressure dams and **b** protective dams; *16*—construction of ponds for waste, biological materials, etc. *Melioration: 17*—erosion from irrigation, soil silification, changes in the micro-landscape, secondary soil salinization, and suffusion-subsidence phenomena that are associated with irrigation; *18*—compaction and mineralization of peat, accelerated deflation, degradation of water courses, etc. *Agricultural activities*: *19*—accelerated erosion on tilled lands; *20*—deflation

human-caused landslides, collapses and other processes were not considered in these calculations.

Other regions in Ukraine where landscape changes from mining occurred include the Nikopol manganese ore basin and the Cis-Carpathian sulfur- and salt-bearing regions. In contrast, mining activities are less widespread in the Dnieper brown coal basin, while the Bakhmut and Dokuchayevsk mine-industrial districts here are gravitational, and karst processes are intensifying. Indeed, the rate of karst processes in open pit mines for sulfur mining in the Cis-Carpathians exceed the rate of natural karst formation by an order of magnitude.

Closely related to industrial development is the construction and operation of hydro-power stations, which has caused major changes in the natural landscape, such as along the cascade of reservoirs along the Dnieper basin, i.e., Kiev, Kanev, Kremenchug, Dneprodzerzhinsk, Dnieper, and Kakhovka. In addition to the flooding of large areas in the Dnieper's flood plains and even parts of its first terrace, these reservoirs caused gravitational, karst and other processes along their shores and the Dnieper River's banks. In general, the highest intensity of processes along the shores occurred within the first 5–10 years after the flooding of the reservoirs. The influence zone of the reservoirs also indicate sub-flooding, erosional and other processes.

Water consumption greatly increased throughout Ukraine during the 20th century, necessitating a high degree of water regulation such as the construction of ponds and reservoirs along small rivers and channels. Many anthropogenically induced processes have occurred along the shores and banks of these ponds and channels, including collapses, landslides, subsidence events, and karst formation. Sub-flooding processes are more common in the upper parts of the reservoirs and ponds, while superfluous drainage and deflation processes are more widespread downward from dams. Ultimately, the high level of regulation of surface drainage patterns has resulted in the degradation of many small rivers and creeks, changes in the hydrographic network pattern, and a reduction in the overall length of rivers and streams.

Concomitant with landscape changes that were designed to secure a constant water supply, major landscape changes have occurred throughout Ukraine over many decades to reduce water via drainage and irrigational water melioration. Before 1990, approximately 3.0 million ha of wetlands were drained in Ukraine, which changed the natural landscape of these former wetlands because of leveling, smoothing, and the construction of open drainage channels and drainage pipes. Furthermore, sub-flooding and irrigational erosion, karst, activation, and gravitational processes have become widespread because of broad-scale irrigation in southern Ukraine (more than 2.5 million ha were irrigated prior to 1990).

The building and operation of transportation infrastructure, especially railroad lines and roads, also substantially influenced the landscape morphology and triggered landscape-changing processes. For example, approximately 300 active landslides occurred along the railroads of Ukraine in 1995 compared to only 70 landslides in 1950. Active landslides can endanger regular railroad operations and are particularly common in the Carpathians and the mountainous part of Crimea, particularly near Sevastopol. Furthermore, the deformation of weak ground because of the weight of railroad embankments has occurred in Ukrainian Polesye and the Cis-Carpathians, and sub-flooding processes have been observed along numerous segments of railroads and artificial embankments.

During the construction of highways, topographic features are leveled via the movement of large amounts of soil, the removal of solid rocks, and so on. These activities often trigger landslides, karst processes, and subsidence. Numerous landslides have occurred along Ukrainian motorways, especially in mountainous regions. Furthermore, numerous landslides, crumbles, and karst forms are present along some of the motorways in the plains of Ukraine.

Similar landscape changes occur from the construction and operation of pipelines, which can cause landslides, crumbles, and the deformation of river channels. The establishment of construction corridors for pipelines and access roads and the development of trenches and tailings for unwanted rocks all transform the landscape and create instability. Depending on the diameter of the pipelines, the widths of their construction corridors range from 20 to 45 m, so the influence upon the landscape and its stability occurs over a much greater area when the pipelines are larger. For example, our data show a substantial number of old and recent landslides that were caused by the construction of the gas pipeline Urengoy-Uzhgorod in the beginning of the 1980s in the Carpathian Mountains. In the plains region of Ukraine, the construction and operation of major pipelines has led to higher rates of gravitational and karst processes.

Agriculture, particularly farming, changes landscapes by causing both sheet and linear erosion along with soil deflation. However, at the end of the 1990s, the areas of ploughed lands that were subjected to sheet flooding, along with this phenomenon's intensity, somewhat decreased because of the abandonment of agricultural land. The intensity of linear erosion on agricultural sites, on the other hand, has increased.

Forest logging in the mountain areas of the Carpathians and Crimea has intensified mudflows, especially during the late 1990s and early 2000s, with the frequency of mudflows increasing from every 5–10 years in the middle of the 20th century to every 2–3 years at the end of the 1990s. The number of mudflow basins has also increased.

Urban development is another powerful factor that changes the natural landscape. At the beginning of the 2000s, 19 large industrial-urban agglomerations existed in Ukraine. Considering the geomorphologic conditions where these agglomerations are located, we suggest the following types of urbanized territories: "near-river", which are located within the limits of large river valleys (Kiev, Dnepropetrovsk, Dneprodzerzhinsk); "seaside" (Odessa, Mariupol, Nykolayev); "watershed" (Gorlovka-Enakievo, Donetsk-Makeevka); and "watershed-slope" (Lvov, Kharkov, Krivoi Rog). New landscape formations in cities include artificial reservoirs and terraces, trenches, tunnels, pits, ditches, rock disposals, small hills, quarries, depression funnels, embankments, collapses, dumps, and others. Negative technogenically induced geomorphologic processes in the cities of Ukraine include sub-flooding, technogenic mudflows, and land subsidence. Accelerated sheet flooding and linear erosion, the accumulation of technogenic grounds, the activation of gravitational processes, and some cases of swamping or, on the contrary, the over-drainage of territories remain common in all urbanized territories. Indeed, 327 towns and settlements of urban type in late 1990s that existed in Ukraine needed protection against different dangerous landscape-changing processes (Kysel'ov 2000).

In summary, major landscape changes occurred across the entire Ukraine due to new industrial technologies, construction of large economic enterprises, transportation infrastructure, and the use of natural resources; as well as interactions between natural endogenous and exogenous processes.

2 Natural Landscapes and Recent Changes

Ukraine has a great variety of natural landscapes because the country contains parts of four landscape zones and two mountain regions. An overview of the landscapes of Ukraine is presented in the *Landscapes map*, which was created in 2004 at a scale of 1:2.5 million (Fig. 2) (National Atlas of Ukraine 2007).

The map shows the landscape structure of Ukraine, where taxonomic variety is represented by two classes of landscapes (plain and mountain), 6 types, 38 genera, and 157 types of landscapes, for which area summaries are presented in Table 1.

While the map and statistics highlight the high level of landscape diversity in Ukraine, even more spatial and regional variety has been formed by natural processes and human interventions.

The intensity of the interactions among people, society and nature during the 20th century depended partly on socio-economic factors and the development level

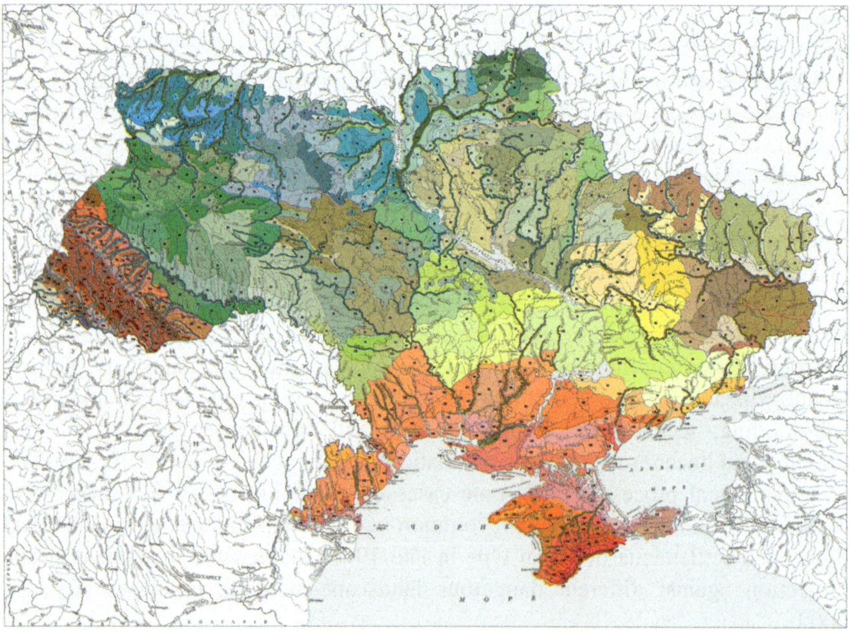

Fig. 2 Natural landscapes of Ukraine

Table 1 Landscape variety of Ukraine

Natural zones	Area, $10^3 \times km^2$	Percent area of Ukraine	Landscape types	
			Number	% of total
I. Mixed forest zone	91.5	15.2	22	14.6
II. Broad-leaf forest zone	43.7	7.2	12	7.9
III. Forest-steppe zone	190.6	31.6	28	18.5
IV. Steppe zone	238.1	39.4	57	37.7
V. Mountain regions	39.8	6.6	32	21.2
Total in Ukraine	603.7	100	151[a]	100

[a]Additionally, six types of floodplain regions are present: 3 in the mountains and 3 in the valleys of lowland rivers

of productive forces. Some of these interactions resulted in the deterioration of the environment and caused management problems and changes in natural landscape properties.

Landscapes always change, and all the landscape associations in Ukraine have been and will be subjected to various types of natural changes according to their natural features, genesis, morphology, and recent state. However, human activities have caused changes in landscape components that have resulted in the disappearance of natural landscapes in many of Ukraine's regions by the late 1990s, depending on their natural conditions, resources and level of socio-economic development, especially along the "Right Bank" of Ukraine, which has experienced substantial landscape changes from human activities, reaching 92–95 % by the late 1990s (Denisik 1998). In the following, we present some examples of anthropogenic changes of natural landscapes.

The most substantial changes in Ukraine's natural landscapes were caused by technogenic and geochemical activities such as chemical pollution, including emissions from metallurgy works and power plants, fuel and power stations, military complexes, auto transportation, and crop fertilization. Heavy metals are particularly concentrated near ferrous and non-ferrous metallurgical enterprises (35 %), thermal and other power stations (27 %), petrochemical enterprises (16 %), highways (13 %), and enterprises that create building materials (8 %) (Bokov and Lushchik 1998). However, heavy metals from the wastes of various industrial enterprises are also spread by water and wind, contaminating both adjacent and remote landscapes.

One technogenic-geochemical process that is particularly dangerous for the environment is mining. The main mineral coal deposits are located in the Donets and Lvov-Volyn Basins, and brown coal deposits are located in the Dnieper Basin. These basins differ considerably, but their shared feature from the viewpoint of anthropogenic landscape changes is that these mines all have typical mining landscapes (Denisik 1998). The mining industry in these basins has resulted in the destruction of large areas of agriculture; once areas are mined, technogenic waste grounds remain. Furthermore, major land loss occurs because of safety zones around mining tailings (the first zone is 200 m wide, the second 500 m). Within these zones, the air is polluted, the soils are salinized and the water table is often high, rendering them unsuitable for future agricultural use. Unfortunately, large portions of mining landscapes have high concentrations of processing waste, which contain large amounts of toxic elements and pollute their surroundings.

Among all types of mining, the relatively new practice of strip mining has the greatest effects on natural landscapes. Under strip mining, layers of rocks are removed and piled as waste at heights of several tens to even hundreds of meters. During this process, toxic rocks that are harmful to plants frequently accumulate on the disposal surface, and the landscape geomorphology is transformed. In Ukraine, strip mines are particularly common in the Donetsk and Krivoy Rog regions, where a considerable amount of different raw material deposits are exploited in this manner.

Because of various forms of mining, the contemporary environment of Donbas is highly disturbed. The entire Donbas region is covered with a dense network of industrial sites, among which metallurgical and coke chemical enterprises and coal mines are the most common. A characteristic feature of Donbas' landscapes is barren rock disposals, which represent 17 % of all the barren lands of Ukraine. Coal extraction in western Donbas is conducted by underground mining, which has caused large damage to natural landscapes, induced landscape transformation, destroyed soil and vegetative covers, and affected natural hydrological and hydrochemical processes. These phenomena are especially prominent in the Samara River basin. Here, coal extraction has lowered the river by 3–5 m below the groundwater level, hence replacing floodplain forests, tilled soils, and hay meadows with open water bodies and swamps. Prior multigrass-feather grass and meadow steppes were not preserved, and the only remainders of natural vegetation are small reserves on "unsuitable" lands (ravines, rock exposures, etc.). Examples of what was previously magnificent steppe vegetation have persevered on "unsuitable" lands near the town of Khartsyzsk, but this instance is a rare exception.

In terms of necessary steps for environmental protection, the prevention and mitigation of the negative consequences of the technogenic migration of chemical elements is the most important issue for mining landscapes because these chemical elements deteriorate environmental conditions over large areas. More data are required regarding the intensity of chemical elements and their technogenic migration, along with the resilience of natural systems to pollution and their self-clearing potentialities.

Mining activities represent only one form of human land use, albeit a very stark one. The forms of land use in Ukraine are diverse and can be roughly ranked as follows: reserve, unused, commercial hunting industry, forestry, recreational, meadows and pastures, agricultural (including melioration), water management, settlements, roads, industrial sites, and mining cites.

The present landscapes of Ukraine have different levels of economic transformation, and we should describe landscape conditions in terms of their "virginity", i.e., the presence of natural or conventionally natural, unchanged and partially changed land cover types such as forests, wetlands, haymaking meadows, pastures, sand dunes, ravines, etc. Based on their percentage in a given landscape, we can rank modern landscapes from very high human influence (>90 % of non-natural land cover) to very low (<40 %). We conducted such an evaluation of the level of landscape virginity in each landscape of Ukraine based on the map *Ukraine Landscapes* map (1:2.5 × 10^6 scale). We verified our analysis based on statistical data from the *Ecology stabilizing lands* map (1:5 × 10^6) (National Atlas of Ukraine 2007).

Ukraine's landscapes differ greatly in terms of their level of virginity, or lack of anthropogenic land cover (Fig. 3). The almost natural, conventionally unchanged landscapes include water areas in large and intermediate rivers, swamps, moving sand and barrier spits, floodplain forests, and forest landscapes. The mountain landscapes in the Ukrainian Carpathians and cuesta-yaila massifs of Crimea have the greatest level of virginity. A very low level of anthropogenic transformation is also characteristic of the sandy forested terraces, ravine systems, alluvial-outwash plains, moraine-outwash plains, significantly forested and swamped lowlands in mixed forests, and broad-leaf forest zones in Ukraine. Mountain slopes and foothills mostly have a low index of anthropogenic transformation because of widespread forests and ravine systems, including the landscapes of Minor Polesye, most of the Novgorod-Siversky Polesye territories, the northern part of Dnieper Heights (i.e., the watershed between the Teterev and Irpin Rivers), the watersheds of southern Bug and Ros, the Podolian Tovtry, the western part of the Transcarpathians, and the northern slope of the outer ridge of the Crimean Mountains.

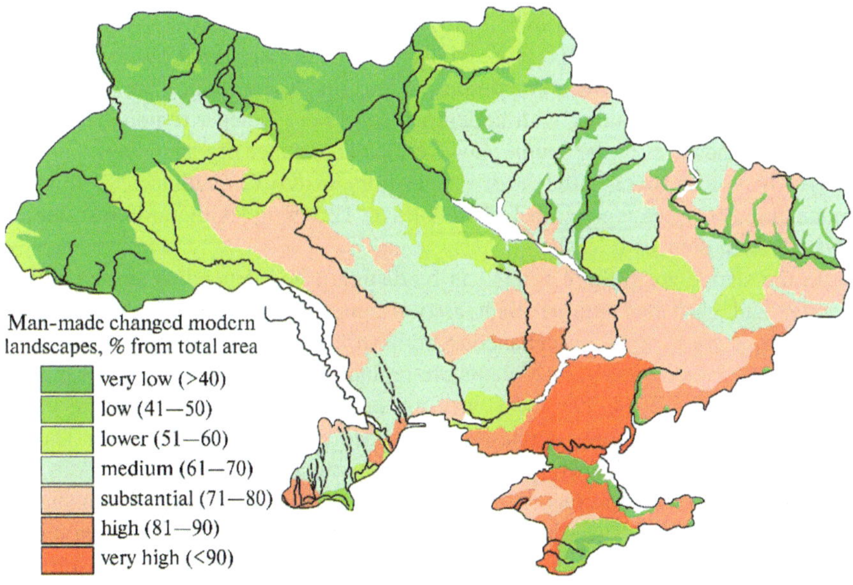

Fig. 3 Landscapes of Ukraine classified according to the percentage of anthropogenic land cover

In contrast, a medium level of transformation occurs in the landscapes of Podolian Heights; the loess islands of Zhitomir, Chernigov and Novgorod-Siversky Polesye; the high Dniester terraces; the landscapes of the northern steppe zone; the Near-Azov Height and Donets Ridge; the Poltava Plateau; and fragments of spurs along the Middle-Russian Hills.

Finally, the landscapes of the Ridge Bug region, the Dnieper-southern Bug, and Uday-Psiol watersheds, and the Dnieper right-bank terraces have a substantial transformation level. High and very high (up to 90 %) transformation levels characterize the overwhelming majority of the landscapes in the steppe Black Sea area, steppe Crimea, and parts of the Vorskla-Orel-Samara-Seversky Donets watershed.

3 Regional Flora and Its Interaction with Human Activity

Three vegetation zones are typical in Ukraine: broad-leaf forests, forest steppes, and steppe zones. Each of these zones are widespread, and a sub-Mediterranean zone occurs as well. The floral and cenotic complexes from different origin centers determine the plant communities in these zones. Boreal complexes of pine forests and oligotrophic sphagnum swamps are common in the north (Polesye), which is typical of taigas. Richer turf-podzol soils, however, contain mixed coniferous/broad-leaf forests of complex structure with boreal and nemoral species. The richest northern Ukraine turf-podzol and grey forest soils, on the other hand, have broad-leaf forests that are dominated by typical nemoral species. At the extreme margin, the highest landscape plots with grey soils are occupied by beech forests, akin to Central European communities, while the lower slopes are covered by hornbeam-oak forest, along with linden-oak across the Dnieper region. The role of oak forests increases to the south. However, the abundance of forests varies greatly. Currently, forest covers 40.2 % of the territory in the Carpathians, 32 % in the mountainous Crimea region, 26.1 % in Polesye, 12.3 % in Ukrainian forest steppes, and 3.8 % in steppes (Golubets 1997).

In the steppe zone, three bands can be distinguished depending on the climate aridity, and each has unique plant associations. Feather grass-tipchak (*Festuka valesiata*) steppes with miscellaneous herbs prevail in the northern part of the steppe zone; southwards, these plants are replaced by tipchak-feather and then by grass or desert steppes. The most northern steppes, which are part of the forest-steppe zone, are meadow steppes. The steppes extend from the lower Danube through the Crimean Peninsula to the eastern boundary of Ukraine.

In addition to climatic factors and soils, elevation plays a major role in shaping plant communities. In the Ukrainian Carpathians and mountainous Crimea region, vegetation is present in distinct altitudinal belts because of increasing moisture and decreasing temperature. For example, in the Carpathian piedmont, a belt of mainly beech and oak forests at elevations above 550 m is replaced by a lower forest belt. Similarly, beech forests dominate on southwestern slopes, while fir-beech and fir-beech-pine forests occupy the northeastern slopes. The upper forest belt (mainly pine forest) is located at over 1200–1300 m altitude, while alpine mossy forests are present above 1500 m.

Different altitudinal belts also formed in the mountainous Crimea region and on northern and southern slopes. The lowest altitudes (150–400 m) and northern slopes contain a forest-steppe belt with pubescent oak forests and meadows that transition to pubescent and rock oak forests at higher altitudes. A belt of beech forests with hornbeam is located at altitudes above 750 m. The southern slope at medium altitudes (400–450 m) contains a belt of pubescent oak forests with some evergreen trees and shrubs. Furthermore, in locations there is an open forest of *Juniperus excelsa* Bieb. or pine *Pinus stankewiczii* (Sukacz.). A belt of coniferous forests (*Pinus pallasiana* D. Don) is located at higher altitudes, while even higher altitudes (>1000 m) contain pine forests (*Pinus sosnowskyi* Nakai). The flat tops of the main ridge of the Crimean Mountains are occupied by meadow steppes. The intrazonal halophyte vegetation is common in Ukraine's forest-steppe and steppe zones, along with zonal aquatic and coastal-aquatic vegetation (Shelyag-Sosonko and Andrienko 1985).

Ukraine is characterized by great biological diversity. Although the country covers only 6 % of the European territory, Ukraine contains 37 % of its vascular plant species. The species diversity of Ukrainian biota during the 20th century was over 70,000, and that of its flora was over 25,000, including vascular plants (5100), mosses (800), lichens (1000), algae (4000), and fungi and myxomyceta (over 15,000) (Davidok et al. 1997). Amongst the 5100 vascular plant species, 4550 were natural. The obligate species in different vegetation types included steppes (880), forests (850), swamps (340), meadows (290), halophytes (210), and aquatic and coastal-aquatic 200 species. In terms of endemic species richness, forests were second to steppes, but their functional importance was higher than that of all the other vegetation types.

Intensive anthropogenic influence has degraded many forest ecosystems. Most importantly, the area of forests has been cut in half as agriculture has replaced forests, decreasing the phytomass per hectare by a factor of 7–9 on average (Golubets 1997). Similarly, oxygen production, carbon dioxide adsorption and annual biomass growth decreased by a factor of almost 1.5, as did the use of nutrients for plant growth. Thus, soils have become depleted and the microelements that are necessary for the well-being of humans and animals have been washed out. Even when forests regrow, differences remain because secondary forests are distinguished by lower rates of photosynthesis, poorer fauna and flora, and lower rates of soil-forming processes. Finally, biogeochemical cycles and the energetic and hydrological functions of the initial forest ecosystems degraded even more as forests were lost. However, only 20 % of the initial forest cover in the mountains still provides 40–45 % of the steady river drainage, and 40–50 % of the forest cover provides 80–90 % of the drainage and almost entirely stops soil erosion and the wash-out of microelements (Golubets 1997).

Swamps cover approximately 2 % of Ukraine (approximately 1.1–1.3 million ha), and peat is present over ~1.2 % of its area. Major efforts were made to drain these swamps, but large-scale drainage has not been beneficial either ecologically or economically for Ukraine. Drainage efforts started as early as 1873 in the swamps of what was then Volyn Gubernia and adjacent regions of Belarus. These works were continued during the entire 20th century. By the end of the 1980s, approximately 50–55 % of all swamps had been drained, including large eutrophic swamps such as Zamglay, Vidra, and Trubezh, and the total area of swamps was cut in half. Swamps were transformed into agricultural areas; eutrophic swamps, which are potentially the most productive, were reduced the most. However, inadequate farming methods have limited the economic benefits of these activities.

During the 1990s, melioration works almost entirely ceased because of the economic crisis, and some swamp vegetation regrew during this time. Currently, approximately 15 % of the swamp areas of Ukraine are protected. Furthermore, the restoration of some former swamps has been proposed, and the remaining natural swamps should be used for ecotourism, hunting and fodder production and to protect drinking water supplies. In light of today's understanding of the environmental role of swamps, we must increase their area to 1.3–1.4 % of Ukraine's area and "renaturalize" the locations that were incompletely drained and where the vegetative cover has not yet been fully destroyed.

Meadows currently occupy approximately 5.5 million ha (9 % of Ukrainian territory). These areas are mainly secondary formations that have arisen from human economic activity. Non-anthropogenic meadows are only found in the sub-alpine and alpine belts of the Carpathians.

Since the beginning of the 20th century, the area of meadows in Ukraine has been reduced by a factor of 1.5. The rate of loss of meadows was particularly high during the 1950–1970s, when the slogan "meadows are our virgin soil" was proclaimed and practically all dry meadows were ploughed. Furthermore, approximately 30 % of the area of flood plains along the medium and large rivers of Polesye were drained and ploughed. Additionally, by the 1990s, a series of water reservoirs almost completely flooded valuable flood meadows in the Dnieper area, approximately 20 % of the Southern Bug area, etc. The drainage and watering of meadows, along with excessive pasturing and hay cutting, all change these areas' structure and floral composition. Ultimately, the drainage of large meadow-swamps had drastic effects because this process changed the groundwater levels, resulting in undesirable changes in species composition and decreases in biological productivity. Superfluous hay cutting also had similar consequences. For example, a cutting cycle of four times per year greatly decreases the amount of valuable species such as *Alopecurus pratensis* L. in the herbal cover. Overgrazing has a similar effect, which creates swamping in wet meadows and desiccation in dry meadows. The species diversity decreased, the share of common and weedy species increased, and the productivity decreased more than twofold under all these scenarios.

Steppe vegetation. Today, steppe vegetation covers less than 1 % of Ukraine's territory, and a very small part (approximately 0.1–0.2 %) is natural unaltered steppes, which are mainly found in natural reserves. The rest of the steppe vegetation is preserved on the slopes of ravines and river valleys. Unfortunately, the steppe's rich biological diversity has practically been lost. The steppe ecosystems have largely been destroyed or severely transformed because of superfluous ploughing in southern Ukraine, which exceeded 85 % of the total area in some regions.

4 Assessment of Changes in Plant Productivity and CO_2 Fluxes from 1990 to 2000

One of the fundamental indicators of climate change, which is of practical importance for human society, is a change in the biological productivity of terrestrial ecosystems. Spatial variations in the net primary productivity (NPP) of terrestrial ecosystems can be very high, ranging from approximately 1000 gC m^{-2} $year^{-1}$ for evergreen tropical rain forests to less than 30 gC/m^{-2} $year^{-1}$ for deserts (Voloshchuk et al. 2002). With higher atmospheric CO_2 concentrations and resulting global climate changes, the NPP values can further widely fluctuate over different areas of the globe (Nilsson et al. 2002). Therefore, understanding regional changes in the carbon cycle requires more detailed analysis of the processes that occur on the Earth's surface.

Important information that can be used to assess these processes is provided by satellite remote sensing of the Earth (RSE). MODIS (Moderate Resolution Imaging Spectroradiometer), which launched in 1999, allows weekly global assessments.

The NASA MODIS NPP product was designed to accurately and continuously measure the productivity of ground vegetation. This product benefits both basic and applied science. The value for basic science is to determine the seasonal dynamics of regional carbon dioxide budgets within the global carbon cycle. Currently, global carbon cycle models are combined with global climate models to create a unified Earth system model that can correctly describe the dynamics of the atmosphere, biosphere, and oceans and their interactions. The practical significance of such models is in their ability to evaluate the dynamics and projections of economically and socially important vegetation products (e.g., crops) and to inform predictions of future climate change.

Several approaches can be used to assess biophysical plant parameters with satellite data. We used coarse spatial resolution spectroradiometer MODIS images to evaluate and map carbon dioxide fluxes. The territory of our study was the western part of Ukraine, which is covered by coniferous forests, deciduous forests, grass cover and agriculture. We chose the MODIS image that was recorded on the 14th of June 2000 with a spatial resolution of 1 km for the analysis because this day had the lowest cloud cover over Ukraine. Furthermore, the end of June coincides with the peak of the vegetation growing season. We were also able to compare the MODIS data and analysis of Landsat ETM images for 2000 (Lyalкo 2007; Lyalкo et al. 2007) (Fig. 4). We compiled a map of the geographic distribution of CO_2 fluxes by using MODIS data and the PRI and NDVI indices for western Ukraine (Movchan 2008).

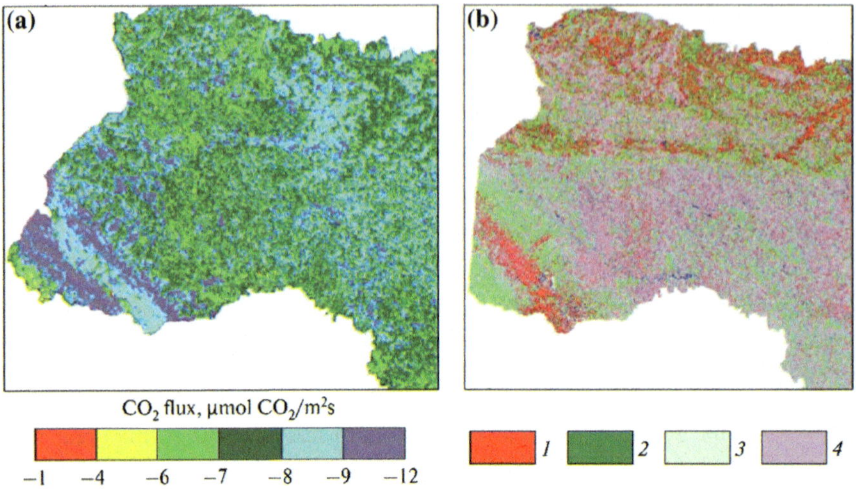

Fig. 4 CO_2 absorption rate by vegetation, as of June 03, 2000 (according to the MODIS images). **a** and **b** Types of vegetation cover as of 2000 (according to the Landsat ETM images) for the western part of the Ukraine (*1*—coniferous forest, *2*—deciduous forest, *3*—grassland, *4*—agricultural)

The results show that deciduous forests absorbed the most CO_2 (approximately 10 mmol CO_2 m^{-2} s^{-1}). Somewhat lower absorption rates (approximately 7–8 mmol CO_2 m^{-2} s^{-1}) were characteristic of coniferous forests, which can be explained by the fact that June is the peak of the vegetation season for deciduous forests. However, the vegetation period for deciduous forests lasts only during the warm season, while coniferous forests absorb CO_2 throughout the year, albeit at lower rates. Therefore, coniferous forests contribute more to reductions in the CO_2 concentration in the atmosphere. The grasslands had the lowest absorption rates (from 0 to 4 mmol CO_2 m^{-2} s^{-1}).

Agricultural areas differ considerably in their ability to absorb CO_2, making average values less meaningful. The agricultural areas in northern Ukraine have high absorption rates, but the absorption rates in the southern part of Ukraine, where grain crops dominate, are nearly as low as those of grasslands.

We calculated the amount of carbon annually absorbed by vegetation by comparing the productivity of different vegetation types with changes in vegetation cover over the western part of Ukraine (Lyalко 2007; Lyalко et al. 2007) by using the LANDSAT TM/ETM for 1990–2000 (Movchan 2008). These results are shown in Tables 2 and 3.

According to our data for the western part of Ukraine, the changes in the vegetation cover areas decreased the amount of absorbed carbon in 2000 by one thousand tons compared to the value in 1990. Thus, a larger fraction of unabsorbed carbon remained in the atmosphere, contributing to a stronger "greenhouse effect".

Table 2 Changes in vegetated areas in the western part of Ukraine based on analyses of LANDSAT TM/ETM images for 1990–2000

Types of vegetation	Vegetation cover areas (km^2)		
	1990	2000	Difference
Carpathian region			
Coniferous forests	6671.5	4669.9	−2001.6
Deciduous forests	6128.9	8515.8	2386.9
Herbage	4300.3	4414.3	114.0
Agricultures	9994.1	9086.8	−907.3
Polissya			
Coniferous forests	17688.5	17524.7	−163.8
Deciduous forests	13128.6	10233.0	2895.6
Herbage	20502.2	19702.2	−800.0
Agricultures	27774.1	27481.5	−292.6
Southwestern region			
Coniferous forests	6455.1	5879.5	−575.6
Deciduous forests	25417.5	36217.7	10800.2
Herbage	15732.7	12694.8	−3037.9
Agricultures	63738.1	55407.1	−8331.0

Table 3 Changes in the annual total carbon that was absorbed by vegetation because of changes in the vegetated areas in the western part of Ukraine from 1990 to 2000

Types of vegetation	Amount of absorbed carbon (kg C year^{-1})		
	1990	2000	Difference
Carpathian region			
Coniferous forests	2400.07×10^6	1679.996×10^6	−720.08
Deciduous forests	2250.84×10^6	3127.43×10^6	876.59
Herbage	821.36×10^6	843.13×10^6	21.77
Agricultures	3255.58×10^6	2960.03×10^6	−295.55
Sub-total			−117.27
Polissya			
Coniferous forests	6363.44×10^6	6304.51×10^6	−58.93
Deciduous forests	4821.48×10^6	3758.06×10^6	−1063.42
Herbage	3915.92×10^6	3763.12×10^6	−152.8
Agricultures	9047.41×10^6	8952.1×10^6	−95.31
Sub-total			−1370.46
Southwestern region			
Coniferous forests	2322.22×10^6	2115.15×10^6	−207.07
Deciduous Forest	9334.58×10^6	13300.95×10^6	3966.37
Herbage	3004.95×10^6	2424.71×10^6	−580.24
Agricultures	20762.69×10^6	18048.86×10^6	−2713.83
Sub-total			537.23
Total for all regions			−950.36

5 Water Resources

Ukraine's water resources consist of river drainage and groundwater. The river drainage of Ukraine itself has an average annual amount of approximately 52.4 km^3, and together with inflows from adjacent countries, reaches 87.1 km^3 or even 209.8 km^3 when accounting for the Danube's drainage through the Kiliya Channel.

The estimated resources of Ukraine's groundwater are 22.5 km^3 per year, only 7 km^3 of which is not hydraulically connected with river drainage (Yatsyk 1997, 2001). Therefore, the total annual water resources are estimated to be as large as 94.1 km^3 on average. In a dry year, this value is approximately 77.2 km^3 during dry years and only 59.4 km^3 during very dry years. On average, the local drainage yields 86,800 m^3 per year from 1 km^2 of the territory and 1000 m^3 per year per capita. During very dry years, these figures are equal to 49,200 and 0.610, respectively, which indicates that Ukraine is insufficiently supplied with fresh water. Furthermore, the distribution of river networks across the country is very non-uniform. Unfortunately, the fewest water resources are found in the areas with the highest concentration of water consumers: Donbas, the Krivoy Rog region, Crimea, and the southern regions.

The main characteristic feature of the water resources and river runoff is heterogeneity within and among years. The territory of Ukraine can be divided into 16 areas depending on the features of the annual river runoff distribution. The general feature of all these areas is that the majority of the annual runoff occurs during spring floods, ranging from 60 to 70 % in the north and northeast to 80–90 % in southern Ukraine.

Groundwater resources are also very heterogeneously distributed: 65 % of them are concentrated in the Dnieper-Donets and Volyn-Podolia artesian basins in northern and northwestern Ukraine. The Black Sea artesian basin and other hydrogeological districts have less favorable conditions for the formation of groundwater. The largest volume of groundwater per capita (5.54 m^3/d) is in the Chernigov region, and the smallest (0.28–0.43 m^3/d) is in the Odessa, Dnipropetrovsk, Kirovograd, Donetsk, Nykolayev, Zhitomir and Vinnitsia regions.

The largest total amount of groundwater is in the Dnieper (12 %) and Dniester (9 %) basins. The basins of the rivers in the Azov Sea area (4.6 %) and Dniester-Southern Bug watershed (0.5 %) are a part of the remaining 18 %. The prospected level of available groundwater resources changes from 90 % in the basins of the Crimean rivers to 14 % in the basins of the rivers that flow into the Sea of Azov. Only 20 % of these resources have been prospected in the Dnieper's basin, and 27 %, 30 % and above 49 % have been prospected in the Dniester, Southern Bug and Seversky Donets basins, respectively. In total, 371 groundwater deposits from 977 sites have been prospected and authorized in Ukraine. The total prospected exploitation resources of groundwater is 5.7 billion m^3 per year (15.7 million m^3/day), or 25 % of the estimated resources. In 2003, the intake of underground water was 2.6 billion m^3 per year (7.1 million m^3/day), which was

11 % of the estimated and 45 % of the exploited resources, respectively. Thus, reserves of fresh groundwater are still available to improve the drinking water supply in Ukraine.

In 2002–2003, groundwater sources provided approximately 14 % of the total water consumption. Over 110,000 wells have been drilled for the intake of underground water. Furthermore, more than 1.9 million shaft wells, mainly in the countryside, and over 2000 springs have been exploited for water intake from the first aquifer (soil) horizon, which is of great importance for the water supply in some areas (mainly in the mountainous Crimea region).

Groundwater is a major source of drinking water to the populations of cities and other settlements in the Lugansk, Volyn, Trans-Carpathian, Zhitomir, Kirovograd, Rivne, Poltava, Sumy, Ternopil, Kherson, Khmelnitsky, and Chernivtsy regions and in Crimea. Approximately 50–81 % of the local groundwater is used for these purposes. Agricultural water supplies also use groundwater. Of the entire intake volume of groundwater, 30 % is used for household-drinking water supply, 42 % for agriculture, and 28 % for technological water supply. The largest groundwater intake occurs in the Lugansk (494 million m^3 per year), Donetsk (473), Lviv (204), Dnepropetrovsk (177) and Kiev (135) regions, which comprise 58 % of the total water consumption in Ukraine. As a whole, only 25 % of the urban water supply in Ukraine is provided by groundwater sources, whereas the use of groundwater in the majority of European countries reaches 90 % to meet the needs for high-quality drinking water.

Lake waters are also included in water resources. Over 20,000 lakes are present in Ukraine, and 7000 of them have a surface area of 0.1 km^2 or more (Shyklomanov 1986). However, lake water resources cannot be a reliable source of water supply because most of these lakes are small and their levels are not steady. Furthermore, numerous coastal lakes have salty or brackish water and therefore cannot be sources of fresh water. Freshwater lakes can thus serve as sources for drinking water only at very local scales.

As previously mentioned, river runoff is very non-uniform in time and space. Numerous reservoirs, channels and water pipes have been constructed in Ukraine, many of them long ago, to regulate and redistribute river runoff. Before 1950, their total area did not exceed 100,000 ha, and their total volume was 1.4 km^3, which allowed them to control only 3 % of the annual drainage. However, by 1985–1986, the area of these reservoirs had increased almost five-fold, and their total volume increased by a factor of 8.5 even without accounting for large reservoirs along the Dnieper and Dniester Rivers. When these reservoirs are included, the usable volume of all the reservoirs (without ponds) is equal to 50.5 % of the average local drainage, and their total volume exceeds this value.

However, the distribution of ponds and reservoirs in the basins of large rivers is very heterogeneous. Reservoirs are relatively numerous in the basins of the southern Bug and Seversky Donets regions and in the forest-steppe and steppe zones of the Dnieper's inflows basins.

In terms of irrigation, the area of irrigated lands by 1989 exceeded 2.6 million ha, and that of drained ones exceeded 3.2 million ha. The bulk of land

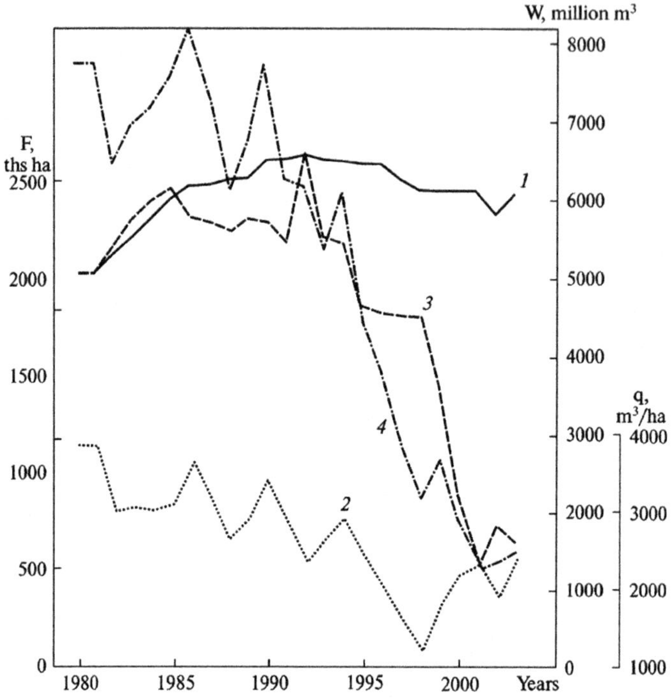

Fig. 5 Dynamics of irrigated lands and water volume that were used for irrigation

reclamation work was performed in 1976–1985, when 42 % of irrigational and 43 % of drainage systems were put into operation (Kirkor 1907). As the areas of irrigated lands increased, the volumes of water that was supplied to fields increased (Fig. 5).

The volume of fresh water from natural sources during 1960–1985 increased by a factor of 2.26. The annual increase in the water volume that was used was 800×10^6 m^3 on average. The largest portion of this gain was predetermined by the sharp increase in water consumption for agriculture in 1960–1985, which reached 500×10^6 m^3 annually. During 1960–1970, the annual increase in water consumption for agriculture was equal to 390×10^6 m^3, while that for 1971–1985 was equal to 567×10^6 m^3.

In agriculture, water resources are used for the needs of rural population, animal breeding, and crop production, for water reclamation-irrigation and watering, and for the drainage of swamps and over-moisturized lands. In total, agriculture consumes 35–40 % of the water that is used in the country, but this portion decreased to 26–30 % in 1997–2003. The irrecoverable (i.e., non-recyclable) water consumption comprised 75–97 % of the used water volume. The irrecoverable water consumption for agriculture reached 70–74 % of all the irrecoverable water consumption in Ukraine, but this percentage decreased to 40–33 % in 1997–2003.

In 1980–1994, agriculture used 9–12 km^3 of water per year. During 1995–2003, this volume decreased to 2–3 km^3 per year. If we compare the volumes of water intake in 1960 and 1988, the total water intake increased by a factor of 2.3 and that for agriculture by a factor of 4.9, but 1984 saw a reduction in the use of water for agriculture. The same tendency was observed in the volume of water that was used for irrigation: 5.25 km^3 from 1985 to 1999 (from 7.58 to 2.33 km^3). The irrigated (actually flushed) areas decreased by 660,000 ha, but specific water consumption has decreased by more than a factor of 2.3 (Fig. 6).

The largest number of water consumers is located in small river basins. Their runoff decreases as a result of the direct water intake from the channels and aquifer horizons that are hydraulically connected with these rivers. Very often, the volume losses reach 30–50 %. An analysis of the water use structure for individual rivers shows that the economic-household services in Polesye consume 3–25 % of the total amount of the river drainage, while those in forest-steppe regions consume 10–20 % and those in steppes consume 20–40 %. Almost everywhere, small rivers are a source of agricultural water. In some rivers, over 40 % of the water resources that are consumed in the economy are spent for household needs (Mihalescu 1983).

The water quality in nature is mainly controlled by hydrobionts, which depend on the hydrological and hydrochemical conditions of a water body. If humans create the necessary conditions for the normal existence of water organisms, the water will have optimal quality, in turn enabling the water to be used continuously (Shvets 1969).

Fig. 6 Dynamics of the water consumption characteristics in the agricultural sector

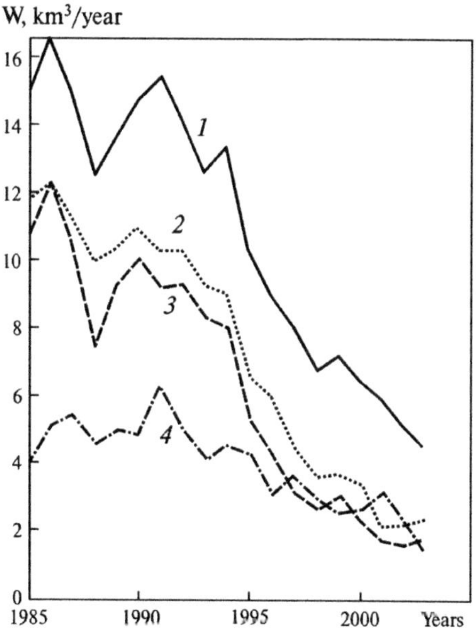

The water quality is characterized by the presence of mineral and organic substances. The pollution of water bodies, including rivers, is divided into biological and anthropogenic components. The biological pollution of rivers occurs during the natural growth of biomasses of hydrobionts, mainly the death and disintegration of hydrophytes and organic substances of autochtonous (formed in the water body) and allochtonous origin (brought from outside). The anthropogenic pollution of water bodies is caused by human economic activity.

One can estimate the contributions of individual branches of the economy to the pollution of surface water and, consequently, the quality of the volumes of waste water in the rivers and other water bodies. Mostly, these environments are polluted by industries, which are responsible for more than half of all waste waters: from 63.8 % in 1989 to 53.5 % in 1998. The second largest contribution to pollution originates from municipal services. Their contribution to the total amount of waste water from 1989 to 1998 increased continuously from 19.3 to 32.6 %. In 1999, this contribution decreased slightly and remained constant until 2004. However, the absolute volumes of waste water decreased from 4.1 km^3 in 1992 to 2.9 km^3 in 2003. The fraction of agricultural waste water during 1989–1997 varied between 15 and 18 % and thereafter decreased to 9–10 %. Other branches of the economy produce no more than 0.5 % of the entire volume of waste water.

In 1990, the discharge volume of waste water without purification was 0.47×10^9 m^3, whereas this value had risen to 0.8×10^9 m^3, or approximately 1.7 times greater, by 2003, which occurred under the condition of almost the same reduction in the volume of water consumption (Yatsyk 2001). A steady tendency of less efficient purification facility operations occurred over the last two decades, which was controlled by equipment wearing out, these facilities' low technological levels and significant energy dependence, and the presence of new chemicals in polluting substances, for which the necessary reagents for water clearing are absent.

Surface water pollution is regularly monitored at 251 points for 195 sites at 101 rivers, 15 water reservoirs, seven lakes, and one channel. Almost all these monitored water objects belong to polluted and highly polluted objects. The most polluted rivers are the Goryn, Desna, Sula, Teterev, Vorskla, Unava, Samara, and Ingul Rivers (in the Dnieper River Basin); the Seversky Donets, Udy, Kazionny Torets, Bakhmut, Lugan, and Bilenka Rivers (in the Seversky Donets River Basin); the Dniester, Tiasmenitsa, Opor, and Striy Rivers (in the Dniester River Basin); the Kalmius, Kalchik, Bulavinka, and Molochnaya Rivers (in the Peri-Azovian Basin); the western Bug and its inflows the Poltva and Luga, the Danube, Latoritsa and Vitcha Rivers (in the Danube River Basin); and the southern Bug River.

The water reservoirs in the Dnieper Cascade, especially the Kiev and Kanev reservoirs, are also highly polluted. For example, 30–240 cases of very significant pollution were observed in 1998 alone. Similarly, approximately 900 cases of highly polluted surface water from 14 pollutant components were registered on 64 % of monitored objects (Ramad 1981). However, observations showed a reduction in some substances. In the basin of the Dnieper River in 1993–1999, drops of petroleum decreased by a factor of 2.7, ammonium nitrogen by a factor of 12, phenols by a factor of 2.2, and fats by a factor of almost 6. Similarly, iron and copper became less abundant, but the same cannot be said for other heavy metals, and chlorides, sulfates, nitrates, zinc and chromium remained constant.

Petroleum products, ammonium nitrogen, phenols, fats, nickel and some other materials in the Dniester basin decreased with a practically constant volume of waste water, but the iron discharge increased. Changes in the Seversky Donets River basin were less substantial than those in the Dnieper basin. The volume of waste water in this location decreased by a factor of 1.5, sulfates by a factor of 1.3, chlorides by a factor of 2, ammonium nitrogen by factor of greater than 3, and phenols by a factor of approximately 4 (very non-uniformly). Sharp reductions in the amount of fats, zinc, and nickel were observed. However, the discharge of nitrates more than doubled. In the southern Bug basin, the discharge of waste water was lower than that in the basins of other large rivers in Ukraine. The volume of waste water changed from 198×10^6 m^3 in 1993 and 1997 to 149×10^6 m^3 in 2003. The total discharge volume of waste water in this basin remained virtually unchanged. Although the amount of polluting substances generally decreased (in particular, petroleum, chlorides, and ammonium nitrogen by a factor of 2), the amount of sulfates, nitrates and fats increased and the amount of heavy metals did not change.

In summary, Ukraine's water resources are relatively poor and cannot provide the population and branches of the economy a long-term water supply of appropriate quality at the existing levels of industrial technologies and agricultural production. Future economic development will likely require increasing volumes of water. This demand has been partially satisfied by the control of river drainage and by intra- and inter-basin transfer. However, the insufficient purification of municipal, industrial and other waste water and the reduction in the natural clearing abilities of rivers because of their unreasonable control levels has resulted in the increasing pollution of surface waters. The inefficient use of natural resources, out-of-date technologies, and a lack of effective legislative, normative, economic, informational and organizational levers have influenced the production and dumping of large waste volumes, much of them toxic, and the pollution of surface and groundwater. Unfortunately, agricultural production has made a major contribution to the pollution of surface and groundwater because of the inflated use of mineral fertilizers and pesticides and soil degradation.

6 Impact of Flow Regulation on River Deltas in the Black Sea Basin

River deltas play an important economic and social role, especially in connection with water use intensification for power, irrigation, drinking and other purposes. However, excessive use of water resources since the 19th century has decreased the water and sediment inflow into the deltas of the Black Sea basin and changed their landscapes.

The river deltas of the Black Sea, similar to deltas around the world, are dynamic systems that respond to natural and anthropogenic changes. Rivers channel water and fluvial constituents from the upstream watershed to the delta, providing freshwater, sediment, and nutrients. Considering the watershed and offshore ocean or sea as a part of the larger delta system is very important because of these tight connections between upstream areas and deltas via river networks and the delta's location on the coast.

Anthropogenic changes in upstream areas and deltas play an important role in the deltas of the Black Sea. Several of the larger threats to this sea's deltas can be loosely classified as those that directly affect wetlands and other delta ecosystems, those that directly affect human communities in the delta, and those that affect the land surface elevation of the delta or relative sea level rise. Relative sea level rises can, in turn, affect both coastal ecosystems and humans.

Delta wetlands are threatened by local development, particularly the conversion of wetlands for agricultural use (Coleman et al. 2008). Upstream development and dam construction can interfere with the flood pulses that deliver freshwater, sediment, and nutrients to wetlands ecosystems (Vörösmarty et al. 2003). Development and population growth on the delta may also result in an increased reliance on groundwater extraction for agricultural, industrial, and municipal use (Wada et al. 2012). Groundwater extraction can increase land subsidence rates by reducing pore-water pressure in buried sediment (Syvitski et al. 2009). Groundwater extraction that exceeds the recharge rates can result in the drawdown of water tables and can cause saltwater intrusion near the coast, which can threaten the freshwater supply for humans and wetlands and reduce agricultural yields (Zhang et al. 2013).

Under natural conditions, low land subsidence rates would be compensated by the deposition of new sediment on the delta plain during flood events. In anthropogenically influenced deltas, however, delta-building processes are disrupted in several ways. In upstream areas, dams and reservoirs trap sediment upstream, resulting in lower sediment concentrations in the river water that reaches the delta (Vörösmarty et al. 2003). Additionally, artificial levees, dikes, and engineered river channels, which serve to protect human communities from floods, prevent the deposition of sediment on the delta. Instead, fluvial sediments are transported offshore to sub-aqueous deltas or farther out to sea (Syvitski et al. 2005).

While some level of human influence can be detected in most deltas, determining which processes are most important for a given system is critical. The deltas in the Black Sea are only slightly urbanized, but growing populations along the coast and upstream are causing changes to physical processes that must be better understood to ensure the long-term sustainability of the delta as a home to both people and ecosystems.

We monitored changes in the environmental state of deltas, including both the long-term and seasonal dynamics of flooding and drying processes on the landscape by using Landsat satellite imagery and ground-based observations in the deltas of the Black Sea Basin (Starodubtsev and Bogdanets 2012; Starodubtsev 2013).

Landscape processes in the Black Sea basin differ among the deltas of rivers that flow into the deep tide-free sea (Danube, Kizilirmak, and Rioni), into the shallow sea (Don, Kuban), and into large estuaries (Dnieper and Dniester) (Fig. 7).

Fig. 7 Major watersheds of the Black Sea basin (river basins: *1*—Danube, *2*—Dniester, *3*—Dnieper, *4*—Don, *5*—Kuban', *6*—Rioni, *7*—Yesilirmak, *8*—Kizilirmak)

In the Danube River delta, the decreased inflow of water and sediment has changed the seasonal and long-term hydrological regimes as a result of flow regulation and economic activity in the basin. Therefore, approximately 800–900 ha of the water surface in the delta is transformed into wetlands each year because of siltation of reservoirs and economic activity. At the same time, abrasion processes along the sea coast and some renaturalization of previously cultivated wetlands have reduced the area of terrestrial ecosystems at a rate of 400–500 ha per year (Fig. 8; Table 4).

In the rivers Don and Kuban, which flow into the shallow Sea of Azov, are no longer growing into the sea because of a reduction in sediment runoff. Furthermore, water onset and strong winds are causing major transformations in the delta landscape, and intensive industrial construction and aquaculture (the Don River) and the development of rice (the Kuban' River) have reduced the wetland and floodplain areas.

The rivers that flow into the long Dnieper and Dniester estuaries also have such low sediments loads that their deltas ("limans") are no longer growing. Furthermore, the overgrowth of delta lakes and canals has occurred at a rate of 150–160 ha/year in the Dnieper delta and 30–40 ha/year in the Dniester delta. Algal blooms are common in the coastal water, and the water quality has widely deteriorated (Table 5).

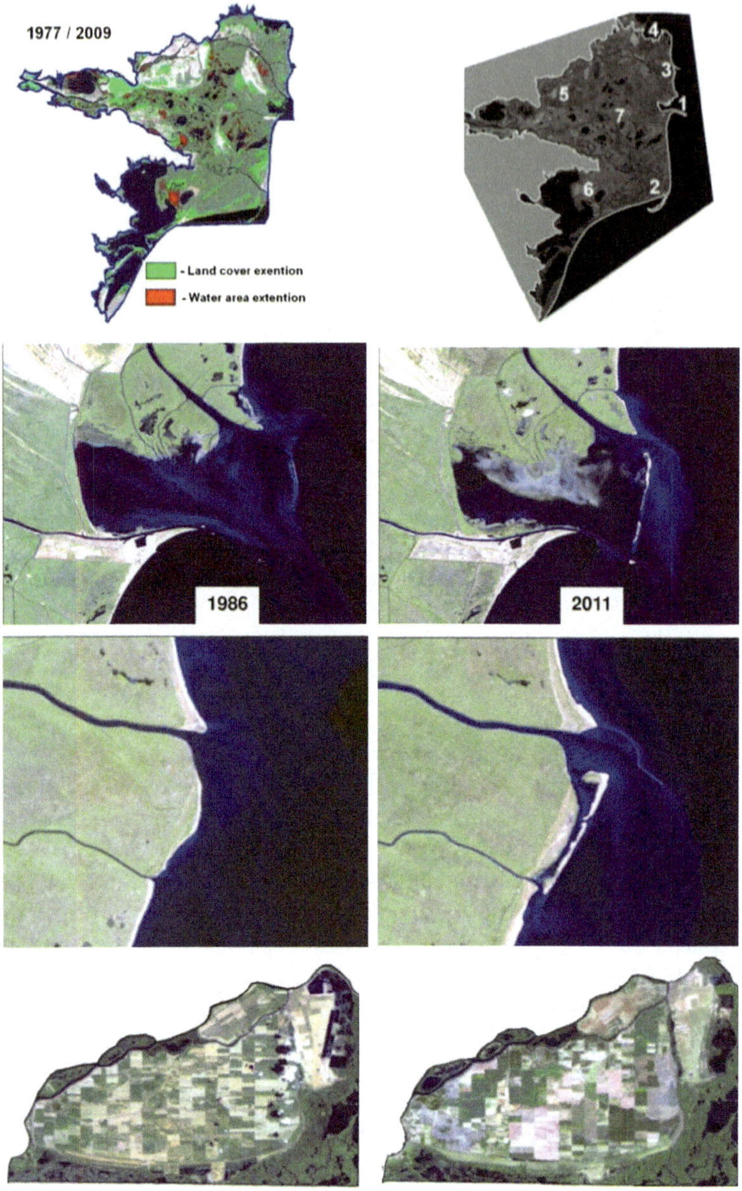

Fig. 8 Land cover change detection in the Danube delta from 1977 to 2009, and the key sections of the Delta (section 1 is the area between the Kiliya and Sulina river canals; Section 2 is near the St. George canal; Section 3 is near the Bystroe canal; Section 4 is the region around the Ochakov canal; Section 5 is agricultural lands; Section 6 is the area near the Rezim lake; and Section 7 is near the Sulina canal), followed by examples of changes as depicted in Landsat satellite image for key section 1 (2nd row), section 3 (3rd row), and section 5 (4th row)

Table 4 Land cover area (in ha) and changes in key sections of the Danube delta

Type of surface	1986	2011	Changes (ha)
Key section 1			
Deep water	19885.1	18422.1	−1463.0
Shallow water	599.4	1271.4	+673.0
Reedbeds and cattail with some sandbars and buildings	14167.0	14957.1	+790.1
Total key-1 area	34651.6	34651.5	
Key section 2			
Deep water	10341.5	11454.9	+1113.4
Shallow water (plume)	6527.3	8428.5	+1901.2
Slightly overgrown water area	3150.7	226.2	−2924.5
Overgrown water area	–	226.4	+226.4
Submerged marshes	383.7	–	−383.7
Marshes with reedbeds and cattail	7589.0	5238.0	−2351.0
Dense reedbeds and cattail with sand	2564.6	4982.9	+2418.3
Total area	30556.8	30556.9	
Key section 3			
Deep water	3276.9	3011.7	−265.2
Shallow water	596.2	701.1	+104.9
Overgrown water area	58.4	59.3	+0.9
Sandbars	49.4	78.9	+29.5
Marshes	1961.7	1306.2	−655.5
Dense reedbeds and cattail with bushes (riverine levees)	1490.0	2275.4	+785.4
Total area	7432.6	7432.6	
Key section 4			
Deep water	8195.1	9771.5	+1576.4
Shallow water	4526.6	2778.8	−1747.8
Overgrown water area	1367.2	1070.6	−296.6
Submerged marshes	4066.8	2262.1	−1804.7
Marshes with reedbeds, cattail and sandbars	8171.5	6839.4	−1332.1
Dense reedbeds and cattail with trees and bushes (riverine levees)	1317.9	4922.7	+3604.8
Total area	27645.1	27645.1	
Key section 5			
Water area	5281.2	3815.8	−1465.4
Marshes	13817.3	4632.3	−9185.0
Waterlogged arable land	8080.8	14196.7	+6115.9
Wet fields (forage crops)	12712.9	21290.5	+8577.6
Dry fields (crops)	12164.1	8099.5	−4064.6
Built-up areas	1079.5	1100.5	+21.0
Total area	53135.8	53135.3	

(continued)

Table 4 (continued)

Type of surface	1986	2011	Changes (ha)
Key section 6			
Water area	18521.2	16030.9	−2500.3
Water, strongly overgrown with macrophytes	2203.5	2560.6	+357.1
Marshes	6059.1	5679.8	−379.3
Reedbeds and cattail with sand plots	13137.5	10707.6	−2429.9
Uplands and drying wetlands	4942.5	9884.5	+4942.0
Total area	44863.7	44863.4	
Key section 7			
Water of lakes	743.4	672.8	−70.6
Shallow waters (slightly overgrown)	415.5	–	−415.5
Water, overgrown with macrophytes	861.1	453.2	−407.9
Marshes and water	1403.5	964.9	−438.6
Reedbeds and cattail	1715.7	2344.6	+628.9
Uplands and levees	278.6	961.2	+682.6
Total area	5417.8	5396.5	(−27.3)

Table 5 Land cover changes in deltas of the Dnieper and Dniester Rivers

Type of surface	Area (ha)			
	Dnieper—delta		Dniester—delta	
	13.07.1986	16.08.2010	15.07.1975	01.08.2011
Water area	11605.2	11567.2	16343.1	17166.1
Shallow water	2239.8	1563.8	1490.8	1063.4
Water, overgrown with floating plants, and fens	4709.7	1334.4	2410.2	1187.8
Settlements and anthropogenic-changed territories	1865.3	3180.7	–	–
Overflow lands ("plavni") with shrubs and forests	9052.9	9509.9	–	–
Overflow lands ("plavni") with reeds and cuttails	7910.1	10227.1	12579.9	11372.2
Swamps overgrown by reeds	–	–	4172.0	4045.9
Waterlogged and upland meadows and settlements	–	–	8201.7	10362.3
Total area	37383.0	37383.0	45197.7	45197.7

7 Conclusions

Socio-economic transformations after the breakup of the Soviet Union, together with the effects of climate change, have been pronounced in Ukraine. Numerous recent publications have highlighted the relationship between climatic changes and terrestrial ecosystems (air, water, soil, and biota) on both the global and regional scale in regions of temperate climatic conditions that are similar to the object of our research (Groisman and Ivanov 2009). Unfortunately, this relationship is not often assessed systematically by accounting for all the associations and feedbacks of different physical and genetic processes in a changing environment.

Based on our understanding of the ongoing and projected environmental changes in Eastern Europe and particularly in Ukraine, we make the following recommendations for future studies. According to projections of warmer and dryer climate in the region from global warming, we suggest that special attention should be given to hydrometeorological modeling at the regional scale, and results should be implemented into optimal land management and the development of mitigation and adaptation measures. In particular, we suggest others to consider plans for reforestation. Such plans exist in each country in non-boreal Eastern Europe, but the implementation of these plans has been slow and should be given higher priority.

An analysis of available data shows close relationships among changes in Eastern Europe's terrestrial systems (air, water, soil, and biota). Generally, this part of the world has experienced a less continental climate during the last century, which manifests as lower-amplitude seasonal cycles of the surface air temperature and temperature increases during the winter. Furthermore, fewer differences are present in annual precipitation patterns, particularly between the northern and southern regions. The annual precipitation over most of Eastern Europe has increased during the past 50 years but sometimes decreased during spring and autumn months. The latest global climate models for Ukraine project the following major trends in the regional climate: 2.0 ± 0.5 °C increase in the mean annual surface air temperature by the end of the 21st century; 10–15 % increase in precipitation totals, with a 1.5–2.0 °C increase in the global surface air temperature; and a precipitation decrease in the southern and southeastern regions of Ukraine with further global warming above 2–3 °C because of a northern periphery shift in the subtropical anticyclone zone to these regions.

More research is clearly required given these projected changes and the existence of considerable changes in the functioning and disposal of ecosystems because of global climate changes since the mid-1980s. Model projections and assessments of the impacts of climate changes on individual species, crop yields, forestry, and the carbon cycle do exist, but models that can project the functioning and changes of natural ecosystems are not well developed. The main problems involve understanding the complex interactions of different forcing factors on ecosystems and their internal feedbacks. However, such models are urgently needed to preserve the regional biosphere.

In the near future, ecologists should focus on the development of theory and tools to study the Earth's carbon, energy, and water cycles, along with the interactions between the Earth system and human society; in situ and satellite data provide a unique perspective for this approach. Special attention should be paid to the development of effective methods to project changes in land cover, the frequency and intensity of extreme climate events, and agricultural production. Additionally, recently created geodynamic models should be implemented into scientific products to enhance the efficiency of industrial practices and activities that are designed to protect natural resources and the environment.

References

Bokov VA, Lushchik AV (1998) Basis of environmental safety. Simferopol, SONAT, 224 pp (in Russian)

Coleman JM, Huh OK, Braud DeWitt (2008) Wetland loss in world deltas. J Coastal Res 1:1–14

Davidok VP, Movchan YI, Parchuk GV (1997) National report of Ukraine about biodiversity preservation. Kiev, Takie dela, 21 pp (in Russian)

Denisik GI (1998) Anthropogenic landscapes of the right-bank of Ukraine. Vinnitsa, Arbat, 292 pp (in Russian)

Golubets MA (1997) Layer of life. Lvov, Polis, 185 pp (in Russian)

Groisman PY, Ivanov SV (2009) Regional aspects of climate-terrestrial-hydrologic interaction in non-boreal eastern Europe. NATO Science Series. Springer, Berlin, 269 pp

Kirkor OF (1907) Materials about variations of river's water composition: chemical research of Ros' river water in 1904–1905. Kiev, Lubkovsky Press. 124 pp (in Russian)

Kysel'ov YV (2000) River systems degradation of northeast Donetsk ridge. Messenger Lviv Univ Geol Ser 27:126–130 (in Ukrainian)

Lyalкo VI (2007) Estimation of influence natural-anthropogenic changes of CO_2 fluxes in the vegetation-atmosphere system over creation greenhouse effect. Rep Nat Acad Sci Ukr 4: 130–137 (in Ukrainian)

Lyalкo VI, Artemenкo IG, Zholobaк GM (2007) Research of influence CO_2 and CH_4 changes in atmosphere over climate using remote sensing data. Geol J 4:7–16 (in Ukrainian)

Mihalescu A (1983) Umweltsuenden-Katalog. Muenchen, 143 pp

Movchan DM (2008) Possibilities of the use of the satellite MODIS data for estimation both the productivity of various types of vegetation and CO_2 fluxes in the vegetation-atmosphere system in the example of Ukraine's territory. Rep Nat Acad Sci Ukr 8:108–112 (in Ukrainian)

National Atlas of Ukraine (2007) Kiev. Kartografiya, 440 pp (in Ukrainian)

Nilsson S, Shvidenko A, Stolbovoi V (2002) Russian forests and fluxes of major greenhouse gases. In: XI IBERA conference. Krasnoyarsk

Ramad F (1981) Basis of applied ecology: human impact on environment. L. Gidrometeoizdat, 544 pp (in Russian)

Shelyag-Sosonko YR, Andrienko TL (1985) Vegetation of Ukraine. Nature of Ukrainian SSR. The vegetation. Kiev. Naukova dumka, pp 130–201 (in Russian)

Shvets GI (1969) Blue gems of Ukraine. Kiev. Vysshaya shkola, 176 pp (in Russian)

Shyklomanov IA (1986) Water consumption and water supply dynamics in the world. Water Resour 6:119–139 (in Russian)

Starodubtsev VM (2013) Changes in the Danube delta according to remote sensing data by Landsat satellite. Arid Ecosyst 3(4):258–262

Starodubtsev VM, Bogdanets VA (2012) Deltas degradation as a global process. Environment and Ecology in the Mediterranean region. Cambridge Scholars Publishing, Newcastle upon Tyne, pp 299–306

Syvitski JPM, Vörösmarty CJ, Kettner AJ, Green P (2005) Impact of humans on the flux of terrestrial sediment to the global coastal ocean. Science (New York, NY) 308(5720):376–380

Syvitski JPM, Kettner AJ, Overeem I, Hutton EWH, Hannon MT, Brakenridge GR, Day J, Vörösmarty C, Yo Saito, Giosan L, Nicholls RJ (2009) Sinking deltas due to human activities. Nat Geosci 2(10):681–686

Voloshchuk VM, Boychenko SG, Stepanenko SM (2002) Global warming and climate of Ukraine. Kiev. Kyyivsky universytet, 116 pp (in Ukrainian)

Vörösmarty CJ, Meybeck M, Fekete B, Sharma K, Green P, Syvitski JPM (2003) Anthropogenic sediment retention: major global impact from registered river impoundments. Global Planet Change 39(1–2):169–190

Wada Y, van Beek LPH, Bierkens MFP (2012) Nonsustainable groundwater sustaining irrigation: a global assessment. Water Resour Res 48(November 2011):W00 L06

Yatsyk AV (1997) Ecological basis of rational water consumption. Kiev. Genezys, 640 pp (in Russian)

Yatsyk AV (2001) Environmental safety in Ukraine. Kiev. Genezys, 216 pp (in Russian)

Zhang W, Feng H, Zheng J, Hoitink AJF, van der Vegt M, Zhu Yu, Cai H (2013) Numerical simulation and analysis of saltwater intrusion lengths in the pearl river delta. China J Coast Res 287(2):372–382

Forest Changes and Carbon Budgets in the Black Sea Region

M. Ozdogan, P. Olofsson, C.E. Woodcock and A. Baccini

Abstract The temperate forests in the Black Sea region contain some of the last remaining intact forests between southern Europe and West Asia. The collapse of the Soviet Union brought great political and institutional changes to the region that have already impacted these forests, which have experienced long land use and management histories. In this chapter, we review and synthesize research on forest changes and carbon budgets that are associated with decentralization in the Black Sea countries, focusing specifically on Bulgaria, Georgia, Romania, and Ukraine. Our analysis shows that each of these countries followed a different path in forest management, somewhat mimicking their own history of transition from centrally controlled to market-based economies. In Romania and Bulgaria, a period of economic hardship and weakened institutions resulted in large-scale forest changes, but the net effect of these and other historic forest disturbance events has allowed Bulgaria and Romania to remain a terrestrial carbon sink. Although increases in logging could result in net carbon emissions, great potential exists for carbon sequestration as a result of forest expansion on degraded and abandoned farmland, particularly in Romania. Georgia continues to struggle with establishing and enforcing forest management with a suitable mix of private and public uses to meet the growing demand, particularly for energy. To this end, the future of Georgia's forests and its carbon implications remain uncertain and will mostly depend on its relationships with Russia and its ability to exploit its status as an energy corridor. Ukraine also continues to struggle with establishing suitable forest administration and ownership with high rates of illegal logging. However, natural forest regrowth on large tracts of abandoned farmland can sequester unprecedented amounts of carbon, and large-scale afforestation programs can greatly aid this process.

M. Ozdogan (✉)
Department of Forest and Wildlife Ecology, University of Wisconsin-Madison, Madison, WI, USA
e-mail: ozdogan@wisc.edu

P. Olofsson · C.E. Woodcock
Department of Earth and Environment, Boston University, Boston, MA, USA

A. Baccini
Woods Hole Research Center, Falmouth, MA, USA

G. Gutman and V. Radeloff (eds.), *Land-Cover and Land-Use Changes in Eastern Europe after the Collapse of the Soviet Union in 1991*,
DOI 10.1007/978-3-319-42638-9_7

These findings suggest that all these countries could play an important role in the terrestrial carbon budgets of the Black Sea region. This outcome is partly connected to the land use legacy of the Soviet Union: large areas of relatively young and regrowing forests, a result of high forest harvesting rates during the latter half of the 20th century, have tremendous carbon sequestration potential in each country that is reviewed here. At the same time, the effects of this legacy are quickly replaced with land use and forest management decisions that are made today.

1 Introduction

The ecoregion that surrounds the Black Sea between southeastern Europe and Asia Minor holds some of the last remaining intact forests of Europe. Located within the temperate climate zone, the region has considerable topographic and climatic diversity and associated land cover (Fig. 1). While forests only occupy approximately 20 % (approximately 324,000 km^2) of the total area of all the circum-Black Sea countries (excluding Russia), these forests are almost exclusively located in mountainous areas and support a high diversity of fauna and flora, including a large number of woody species and old growth forests. Following the collapse of the Soviet Union, the Black Sea region experienced a significant political and

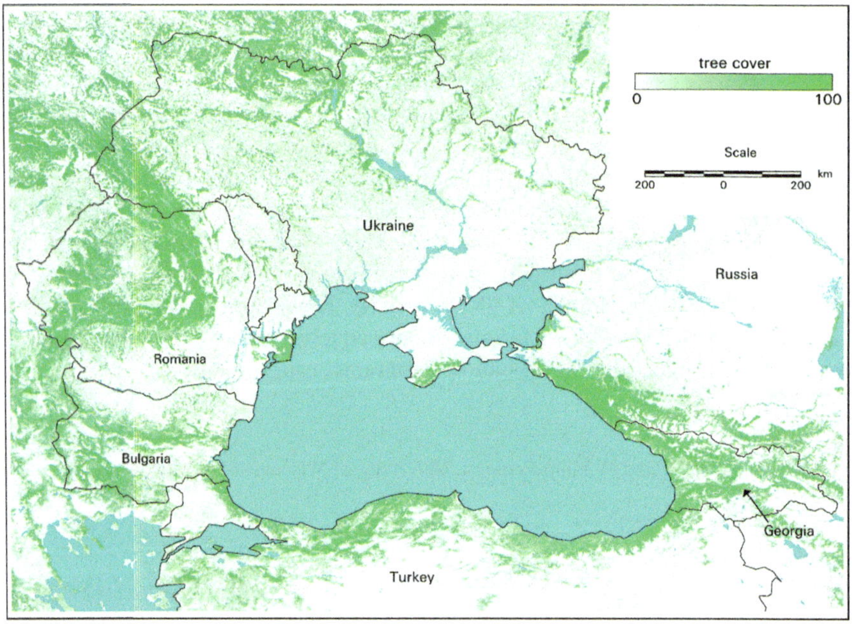

Fig. 1 Distribution of forests in the Black Sea region. The tree percent data are from the Vegetation Continuous Fields product (Hansen et al. 2002)

institutional transformation; until recently, the consequences of this transformation for forest resources remained uncertain. For example, a large portion of forest holdings in both Bulgaria and Romania have been privatized since 1990 as part of larger land ownership reform. The governments of Ukraine and Georgia, on the other hand, have been more reluctant to relinquish control, privatizing only smaller portions of these forested areas under strict conditions.

In this chapter, we review and synthesize research on forest changes and associated carbon budgets in the Black Sea countries. We specifically focus on Bulgaria, Georgia, Romania, and Ukraine. All these countries experienced dramatic changes in socio-economic conditions after the collapse of the Soviet Union, but the changes in forests have followed a different trajectory in each country, leading to different changes in their carbon budgets. We did not include the Russian Federation because this country's forests along the Black Sea coast are rather insignificant compared to the forest resources of the country as a whole, nor did we include Turkey because the dramatic changes that unfolded following the breakup of the Soviet Union did not have a major impact on Turkey's forests.

2 Common Origins, Different Paths: A Review of Institutional Changes in the Forest Sector

The collapse of the Soviet Union brought large institutional transformations to the forestry sector in the Black Sea countries. However, these transformations followed significantly different paths in each of the four countries that we focus on here. For example, while a large portion of forests have been privatized in Bulgaria and Romania since 1990, the restitution of forests—a term that describes the act of returning forests to their pre-Soviet owners—has been slow in Ukraine and Georgia and has occurred under strict conditions. While the economic value of the forests in each country is high and provides an incentive for timber production, the development of legal and institutional mechanisms that govern the management, restitution, and ownership of forests, especially during the early 2000s, resulted in significant changes in forest cover, which will influence the carbon fluxes in the region for decades to come (Olofsson et al. 2011). To this end, we provide a short review of the forest sectors in each of the four Black Sea countries and summarize the changes in forest-related laws and institutions, which directly reflect the amount of changes that are observed in the landscape with remotely sensed measurements.

2.1 Bulgaria

Approximately one third of Bulgaria's land base is forested, with coniferous and broadleaf (deciduous) forests predominating at different altitudes. Bulgaria's forests are one of the most diverse ecological regions in Europe because of the complex topography and their unique location in a climatic transition zone between the European continental, Eurasian steppe, and Mediterranean climatic zones. Intensive afforestation programs and careful utilization have increased the country's forested area in recent years, adding 204 km^2 between 1990 and 2000 (World Bank 2002). At the same time, the rate of creation of new forests has dropped significantly in recent years because of the lack of infrastructure and terrain with steep slopes. Bulgaria's forests also remain threatened by the long-lasting impacts of heavy industry and the increasing resource demands of privatization and development.

Since the breakup of the Soviet Union, Bulgaria's forestry sector has undergone a series of policy and institutional reforms, starting with the 1997 *Forest and Forestland Ownership Restitution Act*, which marked the beginning of the structural reform in the forestry sector (Table 1) (Cellarius 2001). The latest reforms were adopted in 2011, and several important but controversial amendments have followed. One of the most important outcomes of these reforms has been the total restitution of forest areas to private individuals, legal entities and municipalities,

Table 1 Summary of changes in institutional policies regarding the forestry sector in the four Black Sea countries since the breakup of the Soviet Union

Year	Event	Outcome
Bulgaria		
1995	Concession Act (Amended in 1999)	This act governs the terms and procedures for granting concession including forested areas to private parties in the form of "special right to use"
1997	Adoption of the new Forest Act (Law for the Restitution of Property of Forests and Forest Lands)	Provides the legislation for restitution and private ownership of forests first time since 1958. The management, protection, and development of forests are moved under the purview of an interdisciplinary body, the supreme Forest Management Board
2002	Initiation of the development of a National Forestry Policy Strategy	Incorporation of both economic and ecological needs into forest management in the context of larger ownership reform
2005	Introduction of the certification process	Certification process in line with European standards is piloted in small areas
2011	Adoption of new forest law	Defines the principle of sustainable management of all types of forests. The law encourages the certification process in forests. It is innovative and based on the European practices
2012	Amendments to the 2011 Forest Act are adopted	This is considered to be a lobbyist amendment which would allow for cheap and easy construction on protected areas

(continued)

Table 1 (continued)

Year	Event	Outcome
2012	Further amendments to the 2011 Forestry Act	Further amendments following intense public pressure which will lead to the protection and sustainable government of the Bulgarian forests
Georgia		
1991–2000	Reduction of timber and fuel imports from Russia	Intensive timber harvesting activities that are unprecedented in recent past. Increase in illegal fuel wood collection by local population
1995	Initial changes in existing Soviet forest code	To regulate the activities and the utilization of the forest fund, as well as defining forest categories and their utilization
1998	Decree from the president amending the law of timber	Temporarily changes the status of timber exports from the country exports
1999	Georgian Forest Code is adopted	Substantial powers are given to local bodies for the management of local forests, and there is provision for public participation in decision-making processes regarding the management of the state forest estate
Romania		
1987	Law for the conservation protection and development of forests and their rational exploitation	Initial attempts at changing the forest code
1991	Forestry Code	Development of Forest Estates (public and private). Start of the official restitution process, returning 1 ha of forest to each legal heir of pre-WW2 individual owners
2001	The restitution Law 10	Up to 10 ha are being restituted to individuals, up to 30 ha to churches, and all forests are being restituted to communities
2002	Regulation No. 635	Concerns the general management practices, timber harvesting regulations
2003	Founding of National Working Group on Forest Certifications	Initial attempts at the certification process
2008	Forest Code (Law 46/2008)	Second major changes to the forest code since the breakup
2008	Regulation No. 606	Concerns timber harvesting regulations

(continued)

Table 1 (continued)

Year	Event	Outcome
Ukraine		
1991	The Land Code	The start of the land reform in Ukraine. It provides many forms of land ownership, including state, collective, and private. This includes the land of the Forest Fund
1994	Forest Code	Develops provisions of the first basic laws of the national legislation such as the Land Code (1991), the Law on Environmental Protection (1991)
2001	The new Land Code	Considered as a major step towards private ownership of land and development of land markets in Ukraine. The Code contains the roles and provisions needed for private parties to own land, use land and transfer land as they see fit. Expressly indicates what lands belong to the forest fund. Possible forms of ownership are recognized. The state strategy and an order for transfer of the forests for communal and private property is initiated. The term "forest fund" is corrected by excluding forest protection belts and other line forest stands
2006	Forest Code of Ukraine	A number of regulations supporting the concept of the reform and development of the Forest Sector of Ukraine are passed

which, as of 2006, amounted to approximately a quarter of the total forest area (slightly over one million ha). This non-state ownership is further distributed equally between physical persons and other legal entities (46 %) and municipal forests (52 %), with small additional amounts (approximately 2 %) set aside for religious communities. In general, the forests that are associated with individual owners are small; most are less than 1 ha in size. In contrast, the municipal forests are usually several hundreds of hectares, which have important implications for forest harvesting and carbon dynamics.

2.2 Georgia

Approximately 40 % of Georgia's land area is forested, and the majority of its forests are located on high slopes, covering the Greater Caucasus and Lesser Caucasus Mountains. Over 80 % of these forests consist of broadleaf species (almost 50 % beech), while the rest of consist of conifers. Georgia's forests, which are rich in biodiversity, contain more than 4100 of the estimated 6350 species in the entire Caucasus region, including 395 species of woody plants (FAO 2015; Torchinava 2005).

Since independence, Georgia's forests have been under severe pressure from illegal and unsustainable logging and poor management practices. In particular, the intensive timber harvesting activities in Georgia are unprecedented in the country's history. The primary causes for forest degradation in Georgia are twofold. First, timber imports from Russia have almost completely stopped following independence. With the dissolution of the Soviet Union, large volumes of wood that have been imported from Russia for construction and pulp production were no longer available, practically stopping Georgia's processing enterprises (FAO 2015). While reliable economic data on the forestry sector are scarce, one report has estimated that Georgia's imports of timber resources from surrounding regions (mainly Russia) reached 2.5 million m^3 (M m^3) annually, which constituted more than 85 % of the country's requirements (FAO 2005; World Bank 2007). The second factor that has contributed to forest degradation is a sharp reduction in fuel imports, mainly from Russia (Jervalidze 2006). Rural poverty, coupled with a lack of affordable alternatives to wood as a heating fuel, has resulted in widespread illegal harvesting by citizens. While no reliable sources exist regarding the specific volume of illegally harvested wood in Georgia, estimates vary between 2.5 and 8 M m^3 per year (FAO 2005, 2015; Torchinava 2005). Olofsson et al. (2010) performed a remote sensing-based study and estimated that approximately 22,000 ha of forest was illegally logged between 1990 and 2000, but this estimate has a high degree of uncertainty because of the local and small-scale nature of the logging activities.

In terms of the institutional framework, the 1999 Forest Code established the legal basis for the protection, use, and restoration of Georgia's forest resources (Table 1). The Code's primary objectives were the protection and sustainable development and management of forests and a shift in forest management from central planning to market-based management. The World Bank indicated that the Code has been improved since 1999, but further amendments are needed to improve its application and transparency (Torchinava 2005; World Bank 2002, 2009).

The Forest Code allows for multiple forms of forest ownership, including state, municipal, community, church, and private ownership. The code also allows for the long-term leasing of state forests. Licenses to use-rights for state forests are auctioned. The only exception to the license requirement is use for fuelwood by households. Most commercial loggers are private and do not have long-term rights to forested land. Thus, commercial loggers tend to focus on short-term monetary gain rather than medium-term sustainability in their forest operations (World Bank 2002; GOG 2006, FAO 2010).

In 2007, the government of Georgia launched a large-scale reform of the forestry sector to transfer responsibility for forest management and maintenance with long-term leases (up to 50 years) (Egiashvili and Ratiani 2008). However, only a few reforms were implemented, and only 12 long-term leases were auctioned while the Forestry Department was restructured and reorganized (Macharashvili 2009; FAO 2010).

2.3 Romania

Romania has some of Europe's last relatively undisturbed forest ecosystems, and substantial concerns have been expressed regarding unsustainable forest use from forest restitution processes (Ioras and Abrudan 2006; Knorn et al. 2012; Olofsson et al. 2011). As a primary example of a country that chose to restitute its forests, Romania's restitution process included three phases: the first restitution law (18/1991) returned a total of 350,000 ha (Cirelli and Uliescu 1997; Mantescu and Vasile 2009), the second law (1/2000) targeted another 2 million ha, and the third and final law (247/2005) restituted all remaining forests that were privately owned prior to World War II (Table 1). Together, 70 % of all Romanian forestland has been or will be transferred into non-state ownership, doubling the number of individual forest owners from approximately 400,000 in 2000 (Ioras and Abrudan 2006; Abrudan et al. 2009).

Overall, Romania's forest restitution process was fairly complex, which created a period of economic hardship. One outcome of this transition period was greater incentives for new owners to harvest wood rapidly while institutions and forest law enforcement were weak (Nichiforel and Schanz 2011). On the other hand, the relatively high forest harvest rates of the socialist period have declined considerably since the breakup (Turnock 2002), and a considerable portion of the farmland was abandoned in post-socialist Romania, which resulted in forest regrowth (Kuemmerle et al. 2009). Furthermore, Romania became a member of the European Union in 2007, requiring new forest legislation, management practices and a substantial enlargement of its protected area network.

2.4 Ukraine

The total forest area in Ukraine is approximately 100,000 km^2, comprising 17 % of the country's land area. Coniferous forests dominate the landscape, covering almost half the total forested area and more than half the total growing stock. The forests in Ukraine are distributed very unevenly across the country because of climatic conditions and a long history of anthropogenic impacts. Crimea and the Steppe regions to the south have relatively low forest volumes. In contrast, the largest forested areas are concentrated in the northern and western parts of the country, particularly in the Polesia (mixed forests) region and the Ukrainian Carpathians. In terms of tree types, pine is dominant in Polesia and the northern part of the Steppe, oak in the Forest-Steppe and the southern part of the Steppe, and spruce in Crimea and Carpathians. Historically, all these areas have been managed very differently because of varying growing conditions. However, the government of Ukraine set targets during its last round of large-scale assessment to increase the forest area in all regions (World Bank 2006).

Ukrainian forests and forest management practices have several distinctive features compared to the other countries in the region. For example, the country generally has a low percentage of forest cover but a high percentage of reserved forests and a rich history of forest management by a diverse group of organizations (e.g., forests were managed by enterprises, institutions and organizations under more than 50 ministries and departments). Moreover, the majority of Ukrainian forest area is planted and requires careful maintenance (State Forest Resources Agency of Ukraine 2012).

In terms of ownership, the majority of the forests in Ukraine are state property. The forest ownership coincides with the ownership of the land on which the forests are located. Three types of forest land property rights are recognized in Ukraine in compliance with the new Land Code that was adopted in 2001: (a) state ownership, excluding communal and private forests; (b) communal forests, mostly located within the boundaries of settlements; and (c) private forests with stands on plots up to 5 ha within agricultural and farming lands or stands that are grown on private plots (Table 1). Unlike other ex-Soviet countries, the old model of state-owned forest management has prevailed, which has led to small-scale harvesting and processing (Nordberg 2007).

The Ukrainian Forest Code of 2006 stipulated that the use of forests can be temporarily or permanently granted based on decisions of executive power or local government bodies. For example, all forests in state, municipal or private ownership can be temporarily used over short (up to 1 year) or long periods (from 1 to 50 years), which would include the transfer of management rights and responsibilities from public to private institutions. Thus, most timber harvesting is organized by state and private forest harvesting companies.

The forest sector plays an important role in rural areas in Ukraine. With the collapse of the Soviet Union, however, the forest sector suffered a sharp decline, leading to ownership changes and losses in cheap timber supplies and traditional sales markets. Coupled with the loss of cheap raw material (mostly imported from Russia), these changes resulted in the closure of a significant proportion of the large-scale forest industry during the mid-1990s, and recovery has been slow (Nordberg 2007). Furthermore, reforms in the forest sector have been very limited compared to in most other transition countries in Europe.

3 Quantifying Forest Changes and Associated Carbon Fluxes

The short review of the above four Black Sea countries shows how these countries followed divergent paths in developing their forestry sectors in the aftermath of the collapse of socialism and decentralization. While the net effect of these factors in terms of forest changes has not been adequately characterized for some time, literature is beginning to emerge that quantifies changes in forested areas and the ramifications of these changes on regional carbon budgets. Here, we review and synthesize the existing work to provide comprehensive information for the region regarding the rates of timber harvesting and the conversion of forests to other land uses and the effects of conversion on regional carbon budgets following the collapse of the Soviet Union.

In this synthesis, we focus on studies that utilized remotely sensed data to quantify timber harvesting and the conversion rates of forests to other land uses. The reasons for this decision were twofold. First, a substantial portion of the inventory data in these countries only reports the rate of timber harvesting without clear geo-references, which severely limits its use. Although new continuous inventory systems have been introduced with the help of international donors (e.g., the World Bank), this process has been slow and non-transparent. Second, disparate data sources for forest harvests, land-use change, and reforestation complicate comparisons within each country over time and among countries. We acknowledge that remotely sensed studies have other limitations and can be fraught with errors if the harvested plots are small, if most harvesting is in the form of selective logging, and if frequent cloud cover is present over mountainous areas (Olofsson et al. 2009). Fortunately, the existing studies have carefully addressed each of these issues and provide the best consistent picture for differential changes in each country following the breakup of the Soviet Union. These studies have also made extensive use of statistical techniques to identify and adjust for errors during the classification of satellite data (Olofsson et al. 2013, 2014).

We focus on studies that have utilized a book-keeping carbon model that is driven by changes in land use and management to quantify the sources and sinks of carbon that are attributable to changes in and the management of forests (Houghton 2003a, b; Houghton and Hackler 2003). This book-keeping approach requires information on the rates of carbon accumulation (in vegetation and soil) that accompany forest growth. In many studies, these variables were obtained from the literature and from forest inventory data for countries of interest. This model is generally propagated over an annual time step, and unit area changes in vegetation and soil following land-use changes are defined for different types of ecosystems and land uses. For example, when a forest is cleared, some of the initial biomass is left on the site (slash) and some is removed (wood products). Slash that is left is either burned or decays exponentially, releasing carbon into the atmosphere in the same year that it is burned or during subsequent years, respectively. Material that is removed from the site is tracked in pools that decay with time constants of 1 $year^{-1}$ for fuelwood, 0.1 $year^{-1}$ for short-lived wood products, and 0.01 $year^{-1}$ for long-lived wood products. Following harvesting, forests are generally permitted to grow back. In the model, the annual rates of carbon accumulation in regrowing forests define the amount of carbon in each age class, thus allowing the annual uptake and average biomass of forests to be calculated. Ecosystems that are not cleared, abandoned, grazed, or harvested are not included in the analysis (i.e., they are assumed to be unchanged with respect to carbon stocks, Olofsson et al. 2009).

3.1 Bulgaria

Most of Bulgaria's forests are located in its mountains, but the forest area in Bulgaria has steadily increased in recent decades because of intensive reforestation activities. The expansion of the forest area has occurred mainly at the expense of lands that were only marginally suitable for agricultural use. Although Bulgarian forests have traditionally provided a number of economic services, especially in rural areas, forest management has traditionally focused on timber production. However, following the country's European integration, Bulgaria's forest management practices have shifted away from single-use to multi-use, including renewable energy production (wood burning), protection, mitigating climate change, and maximizing ecosystem services (FAO 2011).

Remote sensing-based studies of forest changes in Bulgaria are rare—only a handful of investigations have focused on monitoring forest harvesting. Our own analysis, which involved Landsat data in southwestern Bulgaria, revealed a limited amount of forest cover change during 1990, 2000, and 2010 (Fig. 2). Most of these changes were located in mountainous areas and in coniferous forests.

A detailed review of the literature suggests that a number of institutional and technical factors influence the quantity, quality and location of forest harvests in Bulgaria, which has important implications for carbon release and sequestration.

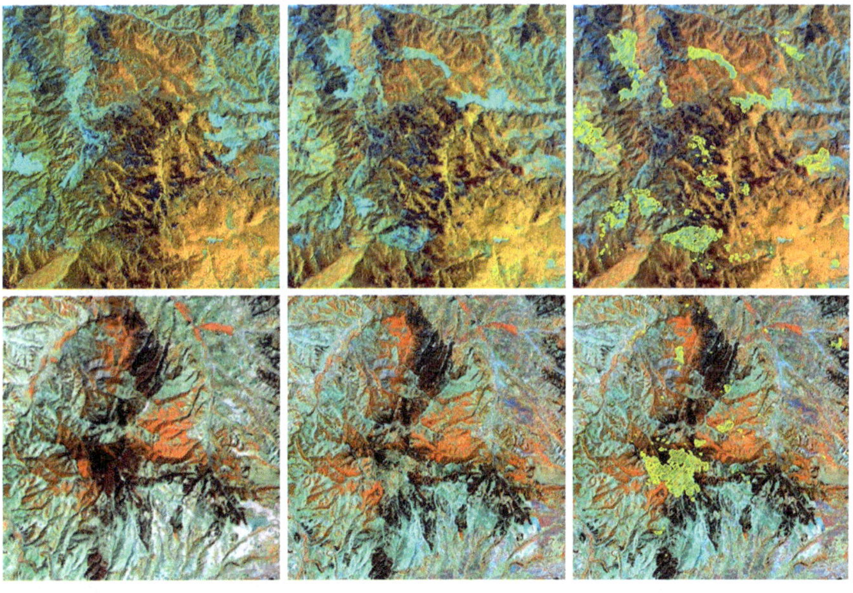

Fig. 2 Changes in forest cover in SW Bulgaria between 2000 and 2010 (*top row*) and between 1990 and 2000 (*bottom row*). In the *right panels* of each row, the pre- (*left panel*) and post-harvest (*center panel*) differences are highlighted in the extracted forest area (*yellow shaded areas*). The size of each window is approximately 25 km

First, the lack of infrastructure, especially in mountainous areas where most of the growing stock occurs, limits the harvesting of mature and overstocked forests. For example, the forest road density is 7.9 m/ha on average, which is low compared to other European countries with broadly similar topographic conditions (e.g., Austria: 36 m/ha, Switzerland: 40 m/ha, France: 26 m/ha, and Germany: 45 m/ha). Forest stock indicators show that Bulgarian forests are maturing because of this non-optimal harvesting practice: during 1965–2000, the growing stock more than doubled from 252 to 526 M m^3 (Trichkov and Dinev 2013). On the other hand, accessible mature wood resources are intensively utilized because of the high demand for saw logs and other large wood products.

Second, timber harvesting is performed through regeneration- and sanitary-cutting and thinning. New stands are established mainly through natural regeneration. Harvesting in the form of clear-cuts is rare and typically only applied in stands with species that require intensive management. Under the latest forest code, the maximum size of clear-cut areas is limited to 5 ha.

Third, the number and extent of illegal forest harvesting activities has dramatically increased in recent years. Although illegal activities take many forms, an important activity is the collection and use of forests for firewood (WWF 2005). These activities usually target the most accessible sites near forest roads, so these sites become easily overexploited. In younger forests, the remaining trees are of inferior quality because the best ones have been harvested. Moreover, the use of timber for firewood has increased, serving as the cheapest form of household heating. In recent years, the number of households that use firewood has doubled, and now 40 % of households use firewood for heating or boiling water.

Lastly, the process of restituting state forests to their former private owners, which is considered to be one of the most important aspects of institutional reform in the forest sector, has been completed, and no further restitution is expected. As of 2010, three quarters of the forests are state owned, while the rest is owned by individuals (10 %), municipalities (12 %), and various other institutions (2 %). One outcome of this distribution is that although municipal forests are often in fairly large blocks, individual ownership is fragmented and blocks are typically less than 1 ha in size. Moreover, private owners tend to know little about sustainable management practices and often work within small and fragmented forest structures.

Unfortunately, the implications of Bulgaria's institutional and economic issues on forest carbon balance are not well known. The forest stock indicators show that as a result of Bulgarian forests have matured because of non-targeted harvesting practices, and evidence strongly suggests that this trend is continuing. The total annual increment in forests is approximately 14.5 M m^3, but only 5–6 M m^3 of wood is harvested (Trichkov and Dinev 2013). Additionally, some evidence exists for the effects of land use and forest disturbance on soil organic carbon. Zhiyanski et al. (2009) found that the overall carbon stock in both the forest floor and soils was highest for high altitude pastures (81 tons/ha), followed by spruce plantations (77 tons/ha), while beech forests and pine plantations had comparatively low carbon stocks. The net effect of conversion from natural pastures and beech forests into coniferous plantations in the central Stara Planina mountain area decreased the soil organic carbon and carbon from the upper mineral soil decades (Zhiyanski et al. 2008, 2009). On the other hand, the large organic carbon storage in the forest floor in spruce plantations compensated the carbon that was lost from mineral soil after the land-use change. While carbon sinks may be saturating in European forests (Nabuurs et al. 2013), including Bulgaria's, most existing studies point towards Bulgaria's forests as an important carbon sink in the Black Sea region (e.g., Zhiyanski et al. 2009).

With respect to changes in carbon sequestration in Bulgarian forests, the future will depend on the direction and use of forest products and changes in land use. For example, the harvested quantities of wood can be increased by 8–10 M m^3, significantly reducing forest waste and dramatically increasing the processing of wood for energy, for instance, by converting from conventional wood products into pellets, chips and briquettes (Trichkov and Dinev 2013). These changes could increase the contribution of biomass-based renewal energy sources in Bulgaria's energy use without significantly increasing forest harvesting. The likely effect of these changes for carbon sequestration would be to allow Bulgarian forests to remain as a carbon sink for some time in the future.

3.2 *Georgia*

Forests play an important role in Georgia's society, both historically and in recent times. This country, which is situated on the Caucasus, is home to diverse fauna and flora, and almost half the country, approximately 30,000 km^2, is covered by forests. Broadleaf species occupy the majority (80 %) of the forests, while approximately 20 % is covered by conifers. Most of these forests are middle-aged (50 %) or mature (22 %) (Metreveli 2002).

The economic and social implications of independence from the Soviet Union in 1991 were dramatic for Georgia. Unemployment increased sharply while economic activity and the population dropped (GOG 2009). However, what remains unclear are the environmental implications of the social and political transitions. Evidence suggests that deforestation in the form of illegal fuel wood harvesting, mostly to mitigate a growing energy crisis, has sharply increased during the transition period (Metreveli 2002). For example, the volume of illegal cuttings for firewood reached its highest level, 106,000 m^3, in 1996 (Oy 2005).

This uncertain period also coincides with a reduction in forest regeneration. For example, the increase in wood volume from regeneration between 1990 and 1997 was 30–40 times lower than during the 1970s and 1980s (Oy 2005). Most of the reduction in regeneration activities was related to forest degradation, which is reflected in the correlation of age groups, reduction in the area of highly productive forest species, and lower canopy cover. The process of forest degradation appears to have continued at least until the end of the 1990s because of the illegal removal and export of timber to foreign countries (Metreveli 2002).

To date, the most comprehensive study on the nature and rate of land use change, deforestation, illegal logging, and associated carbon dynamics was conducted by Olofsson et al. (2010), who used historical land use rates, satellite estimates of forest changes between 1990 and 2000, and a carbon book-keeping model to obtain the rates of land use change to estimate the associated carbon sinks and sources over time. The results of this study suggested that less than one percent (0.82 %) of the forest that was present in 1990 had been lost by 2000 and that much of that loss was concentrated in the western part of the country. The eastern area along the Greater Caucasus range to the north showed no evidence of forest loss. Even the breakaway regions of Abkhazia and South Ossetia, which have not been under full governmental control since the beginning of the 1990s, showed little evidence of forest loss (0.19 and 0.26 %, respectively).

The results, when translated into the amount of carbon with the help of the book-keeping carbon model, showed that Georgian forests are a carbon sink of approximately 0.3 Tg C/y (0.35 Tg in 2004; 0.26 in 2010), but these estimates do not include illegal logging. The sink in 2004 was equivalent to 31 % of the anthropogenic emissions. Moreover, Georgia will remain a sink, with the magnitude slowly declining to zero by approximately 2040. These rates were calculated with the best estimates from remote sensing and assuming a forest loss rate that remains constant at 1990–2000 levels. Assuming that the current forestry activities do not change, the illegal logging rate is the main determinant of the magnitude of the future carbon sink. The effects of illegal logging, as quantified by Georgia's Forestry department, resulted in a larger sink, which will not turn into a source until approximately 2060. Finally, further analysis that incorporated various land use change scenarios, in which the observed logging rates were linearly increased and decreased, indicated either an accelerated net carbon release but an increase in the source strength or a continued carbon sink, respectively (Fig. 3).

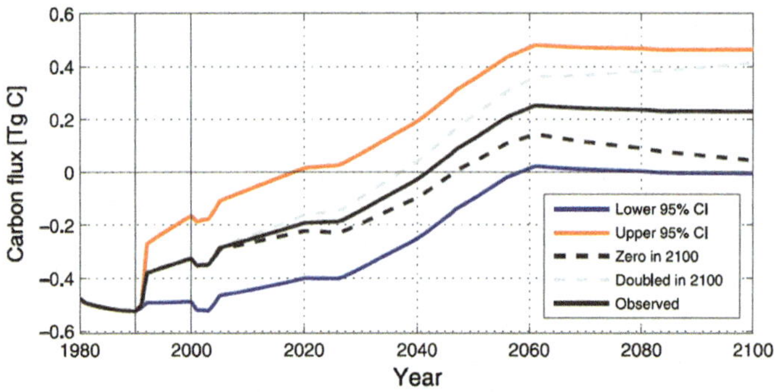

Fig. 3 Net carbon flux for different scenarios of illegal logging in Georgia. "Lower 95 % CI" refers to the lower 95 % confidence interval of the remote sensing estimate. "Upper 95 % CI" refers to the lower 95 % confidence interval of the remote sensing estimate. "Zero in 2100" refers to a linear decrease from the current rate to no illegal logging in 2100. "Doubled in 2100" refers to a linear increase from the current rate to the current rate times two in 2100. "Observed" means the current illegal rate as observed from satellites. The two vertical lines show the time periods that were covered by the remote sensing analysis. From Olofsson et al. (2010). Reprinted with permission from the publisher

The satellite-based assessment of logging rates in Georgia by Olofsson et al. (2010) may have omitted significant amounts of tree removals because most forest harvesting occurs in the form of partial logging. Although a location was identified as logged if half of the trees had been cut, determining the percentage of removal was challenging and resulted in deforestation area estimates with high uncertainty. Nevertheless, unlike in other parts of the former Soviet Union, the rate of deforestation following the collapse was low, with just 0.8 % of the forest having been removed between 1990 and 2000 (Torchinava 2006). Most timber harvesting appears to be for local household use in the form of fuel wood. Although we can assume that the observed change is mainly a result of illegal logging, evidence of large-scale clear-cutting is lacking. Moreover, if recent economic growth continues, illegal logging will likely decrease as the reliance on wood for fuel declines (Olofsson et al. 2010).

3.3 Romania

Roughly a quarter of Romania is covered with timber-rich and generally well-managed forests. However, despite the country's long tradition and advanced technical capacity in forest management, many Romanian agencies that are responsible for forest management have struggled to adapt to changing ownership and the utilization of forests and have not easily embraced multipurpose management (Dembner 1994). As described earlier, the most important event that underlies the changes in forest management in Romania is forest restitution. Similar to other Eastern European countries, Romania embarked on a process of restitution soon after the restoration of democracy. While the subject of forest restitution in Romania has been addressed in the literature for some time, many of these investigations approached the topic from economic and social perspectives (see, for example, Surd and Turnock 2000; Tickle and Clarke 2000; Lawrence and Szabo 2005; Lawrence 2008). In recent years, however, research that is based on satellite remote sensing has emerged and carefully quantified the amount of forest changes that are associated with forest restitution. For example, Griffiths et al. (2012) used Landsat data to investigate how three phases of forest restitution affected forest disturbances from both natural disturbances and changing forest management regimes. Their results showed that forest disturbances increased substantially since the collapse of socialism in 1989, with 750 km^2 of disturbed forest (4.5 % of the total studied forest area) (Fig. 4). Furthermore, forest disturbances increased between each consequent restitution law (34, 21 and 32 %). Non-state ownership and the species compositions of restituted forests appear to influence the degree of disturbance.

A second study by Knorn et al. (2012), who also used Landsat data, showed that the forest disturbance rates in Romania (Carpathians) increased sharply in two waves after 1995 and 2005, which were triggered by rapid ownership and institutional changes. In addition to the magnitude of forest changes, this study shows substantial amounts of forest disturbance inside protected areas, even within core

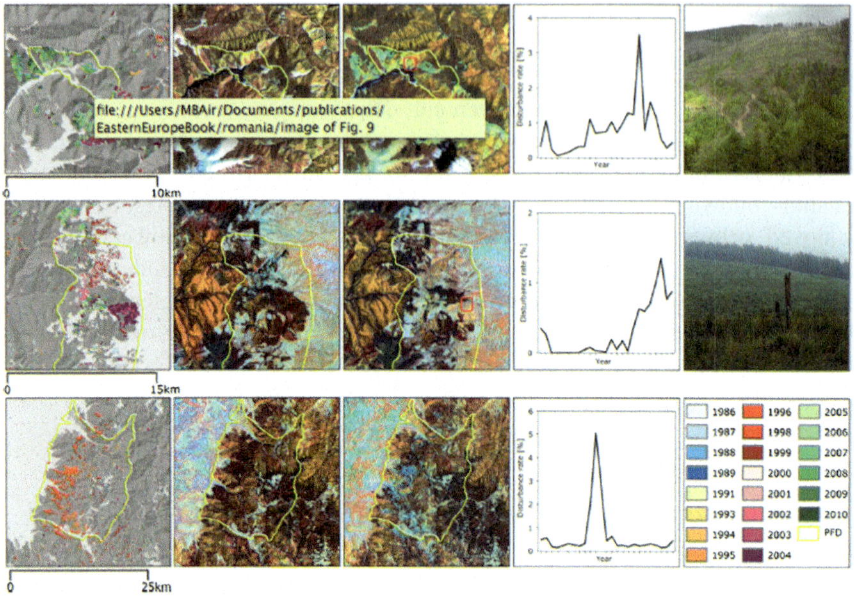

Fig. 4 Three examples of a disturbance map (*left column*) and imagery (RGB = 453) for private forest district areas. The development of the annual disturbance rates for the three PFDs is provided (*fourth column*, *dotted line* indicates the year of legal establishment). The examples show the PFD Papusa–Rucar (*top*), PFD Reghin–Gheorghieni (*middle*), and PFD Targu–Secuiesc (*bottom*). The *red frames* in the imagery indicate the location of two pictures that were taken during a field visit, exemplifying areas of excessive logging (*right*). From Griffiths et al. (2012). Reprinted with permission from the publisher

reserve areas, which suggests that the effectiveness of Romania's protected area network may be decreasing in terms of its ability to safeguard biodiversity.

As with forest disturbance studies that involve satellite data and quantitative models, information regarding the relationship between forest changes and regional carbon budgets has emerged only recently. For example, Olofsson et al. (2011) investigated the implications of forest restitution on the terrestrial carbon balance in Romania by using the aforementioned carbon book-keeping model and satellite-based estimates of forest disturbances. According to both historical and recent logging data, this study shows that high logging rates during socialism resulted in substantial terrestrial carbon emissions, with Romania being a net carbon source until the 1980s. Although the period after the collapse of the Soviet Union was marked by dramatically lower forest harvesting rates, forest restitution has increased forest disturbances in line with other satellite-based investigations. The net effect of both historic and recent forest transition events has allowed Romania to remain a terrestrial carbon sink, offsetting 7.6 ± 2.5 % of anthropogenic carbon emissions (Fig. 5). However, a further increase in logging could result in net emissions from terrestrial ecosystems during the coming decades. Although great

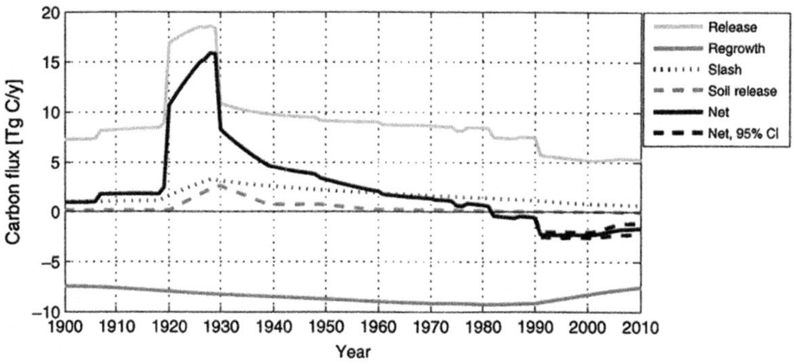

Fig. 5 Terrestrial carbon flux in Romania according to the baseline rates. Because the model only associates the release and uptake of soil carbon with permanent forest loss and gain, the soil carbon flux is close to zero and therefore not plotted (a positive flux equals terrestrial emissions). From Olofsson et al. (2011). Reprinted with permission from the publisher

potential exists for carbon sequestration from forest expansion on degraded land and abandoned farmland, the challenge in Romania is to comply with legal requirements, technical parameters, and environmental needs when managing and harvesting forests, regardless of their condition, size, type, or ownership status.

3.4 *Ukraine*

Compared to the other Eastern European countries that are considered here, Ukraine's forested area is small with respect to the total area of the country. The unique characteristics of Ukrainian forests combined with more tumultuous and slow socioeconomic transition have led to large but not well-quantified changes in forested areas (Nordberg 2007). In other words, the sharp economic decline during the first 10 years of the post-socialist period in Ukraine has had significant negative effects on the forestry sector that were further exacerbated by slow and limited reforms compared to most other transition countries in the region (Nordberg 2007). The net effect of these negative developments purportedly led to substantial changes in forest area, mainly by illegal forest harvesting activities, but quantitative evidence has only recently emerged. On the positive side, the last decade has been marked by more favorable conditions in the forestry sector, including growth, compliance, and environmental considerations.

A few satellite-based assessments of forest change studies in Ukraine have provided a clear picture of the effects of the transition period on forest disturbance rates and associated carbon fluxes. In an important publication that focused on the Ukrainian Carpathians, Kuemmerle et al. (2007) compared post-socialist forest disturbance rates by using satellite data among three countries in the region and found that their forest disturbance rates differed markedly, although the

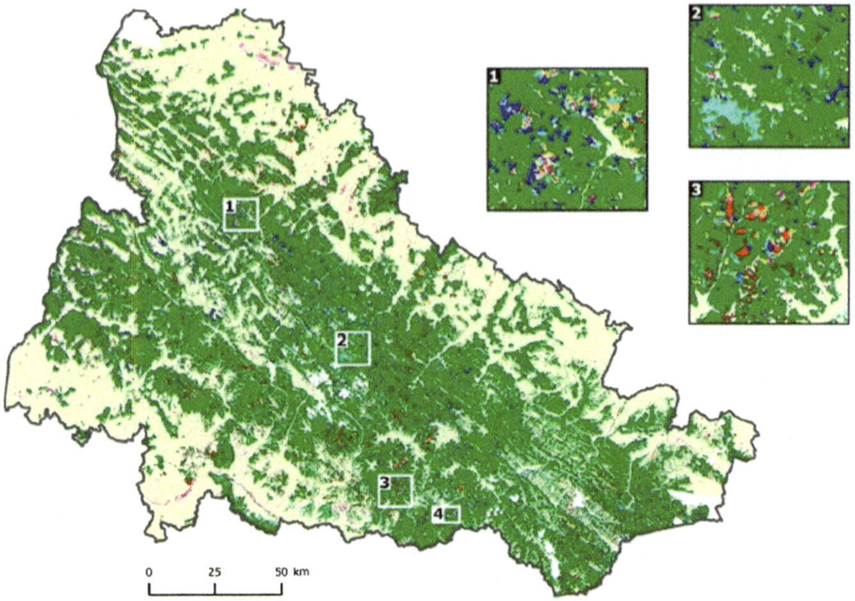

Fig. 6 Forest cover changes between 1988 and 2007 in Ukrainian Carpathians. From Kuemmerle et al. (2009). Reprinted with permission from the publisher

socioeconomic system change increased the harvesting in all countries during 1988–1994. For example, the disturbance rates in Ukraine were 4.5 times higher than those in Poland, and harvests tended to occur at higher elevations. Different harvesting rates between these countries, even under similar environmental conditions, were also identified in econometric analyses (Alix-Garcia et al. 2012). Hence, these differences in the disturbance rates among countries appeared to be most closely related to broad-scale socioeconomic conditions, forest management practices, forest policies, and the strength of institutions.

Kuemmerle et al. (2009) further quantified the extent of illegal logging and reforestation in the Ukrainian Carpathians by comparing satellite-derived forest trends to official statistics and inventory maps between 1988 and 2007 for the entire Ukrainian Carpathians. Although the forest cover modestly increased, primarily in the peripheral areas, forest loss was ubiquitous in the interior Carpathians and remote areas, a continuation of unsustainable forest use from socialist times into the post-socialist period (Fig. 6). The authors estimated that illegal logging was at least as extensive as documented logging during the early 1990s and so-called sanitary clear-cuts represent a major loophole for overharvesting and logging in restricted areas.

Unfortunately, the carbon implications of forest transition during the post-socialist period in Ukraine are harder to quantify because of a lack of observational or model-based studies and the diverse nature of land use changes, which affect the national carbon budgets. For example, large tracts of abandoned farmland areas in the country likely sequester substantial amounts of carbon, while both legal and illegal forms of logging on already limited forested areas contribute to the release of carbon from the land to the atmosphere. In one important study, Kuemmerle et al. (2011) examined how historic and more recent land-use trends affected the net carbon fluxes in western Ukraine and assessed the region's future carbon sequestration potential by using satellite data and a carbon book-keeping model. The results indicated that the region became a carbon sink between 1930 and 1970, which continues today despite intensive logging during socialism. In recent years, the most important contributor to Ukraine's land use-related carbon sink appears to have been the vast amount of abandoned farmland that resulted from the collapse of the Soviet Union.

Despite its common history with other Eastern European countries with respect to decentralization, Ukraine appears to be on a different path in terms of both the fate and management of its forest resources and the carbon implications for large-scale land use transformations. Although the country's forests may be subject to illegal logging and continued struggles with administration and ownership, the carbon picture that is emerging from Ukraine is positive. Additionally, evidence supports continued carbon sequestration in Ukraine. For example, the country's parliament ratified the Kyoto Protocol on the UN Framework Convention on Climate Change in February 2004. Considering the large amounts of unused low-fertility arable lands, Ukraine has great potential for carbon sequestration with the help of large-scale afforestation programs, and the Kyoto ratification certainly improves the possibility that these programs are actually implemented.

4 Conclusions

The bumpy transition from centrally controlled to market-based economies in the four Black Sea countries has clearly resulted in major changes in forest resources. However, the amount, magnitude, and underlying causes of forest changes in each of these countries are markedly different, mimicking the different paths of the socioeconomic and institutional transformations that followed. However, quantitative information regarding the nature and amount of forest changes, particularly in a comparative form, is only recently emerging. Even less is known regarding the effects of these divergent changes on the regional carbon budgets. What is clear, however, is that all these countries could play an important role in the terrestrial carbon budgets of Europe, either through better management of existing forest resources or through the vast areas of abandoned agricultural land that sequester carbon at unprecedented rates, at least in the recent past (Jannsen et al. 2003). To this end, one common outcome of the carbon budget research that was reviewed here is that the land use legacy of the Soviet Union still plays an important role in the current carbon dynamics. For example, carbon sequestration in regrowing forests following the often high rates of forest harvesting during the latter half of the 20th century offsets large amounts of carbon today.

Both Romania and Bulgaria, which have the largest forest area in the Black Sea eco-zone, experienced a rough transition to market-based economies that was marked with a period of economic hardship and weakened institutions. Thus, large-scale forest changes were common in both countries during the early transition period. On the other hand, the net effect of this transition and other historic forest disturbance events has allowed Bulgaria and Romania to remain a terrestrial carbon sink. With their completed forest restitution processes, strengthened institutions, and European integration, the future for forest management in terms of both economic and environmental needs appears bright in both countries. Although increases in logging could result in net emissions from terrestrial ecosystems in the coming decades, great potential exists for carbon sequestration because of forest expansion on degraded and abandoned farmland, particularly in Romania.

On the other hand, both Ukraine and Georgia appear to be on a different path in terms of both the fate and management of their forest resources and, by extension, their carbon implications. Despite the economic and biological importance of its forests, Georgia continues to struggle with establishing and enforcing forest management with a suitable mix of private and public uses to meet the growing demand. To this end, the future of Georgia's forests and its carbon implications remain uncertain and will mostly depend on the country's relationships with Russia and ability to capitalize on its status as an energy corridor from Central Asia to the west.

Ukraine, which has the smallest forest area among the four Black Sea countries that were considered here, continues to struggle with finding the right form of forest administration and ownership. Additionally, evidence supports that its forests are continually subject to illegal logging. However, the carbon picture that is emerging from Ukraine is positive. The large tracts of abandoned farmland areas sequester unprecedented amounts of carbon because of natural regrowth, and great potential exists to increase this factor with the help of large-scale afforestation programs.

References

Abrudan IV, Marinescu V, Ionescu O, Ioras F, Horodnic SA, Sestras R (2009) Developments in the Romanian forestry and its linkages with other sectors. Notulae Botanicae Horti Agrobotanici Cluj-Napoca 37:14–21

Alix-Garcia J, Kuemmerle T, Radeloff VC (2012) Prices, land tenure institutions, and geography: A matching analysis of farmland abandonment in Eastern Europe. Land Econ 88(3):425–443

Cellarius B (2001) Seeing the forest for the trees: local-level resource use and forest restitution in postsocialist Bulgaria. GeoJournal 2001:599–606

Cirelli MT, Uliescu M (1997) Interim report to the government of Romania on forestry legislation, FAO, TCP/ROM/6612, Rome, 1997

Dembner SA (1994) Forestry in countries with economies in transition. Unasylva 179

Egiashvili D, Ratiani K (2008) Land consolidation in the framework of land market development in Georgia. Report presented at the FAO Workshop on Land Tenure and Land Consolidation—Land Banks and Impact Assessment, June 2008, Prague. http://www.fao.org/regional/seur/events/landcons4/docs/Georgia.pdf (last accessed April 2015)

FAO (2005) Country report: Georgia. Global Forest Resources Assessment, Forestry Department. FAO, Rome
FAO (2010) Country report: Georgia. Global Forest Resources Assessment, Forestry Department. Rome: FAO
FAO (2011) State of Europe's forests 2011. Status and Trends in Sustainable Forest Management in Europe
FAO (2015) Forests and the forestry sector. http://www.fao.org/forestry/23747/en/geo/ (accessed April 2015)
GOG (2009) Ministry of economic development of Georgia: population census 2002. Tech. rep., Ministry of Economic Development of Georgia, Department of Statistics 2009. http://www.statistics.ge/
Government of Georgia (GOG) (2006) Ministry of environment protection and natural resources. Concept paper for water resources management policy—draft
Griffiths P, Kuemmerle T, Kennedy RE, Abrudan IV, Knorn J, Hostert P (2012) Using annual time-series of Landsat images to assess the effects of forest restitution in post-socialist Romania. Remote Sens Environ 118:199–214
Hansen MC, DeFries RS, Townshend JRG, Sohlberg RA, Dimiceli C, Carroll M (2002) Towards an operational MODIS continuous field of percent tree cover algorithm: examples using AVHRR and MODIS data. Remote Sens Environ 83:303–319
Houghton RA (2003a) Revised estimates of the annual net flux of carbon to the atmosphere from changes in land use and land management 1850–2000. Tellus 55B:378–390
Houghton RA (2003b) Why are estimates of the terrestrial carbon balance so different? Glob Change Biol 9:500–509
Houghton RA, Hackler JL (2003) Sources and sinks of carbon from land-use change in China. Global Biogeochem Cycles 17:1034. doi:10.1029/2002GB001970
Ioras F, Abrudan IV (2006) The Romanian forestry sector: privatization facts. Int For Rev 8: 361–367
Janssens IA, Freibauer A, Ciais P, Smith P, Nabuurs G-J, Folberth G, Schlamadinger B, Hutjes RWA, Ceulemans R, Schulze E-D, Valentini R, Dolman AJ (2003) Europe's terrestrial biosphere absorbs 7–12 % of European anthropogenic CO_2 emissions. Science 300:1538–1542
Jervalidze L (2006) Georgia: Russian foreign energy policy and implications for Georgia's energy security. GMB Publishing, London
Knorn J, Kuemmerle T, Radeloff WC, Szabo A, Mindrescu M, Keeton WS, Abrudan I, Griffiths P, Gancz V, Hostert P (2012) Forest restitution and protected area effectiveness in post-socialist Romania. Biol Conserv 146(1):204–212
Kuemmerle T, Hostert P, Radeloff VC, Perzanowski K, Kruhlov I (2007) Post-socialist forest disturbance in the Carpathians. Ecol Appl 17(5):1279–1295
Kuemmerle T, Chaskovskyy O, Knorn J, Radeloff VC, Kruhlov I, Keeton WS, Hostert P (2009) Forest cover change and illegal logging in the Ukrainian Carpathians in the transition period from 1988 to 2007. Remote Sens Environ 113:1194–1207
Kuemmerle T, Olofsson P, Chaskovskyy O, Baumann M, Ostapowicz K, Woodcock C, Hougton RA, Hostert P, Keeton W, Radeloff VC (2011) Post-Soviet farmland abandonment, forest recovery, and carbon sequestration in western Ukraine. Glob Change Biol 17(3): 1335–1349
Lawrence A (2008) Chapter 8: forestry in transition: identity and expertise in post-socialist Europe. In: Bocher M, Giessen L, Kleinschmit D (eds) Environmental and forest governance: the role of discourses and expertise. Proceedings of the international conference, Gottingen, 2007. University of Gottingen Publications, Gottingen, Germany
Lawrence A, Szabo A (2005) Forest restitution in Romania: challenging the value systems of foresters and farmers. Environmental Change Institute, University of Oxford, Human Ecology Working Paper 05/01
Macharashvili I (2009) Forestry sector reform in Georgia—policy brief. Caucasus Institute for Peace Democracy and Development. CIPD, Tbilisi

Mantescu L, Vasile M (2009) Property reforms in rural Romania and community based forests. Rom Sociol 7:95–113

Metreveli K (2002) Forest and forest products: country profile: Georgia. Geneva Timber and Forest Discussion Papers, ISSN 1020-7228. United Nations Publications, 2002, 44 p

Nabuurs GJ, Lindner M, Verkerk PJ, Gunia K, Deda P, Michalak R, Grassi G (2013) First signs of carbon sink saturation in European forest biomass. Nat Clim Change 3:792–796

Nichiforel L, Schanz H (2011) Property rights distribution and entrepreneurial rent-seeking in Romanian forestry: a perspective of private forest owners. Eur J Forest Res 130:369–381

Nordberg M (2007) State forest management reforms in three Ex-Soviet Republics. Reforms, reasons and differences. Doctoral thesis. Swedish University of Agricultural Sciences

Olofsson P, Woodcock CE, Baccini A, Houghton RA, Ozdogan M, Gancz V, Blujdea V, Torchinava P, Tufekcioglu A, Baskent EZ (2009) The effects of land use change on terrestrial carbon dynamics in the Black Sea Region, in regional aspects of climate-terrestrial-hydrologic interactions in non-boreal Eastern Europe, NATO science for peace and security series C: environmental security. In: Groisman PY, Ivanov SV (eds) Springer, Netherlands, pp 175–182

Olofsson P, Torchinava P, Woodcock CE, Baccini A, Houghton RA, Ozdogan M, Zhao F, Yang X (2010) Implications of land use change on the national terrestrial carbon budgets of Georgia. Carbon Balance Manage 5(4):1–13

Olofsson P, Kuemmerle T, Griffiths P, Knorn J, Baccini A, Gancz V, Blujdea V, Houghton RA, Abrudan IV, Woodcock CE (2011) Carbon implications of forest restitution in post-socialist Romania. Environ Res Lett 6:045202

Olofsson P, Foody GM, Stehman SV, Woodcock CE (2013) Making better use of accuracy data in land change studies: estimating accuracy and area and quantifying uncertainty using stratified estimation. Remote Sens Environ 129:122–131

Olofsson P, Foody GM, Herold M, Stehman SV, Woodcock CE, Wulder MA (2014) Good practices for estimating area and assessing accuracy of land change. Remote Sens Environ 148:42–57

Oy SI (2005) Ensuring sustainability of forests and livelihoods through improved governance and control of illegal logging for economies in transition. Country Report—Georgia. Prepared for the World Bank. Helsinki

State Forest Resources Agency of Ukraine (2012) Available from http://dklg.kmu.gov.ua/forest/control/en/index

Surd V, Turnock D (2000) Romania's Apuseni mountains: safeguarding a cultural heritage. GeoJournal 50:285–304

Tickle A, Clarke R (2000) Nature and landscape conservation in transition in Central and South-Eastern Europe. Eur Environ 10:211–219

Torchinava P (2005) Country statement—Georgia: policies affecting forest land use and forest products markets. United Nations Economic Commission for Europe (UNECE). http://www.unece.org/timber/docs/other_mtgs/2005krtiny/reports/geo_e.pdf (last accessed April 2015)

Torchinava P (2006) Georgian statistical yearbook of forestry (Forestry Department, Ministry of Environment Protection and Nature Resources of Georgia)

Trichkov L, Dinev D (2013) Potential of forest wood biomass in Bulgaria, market and possibilities for its utilization. J Agric Sci Technol B3:116–124

Turnock D (2002) Ecoregion-based conservation in the Carpathians and the landuse implications. Land Use Policy 19:47–63

World Bank (2002) World development indicators. The World Bank

World Bank (2006) Ukraine: forestry sector note: status and opportunities for development. 2006. The World Bank, 45 p

World Bank (2007) Integrating environment into agriculture and forestry progress and prospects in Eastern Europe and Central Asia volume II: Georgia country review. The World Bank, Washington DC, 21 p

World Bank (2009) Georgia poverty assessment. The World Bank, Washington DC
WWF (2005) Illegal logging in Bulgaria. World Wildlife Fund Publications, 29 p
Zhiyanski M, Kolev K, Hursthouse A, Sokolovska M (2008) Soil organic carbon and nitrogen contents in different mountainous ecosystems from Central Balkan, Bulgaria. J Balkan Ecol 4:397–411
Zhiyanski M, Sokolovska M, Dimitrova T (2009) SOC stock in different forest-related land-uses in central Stara Planina mountains, Bulgaria. Bull Serb Geogr Soc 4:99–107

Land Management and the Impact of the 2010 Extreme Drought Event on the Agricultural and Ecological Systems of European Russia

Tatiana Loboda, Olga Krankina, Igor Savin, Eldar Kurbanov and Joanne Hall

Abstract Extreme heat waves and droughts are common natural disasters in European Russia. The frequency and severity of heat waves have been on the rise in recent decades across Europe—a trend that is projected to continue into the 21st century. These disasters have complex social, economic, and environmental consequences reaching beyond their geographical boundaries. The extreme heat wave of 2010 had global-scale impacts on food security and regional-scale impacts on ecosystem functioning, air quality, and health. The outcomes were exacerbated by the forestry management and crop rotation practices employed in the region. The century-long economic preference for fast-growing conifers resulted in large uniform single-species even-aged pine stands, which are at least 2.5 times more likely to support fire ignition and to spread than dark-coniferous or mixed stands. Although extreme conditions result in fires that burn through all forest types indiscriminately, uniform pine stands encourage rapid fire growth and spread to uncontrollable levels. Similarly, a recent focus on more economically profitable late-spring crops resulted in the long-term depletion of soil moisture from expanded sunflower and corn cropping, which resulted in decreased soil moisture storage across cultivated lands, leaving them vulnerable to even minor droughts. The major drought of 2010 led to widespread crop yield declines and failure; however, only 2 % of the fields with late-spring crops that were cultivated in 5 of 10 years were

---

Olga Krankina: Retired.

T. Loboda (✉) · J. Hall
Department of Geographical Sciences, University of Maryland, College Park, MD, USA
e-mail: loboda@umd.edu

O. Krankina
College of Forestry, Oregon State University, Corvallis, OR, USA

I. Savin
Department of Agriculture, Peoples Friendship University, Moscow, Russia

E. Kurbanov
Centre of Sustainable Forest Management and Remote Sensing,
Volga State University of Technology, Yoshkar-Ola, Mari El, Russia

G. Gutman and V. Radeloff (eds.), *Land-Cover and Land-Use Changes in Eastern Europe after the Collapse of the Soviet Union in 1991*,
DOI 10.1007/978-3-319-42638-9_8

impacted by drought, versus 63 % of comparable fields where late spring crops were planted in 8 of the 10 years.

1 Extreme Drought Conditions During the Summer of 2010

Positioned on a stable continental plate and far inland from the coastal zone of the Atlantic Ocean, European Russia and Eastern Europe primarily experience meteorological natural disasters. Extreme meteorological events, including droughts, floods, windstorms, snowstorms and extremely low or high temperatures, are recurrent disasters with complex social, economic, and environmental consequences. A 30-year analysis of anomalous temperature events over the extratropical northern hemisphere between 22° and 80°N globally has shown that since 2001, at least 3 % of the northern extratropics have experienced summer temperatures 2 standard deviations (σ) above the mean every year over the 1979–2012 period (Gill et al. 2013). Numerous studies have reported a substantial increase in the frequency of heat waves across Europe in the recent past (Della-Marta et al. 2007; Twardosz and Kossowska-Cezak 2013). The trend toward increased frequency and intensity of heat waves is expected to continue into the 21st century (Meehl and Tebaldi 2004; Elguindi et al. 2013).

Although meteorological extremes are common, the magnitude of the 2010 drought was exceptional in comparison with many previous events, thus classifying it as a "mega-heatwave" (Barriopedro et al. 2011). During the summer of 2010, over 30 % of the northern extratropics experienced temperatures 2σ above the long-term mean, approximately 9 % of the region had anomalies above 3σ, and approximately 3 % of the region experienced a greater than 3.5σ temperature anomaly (Gill et al. 2013). Between mid-June and mid-August of 2010, a persistent strong atmospheric ridge resulted in the hottest summer in recorded history in European Russia with multiple consecutive records set for daily high temperatures, exceeding 40 °C over most of the southern part of the region and 37 °C in Moscow (Grumm 2011; Schneidereit et al. 2012). The spatial extent of the drought was so large that it was observed clearly, even at the very coarse (2.5°) scale of the National Center for Environmental Protection (NCEP) Reanalysis dataset (Kalnay et al. 1996) (Fig. 1). Although the zone of anomalously high temperatures extends from 40° to 60°N, outlined by the 2 °C daily temperature anomaly during the summer of 2010 in Fig. 1, the brunt of the heat wave fell on the major grain producing regions of Russia, northeastern Ukraine, and northwestern Kazakhstan.

The record high temperatures were accompanied by below-normal moisture availability, which was driven by the persistent atmospheric blocking structure (Figs. 1b and 2). Anomalously hot summers in the northern extratropics generally coincide with increased specific humidity (Gill et al. 2013), but from the middle of June, the trajectories of the mean daily temperature and precipitable water

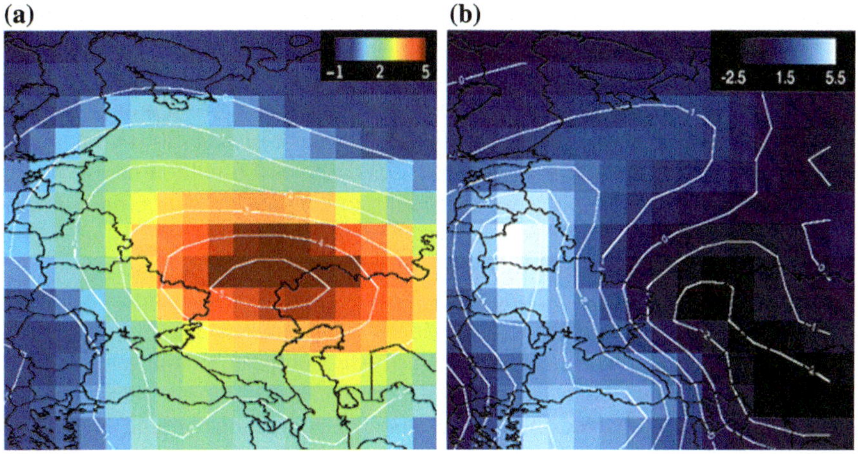

Fig. 1 2010 summer mean daily anomaly in **a** temperature (°C) and **b** precipitable water (kg m^{-2}) from the 2001–2012 base in European Russia (*data source* Kalnay et al. 1996)

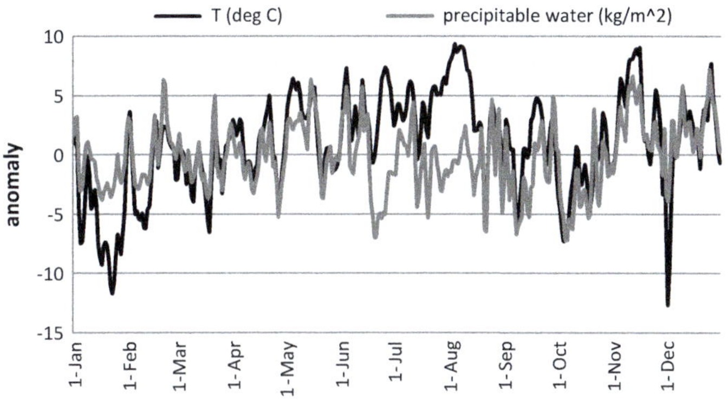

Fig. 2 Daily temperature and precipitation anomalies in 2010 from NCEP reanalysis data over the region, with a mean daily summer T anomaly over 2 °C

anomalies over the drought-affected area diverged (Fig. 2). The temperature anomalies continued to increase in the positive range, whereas the moisture anomalies became strongly negative. The anomaly trends converged again around mid-August, following the dissipation of the persistent blocking event in the region (Witte et al. 2011).

The impacts of this mega-heatwave were broad and were felt at local, regional, continental and global scales. The combined impacts of increased heat and decreased moisture availability led to widespread crop failure within this major global grain-producing region. As a result of the 2010 drought, grain production in Russia dropped by 20–30 % compared to 2009, causing massive perturbations in the global grain markets (Kramer 2010) and fears of food shortages in many grain importing countries (Grumm 2011).

The weakened condition of natural vegetation and highly volatile fire environment supported and promoted the growth of wildfires across European Russia. The 2010 fire season was the largest in satellite record (since 2001), ranging in estimates from ~2.2 million ha (Bondur 2011) to ~4.6 million ha of burned area (see Sect. 4 for our estimates). According to the published data, fire killed 60 people and more than 3500 people lost their homes (Bondur 2011). However, additional damage was caused by high concentrations of tropospheric ozone (O_3), particulate matter (PM_{10}), Carbon Monoxide (CO), Nitric Oxide (NO), and Nitrogen Dioxide (NO_2) produced by fires as they burned across forests, peatlands, and croplands. The population density in European Russia has been historically greater than the average population density of Russia (Stolbovoi and McCallum 2002). However, the concentration of populations around large cities, and particularly in and around Moscow, has grown rapidly over the past decade (Moscow City Government 2015). This growing tendency toward concentrated growth of large urban areas makes larger population groups particularly vulnerable to localized outcomes of extreme events. In 2010, a stagnant anticyclone over the region directed the air toward densely populated areas surrounding Moscow, thus exposing tens of millions of people to extremely high concentrations of atmospheric pollutants (Witte et al. 2011). Point source measurements in Moscow indicate that the maximum allowable concentrations of O_3, PM_{10}, CO, and NO_x were continuously exceeded from late July through late August (Zvyagintsev et al. 2011). At their peak in early August, O_3 concentrations twice exceeded the maximum permissible levels and concentrations of PM_{10} and CO were three to seven times higher (ibid). The extremely high concentrations of CO and PM_{10} were primarily a result of wildfire emissions, which dramatically increased the already high ambient level of anthropogenic pollutants under the stagnant meteorological pattern where the pollutants were allowed to accumulate and recirculate in the atmospheric column (Konovalov et al. 2011; Witte et al. 2011). Satellite observations from various instruments during the period of fire activity reported greatly elevated levels of aerosols in the atmosphere: Moderate Resolution Imaging Spectroradiometer (MODIS) data reveals a 6.8 factor increase in Atmospheric Optical Depth ($AOT_{0.55}$) compared to the 2005–2009 mean (Witte et al. 2011). Over a two-month period, wildfires emitted ~10 Tg CO (nearly 85 % of the annual anthropogenic emissions in the region), which was transported to very densely populated areas and was effectively trapped, causing the pollutants to pool over large urban centers (Konovalov et al. 2011).

Extreme heatwaves put enormous pressure on the health and wellbeing of the population within the affected areas, frequently increasing hospitalizations and deaths (Jones et al. 1982; Semenza et al. 1999; Filleul et al. 2006). The compound effect of the mega-heatwave and the extreme concentration of pollutants resulted in an estimated 55,736 deaths in Russia (Guha-Sapir 2010). A focused study of Moscow attributed additional increased mortality to extreme concentrations of ozone and PM_{10} (Zvyagintsev et al. 2011). According to the reported results, the death toll from the 2010 mega-heatwave in Moscow in July and August exceeded the 2006–2009 statistical mean by 9999 cases, a number far above a standard deviation of 350 for the reported multi-year mean.

In this chapter, we focused on satellite observations of the 2010 drought development and its aftermath on vegetation within the drought-impacted region, as well as on the influence of forestry and agricultural management practices in exacerbating or mitigating these drought impacts.

2 Study Region

We limited the spatial area of the analysis to the full extent of four MODIS tiles (h20v03, h20v04, h21v03, and h21v04 in MODIS grid convention), which cover most of European Russia between 40° and 60°N (Fig. 3). The MODIS land cover product (Friedl et al. 2002) for 2009 was used to assess the impact of drought on

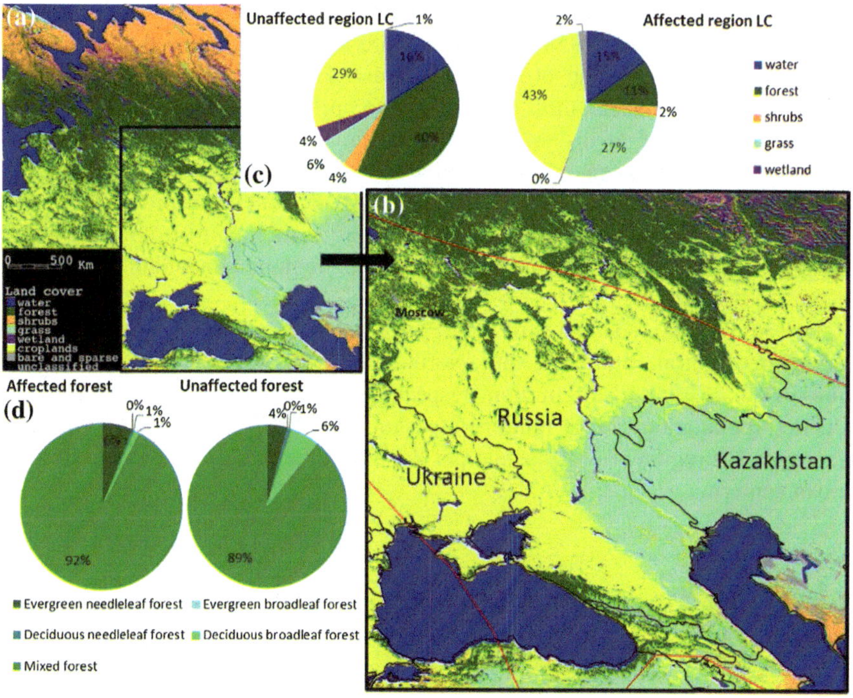

Fig. 3 Study area, land cover and forest types within the regions affected and unaffected by the 2010 drought; **a** location of the study area in European Russia, **b** aggregated land cover types within the 4-MODIS tile focus area—2 °C JJA mean daily temperature anomaly isoline separates, **c** proportionate distribution of aggregated land covers within and outside of the affected areas, **d** proportionate distribution of forest types within and outside the affected areas

vegetation within different land cover categories. Dominant land cover types were developed through aggregation of the International Geosphere-Biosphere Programme (IGBP)-legend version of the MODIS land cover product. We define "forest" as needleleaf and broadleaf evergreen and deciduous as well as mixed forests under the IGBP legend (classes 1–5), shrubs are composed of shrubland and woodland land covers (classes 6–8), grass combines grasslands and open savanna (classes 9 and 10), wetlands remain a separate class (class 1), croplands are defined as a combination of cropland and cropland and natural vegetation mosaic classes (classes 12 and 14), and finally bare and sparse encompass built-up, bare, and permanent snow and ice cover (classes 13, 15, and 16). In this study, we used the 2 °C NCEP Reanalysis summer (JJA) daily temperature anomaly isoline to separate the region into "affected" and "unaffected" areas. Although conditional, this separation allows for better understanding of the magnitude of inflicted vegetation stress as well as a more precise spatio-temporal assessment of conditions preceding and following the drought event.

The study area covers nearly 5 million km^2 across European Russia, south of 60°N, northeastern Ukraine, and northwestern Kazakhstan (Fig. 3b). Croplands and cropland mosaics are the dominant land cover in this area (78 % of all croplands fall within the study area), particularly in the affected regions of Russia and Ukraine. Forests cover ~958,000 km^2 and are found mainly in the northern parts of Russia, with more than 60 % of all forests falling outside the affected area (Fig. 3c). Mixed forests most commonly comprise evergreen needleleaf, with deciduous broadleaf as distant second and third forest types (Fig. 3d). Most of northwestern Kazakhstan is dominated by grasslands, which cover over 1 million km^2. Although wetlands mapped in the MODIS land cover product are found primarily outside the affected regions, in the northeastern part of the focus area, a large (although not well quantified) portion of the region is underlain by peatlands (Konovalov et al. 2011).

Our study area covers most of the densely populated areas of Russia, including the entire Central, Southern, and North Caucasian Federal Districts (population ~38.5, 13.9, and 9.5 million, respectively) and most of the Volga Federal District (population ~29.9 million) (Goskomstat 2010). The mean population density in the region is ~40 people per km^2, with the majority of people living in urban centers.

3 Satellite Observations of Drought Development and Its Aftermath

Daily satellite observations from MODIS allow for the analysis of the impacts of the extreme drought of 2010 on natural and cultivated vegetation within the impacted region. We used 16-day nadir-observed surface reflectance composites (MCD43B4) (Schaaf et al. 2002) and accompanying quality assessment dataset

(MCD43B2) to calculate the Normalized Difference Vegetation Index (NDVI), calculated as the difference in the signal within the Near-Infrared (NIR) and red sections of the electromagnetic spectrum (NIR—red)/(NIR + red) (Tucker et al. 1985) and Normalized Difference Water Index (NDWI), calculated as the difference in the NIR and Short Wave Infrared (SWIR) section of the spectrum (NIR—$SWIR_{1.2}$)/(NIR + $SWIR_{1.2}$) (Gao 1996) to observe the impact of the developing meteorological extreme on vegetation. Although we used only pixels identified by the quality mask as "good quality", winter retrievals of surface reflectance values appear to be strongly impacted by snow and thus exhibit a large amount of variability in the late fall, winter and early spring (October–April) (Figs. 4 and 5). Therefore, we focused our analysis on the full leaf-on period when observations of surface reflectance change are considerably more stable and reliable over the years, as evidenced by the range and standard deviation of values.

As expected, cultivated lands demonstrated a much stronger response to the drought than forests within and outside of the affected region (Fig. 4). The mean amplitude of the NDVI anomaly between June and August on croplands was considerably larger than that of forests. The mean summer NDVI amplitude (2003–2012) over croplands in 2010 exhibited the greatest negative amplitude (−0.06) over the decade. As Fig. 4 shows, the 2010 NDVI was more than one standard deviation below its multi-year mean within and outside of the affected areas.

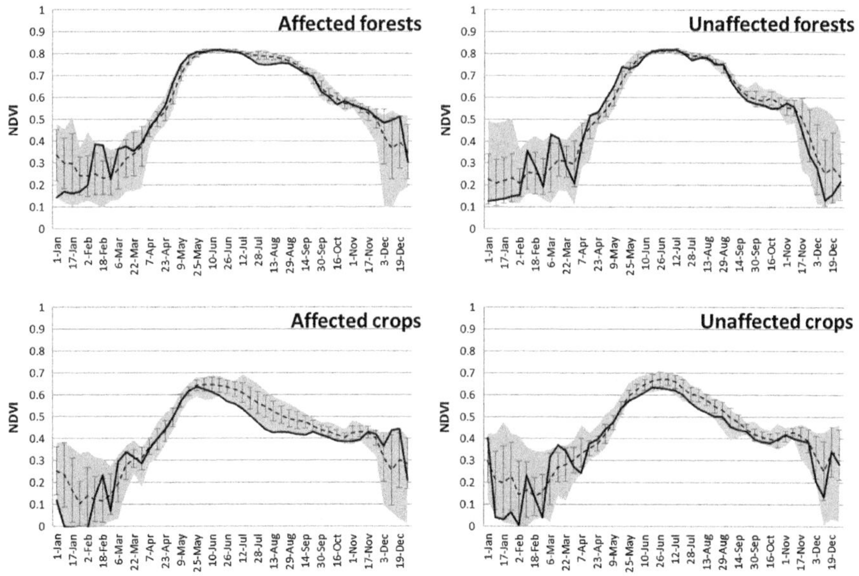

Fig. 4 The multi-year NDVI trend within drought affected (*left*) and unaffected (*right*) areas (as defined by the $+2^{\circ}$ anomaly mean summer temperature in the NCEP reanalysis) within the dominant land cover types: forest (*top*) and croplands (*bottom*). A *solid black line* represents values for the 2010 season, *grey area* represent the range of values between 2003 and 2012, the *short-dashed line* represents the multi-year mean, and the *grey bars* represent a 1 standard deviation spread from the mean between 2003 and 2012

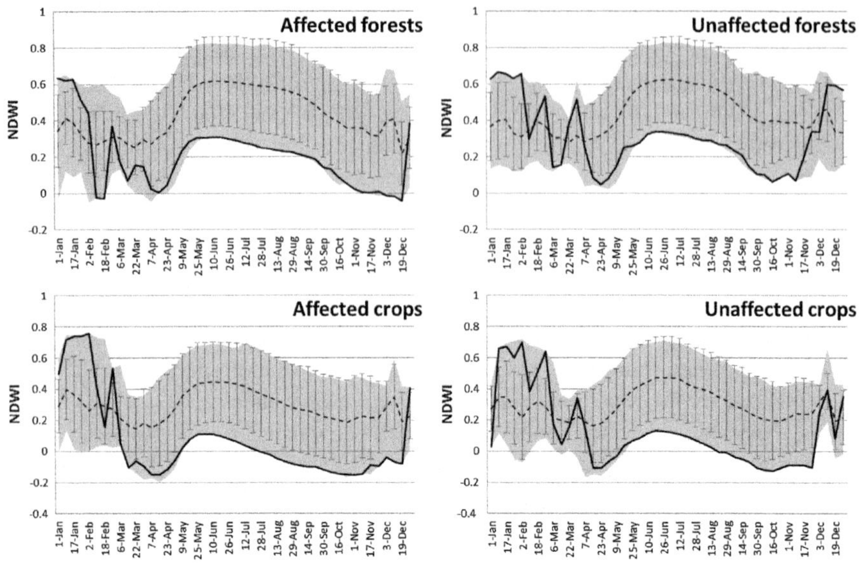

Fig. 5 Multi-year NDWI trend within the drought affected (*left*) and unaffected (*right*) areas (as defined by the +2° anomaly mean summer temperature in the NCEP reanalysis) within the dominant land cover types: forest (*top*) and croplands (*bottom*). The *solid black line* represents values for the 2010 season, the *grey area* represents the range of values from 2003 to 2012, the *short-dashed line* represents the multi-year mean, and the *grey bars* represent a 1 standard deviation spread from the 2003 to 2012 mean

However, within areas affected by drought, 2010 exhibited the lowest NDVI values between 2003 and 2012 and were far below the 1σ from the mean, whereas outside of the affected area, the NDVI was mostly within a 1σ range. A much less significant difference in the NDVI anomaly (−0.01) was noted in 2010 over forests. Overall, the NDVI variability over forests during the full leaf-on period (end of May through mid-August) was very small and the signal was stable between years. However, July and August 2010 NDVI values over the affected forests exhibited a notable decline in forest greenness, with the lowest NDVI values over a 10-year period and well below the 1σ variability. In contrast, in unaffected forests, the NDVI signal remains close to the multi-year mean and, most of the time, is not the lowest on record. Analyses of forest NDVI trajectories between and after the drought (2009–2012) indicate that the 2009 and 2011 summer NDVI values were very close to the multi-year mean without any significant anomaly. Similarly, over affected croplands in 2009 and 2011, there were no strongly anomalous years (with NDVI anomalies of −0.01 and +0.01, respectively). However, 2012 again had one of the lowest summer crop NDVI anomalies (−0.05) and forest NDVI anomalies (−0.01) over drought affected areas in the observed decade.

Even though the magnitude and the temporal trajectory of the NDVI differed between the affected and unaffected croplands, the decrease in surface greenness was evident over all croplands. In contrast, only the affected forests demonstrated a notable decline in greenness between mid-July and mid-August of 2010, corresponding to the exact time of the mega-heatwave without a notable temporal lag. Unaffected forests did not demonstrate a statistically significant difference in summer greenness during the 2010 event. Unlike forests, the unaffected croplands showed a decline in greenness during the 2010 drought from the end of June through the end of September. However, on the affected croplands, the NDVI values for 2010 began to diverge from the mean trend in late May, almost a month before the blocking event was established over European Russia, indicating the early stress on crops that year, which was further exacerbated by the mega-heatwave. By early August, the NDVI on the affected croplands had reached 0.43 (a value similar to mid-September post-harvest conditions) and remained at this low level for the remainder of the vegetative season.

Satellite data indicate that the spring greenup of vegetation in 2011 was delayed. However, this delay was similar to the 2009 greenup trajectory and falls within the range of variability observed between 2003 and 2012. In 2009 and 2011, early season NDVI values over forests presented a slightly negative anomaly, whereas 2010 and 2012 values presented a slightly positive one. Cropped systems show similar greenup trajectories (lower in 2009 and 2011 and higher in 2010 and 2012) but with a lower magnitude of variability from year to year.

In addition to assessing vegetation greenness during the 2010 drought, we analyzed changes in plant moisture content observable from satellite data using the NDWI. Overall, the trends were similar between the NDVI and NDWI signatures, although the range of variability between 2003 and 2012 was several magnitudes higher (Fig. 5). The NDWI signal clearly shows that 2010 was the lowest on record for all crops and the affected forests. However, it offers a less drastic contrast between the affected and unaffected areas than the NDVI. Similarly to the NDVI, the NDWI exhibited the largest difference during the summer of 2010 over cropped areas, with a divergence from the typical trajectory beginning in late May. Although the NDWI in affected forests is still lowest between 2003 and 2010, its amplitude and trajectory were not very different from the 2012 trends. The unaffected forests exhibited the lowest variability of values during the summer, and the 2012 value was slightly below the general trend.

Based on the satellite-derived NDVI and NDWI values, no considerable browning of forests was observed in 2011 following the extreme drought. However, the return of both the NDVI and NDWI values close to the decadal minimum in 2012 may be indicative of forest die-off and an overall reduction in ecosystem productivity at the regional scale. The temporal trajectory of the vegetation indices indicated that vegetation greened up faster during the spring of 2010, indicating a warmer (but not anomalously so) spring. Forests and crops reached peak greenness around mid-May when the curve plateaued. Shortly afterward, at the end of May—beginning of June, i.e., before the establishment of the persistent anticyclone, the crops within the affected area started exhibiting signs of stress. This stress is evident in both vegetation indices but it is expressed more significantly in the NDVI. Despite the obvious decline in crop greenness, no decline was noted in forests until the end of June, i.e., after the mega-heatwave began. The forest NDVI returned to its typical values at the end of August; however, the cropland NDVI collapsed during the mega-heatwave and reached "completed harvest" levels in early August. In general, the NDVI of cropped areas exhibited a gradual decline, with a staged harvesting of crops between late July and early September. The harvest in the South District of Russia begins in the second half of July with the harvesting of winter wheat—a dominant grain crop in this region (FAS/USDA 2007, 2008, 2010a, 2011). The harvest of winter wheat continues through mid-August, moving from the north, in the Volga and Central districts. The harvest of spring wheat, which typically accounts for 60 % of the total wheat cultivation area in Russia, starts in late August and continues into September (FAS/USDA 2007). It is possible that the drought resulted in the unseasonal harvesting of all crops; however, the official statistics indicate that harvesting in the Volga District continued into mid-September (FAS/USDA 2010b). The NDVI and NDWI trajectories in the following year indicate a slightly later (although not outside the observed range) greenup of vegetation following a trajectory very close to the 2009 season.

4 Observations of Fire Activity

The 2010 drought resulted in a record fire season within the MODIS era (Bondur 2011; Witte et al. 2011). Active fire detection from the MODIS instrument characterize fire activity in terms of the number of detected actively burning fires as well as their intensity, expressed through the Fire Radiative Power (FRP) (Giglio et al. 2003). The FRP is a measure of instantaneous energy released by fires as a function of biomass consumption and is thus directly related to fire intensity (Kaufman et al. 1996). Although fire is a common disturbance mechanism in Russian forests, fires in European Russia are rarely large or long-lasting. During the 2010 mega-heatwave, a combination of extreme heat and abnormally low precipitation created highly volatile fire conditions, which resulted in an increased number of fires and in fires of greater intensity (Witte et al. 2011) as well as burned areas of considerably larger size (Bondur 2011).

In the beginning of the fire season until June 22nd, the fire activity within the areas between 45° and 63°N and between 23° and 63°E was not anomalous in either the count of fire detection or the FRP. By June 29th and through August 17th, the daily fire counts were 100–600 occurrences above the 2002–2009 mean, with a mean daily FRP anomaly ranging between 20 and 40 MW, representing a fire season with twice the occurrence and intensity of fires compared to previous years since 2000 (Witte et al. 2011).

Figure 6 shows the comparison in the amount of area burned by aggregated land cover types over European Russia between 2001 and 2011 using the regionally—adjusted burned area algorithm (Loboda et al. 2007) and aggregated MODIS land cover classes (Friedl et al. 2002). The total amount of area burned was actually slightly below the 2001–2011 mean. However, both the spatial patterns (Fig. 7) and

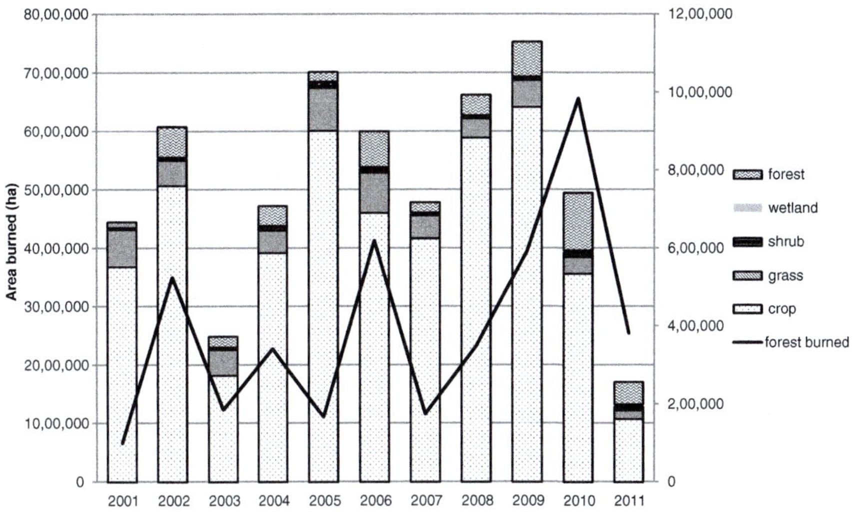

Fig. 6 A multi-year record of the area burned from a regionally adjusted MODIS burned area algorithm within aggregated MODIS land cover classes. The absolute amount of area burned by aggregated land cover types are shown in *bars*. The *black line* separately shows the area of forest burned to highlight the uncharacteristic peak of burning in 2010, which is somewhat muted by the dominance of cropland burning in the bar graphs

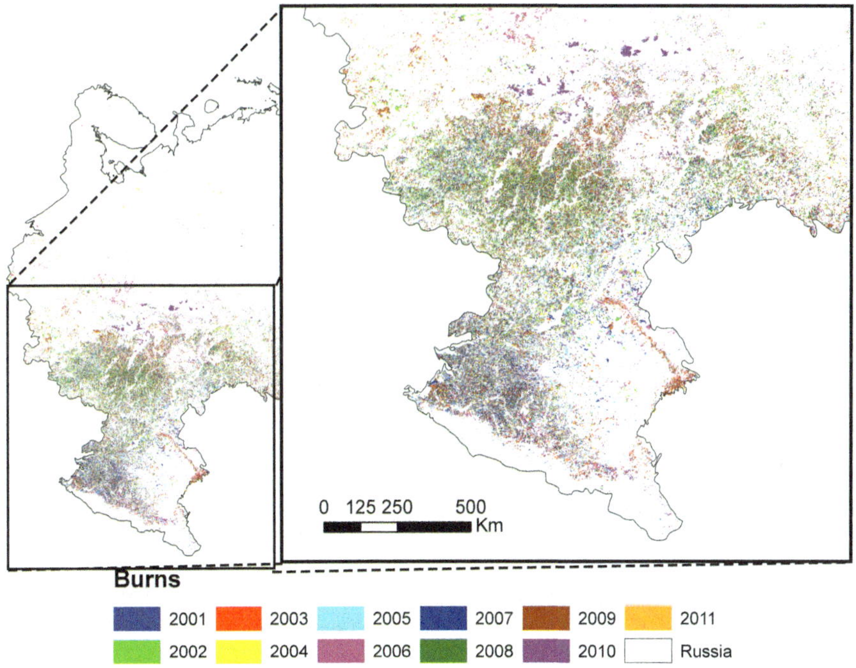

Fig. 7 2001–2011 record of area burned from a regionally adjusted MODIS burned area algorithm in European Russia

analysis of burning as a function of land cover exhibited a distinct shift in the type of burning, which occurred during the mega-heatwave with a considerable decline in cropland burning and a substantial increase in forest burning. In fact, over 22 % of all forests burned between 2001 and 2011 occurred in 2010.

Unlike forests, cropland burning appeared to decrease during the extremely hot summers—2003 and 2010 were two of the three lowest cropland burning years since 2001. However, the lowest amount of cropland burning occurred during 2011. This likely reflects a deviation from the typical crop rotation and management practices caused by the 2010 drought when a sweeping shift in cultivated crops was forced following the collapse of other crops during the main drought event (see Sect. 6).

5 Impacts of Forestry Management Practices on Fire Occurrence and Spread

The absolute majority of Russian forests are federally owned, allowing for widespread implementation of state developed forestry practices nearly everywhere. However, these forestry practices reflect a delicate balance between economic profitability, recreational opportunities, and the long-term sustainability of forest ecosystems and biodiversity (US Forest Service 2013). Although considerably smaller in spatial extent than Siberian forests, forests in European Russia have historically been the major focus of harvest, wood processing and recreational activities (Krankina and Dixon 1992). Forest management favors conifer forests for economic reasons. Forest plantations are created almost exclusively with native conifers, primarily pine (*Pinus sylvestris*). Uniform, single-species, even-aged pine stands in the region are a result of stands replacing fires as well as the selective planting and thinning operations of young forest stands that aim to increase the proportion of conifers (both planted and naturally regenerated) and reduce the proportion of broadleaf species.

Coniferous species have been favored in forestry plantations in the Republic of Mari El for more than a century (Fig. 8) (Romanov et al. 2009). These forestry preferences are representative of comprehensive policies applied across the majority of forests in European Russia during the Soviet era (pre-1992) and are still common today. The results indicate that pine plantations have dominated forest plantations (over 75 %) for nearly a century. Since 1950, the proportion of spruce plantations increased steadily, reaching nearly 50 % by the 1980s, and has remained at that level since. Broadleaf plantations constitute less than 2 % of the total historical range of plantation forests. Oak and poplar plantations were introduced to the forest fund in the mid-1940s. By the mid-1980s, birch plantations were added; however, the total amount of broadleaf plantations remained very low throughout this time. Forest plantations contributed a large proportion to the overall 1,400,300 ha of forest plantations in Mari El (Department of Environmental Protection and

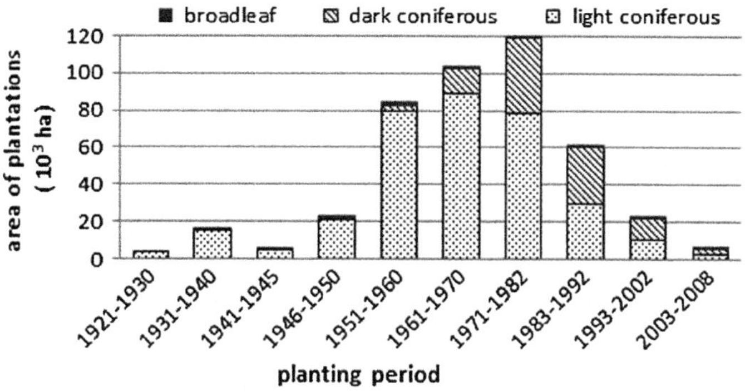

Fig. 8 Dynamics of forest types in forest plantations in the Republic of Mari El, Russia (based on data from Romanov et al. 2009). Light coniferous forests are dominated by pine (97 %), with small larch contribution, dark coniferous forests are 99 % spruce and 1 % cedar, and broadleaf forests are represented by oaks, birch and poplar (67, 19, and 14 %, respectively)

Ecological Safety, Republic of Mari El, 2013). More than 38 % of all coniferous forest stands in Mari El are plantations, with 1450 ha (78 % pine and 15 % spruce) of new plantations added in 2012 (ibid).

This economy-driven preference for conifers created continuous conifer-dominated forest cover that is highly prone to fire and supports rapid fire spread. Previous studies have demonstrated that European pine forests are at least 2.5 times more likely to sustain fire ignition than spruce forests of a similar age (Tanskanen et al. 2005). Moreover, young pine stands are 50 % more likely to burn than mature pine stands (ibid). However, spruce is more susceptible to fire damage, whereas pine can survive repeated fires of moderate intensity, thus making pine a more resilient commercial option (Sannikov and Goldammer 1996). At the same time, pine species are more adaptable to changing environmental conditions than spruce and fir species (Krankina et al. 1997). Fire occurrence also contributed to the management-aided proliferation of pine in the European boreal zone because the naturally reseeding pine is enhanced by fire events through the fire-facilitated release of seeds (Sannikov and Goldammer 1996).

A focused study of areas burned in Mari El (Landsat path 172 row 21), undertaken within the Northern Eurasia Land Cover Dynamics Analysis (NELDA) project (http://www.fsl.orst.edu/nelda/), showed that based on Landsat MSS analysis, the total area burned in a similar catastrophic fire in 1972 was 212,300 ha, which is more that 12 % above the official statistic of 180,000 ha (Fig. 9) (Vorobyov et al. 2012). Our Landsat TM-based assessment of area burned in 2010 within the Mari El Republic reported 100,200 ha of burned area, which also exceeds the officially reported 72,800 ha by nearly 28 %. Approximately one-third (34,228 ha) of the forests that were burned in 2010 were located on pine plantations developed after the 1972 fire. These young dense stands are extremely vulnerable to

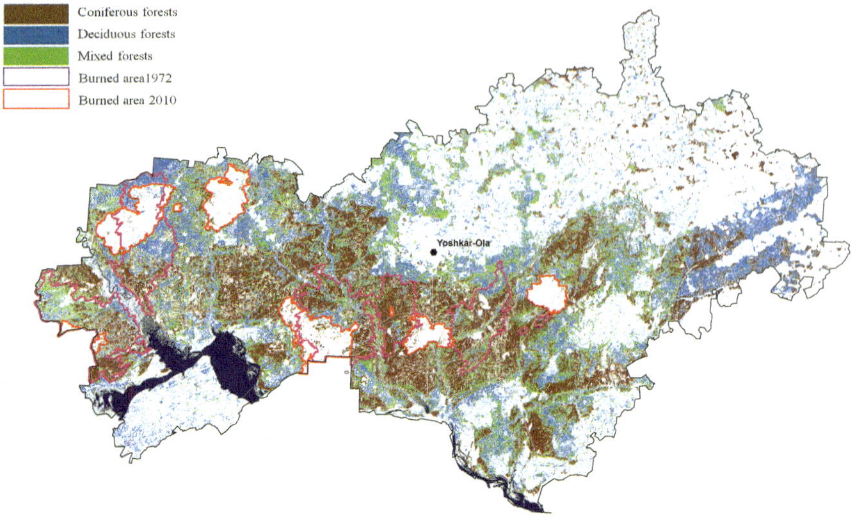

Fig. 9 Distribution of 2010 and 1972 catastrophic fires by forest type in the Republic of Mari El. Forest composition was mapped in 2011 and shows that the overall tree mortality within the burns of 2010 was greater than 90 %

fire as they readily support fire ignition and create a dense fuel matrix to move the fire from the surface into the crowns through ladder fuels of lower branches in young pine stands. Of the total area burned in 2010 in Mari El, the broadleaf forests appeared to burn just as readily as the coniferous stands (49 and 45 %, respectively). The distribution of 2010 burns across different forests types was also approximately equal, with 8 % of all coniferous forests and 11 % of all broadleaf forests in the focus area impacted by fire. It is likely that this is due to the magnitude of the extreme meteorological conditions observed in 2010 (the hottest year on record in many parts of European Russia), which overwhelmed the natural fire resistance of broadleaf and dark coniferous forests.

The long-term forestry preference for coniferous forests in European Russia has created an ecosystem of more drought tolerant forests by preferentially selecting pine over spruce and broadleaf species, which are likely to be better adapted to the increased frequency of droughts. At the same time, these drought resistant species are more likely to incur fire ignition, enhance fire propagation and subsequently increase fire intensity during wildfire events, which are also likely to become more prevalent under the hotter and drier conditions expected in the future. Following the stand replacing fire events, pine species are further promoted through their natural tendency to release seeds after fire and from forestry thinning operations as well as the preferential plantation of pine. It is likely that the fire risk associated with the dominance of pure pine plantations outweighs the potential benefits from their drought resistance as it increases the direct threat to human life, health, and property.

6 Impacts of Crop Management Practices on Drought Related Crop Failure and Persistence of Water Stress

The dependence of different crop types on the availability of moisture is well recognized. Seasonal water consumption for different cultivated crop types within the same region can differ by several factors of magnitude. For instance, the water consumption of spring cereal crops is, on average, 450–650 mm during the vegetative season, whereas water consumption of corn, potatoes, sugar beets and sunflowers is 500–800, 500–700, 550–750, and 600–1000 mm, respectively (FAO 1986).

Under the central management structure of the Soviet Union, all crops in any climatic zone were cultivated within a framework of strictly assigned crop rotations. The crop rotations were scientifically substantiated and adapted to specific geographic conditions. For instance, the primary agricultural zone of Russia is located on the steppe and forest-steppe zones. The mean annual water availability in the vegetative season in this zone is near the minimum amount required by crops, necessitating soil moisture retention for optimizing crop rotation practices involving crop rotations to avoid excessive soil drying.

After the collapse of the USSR, the established crop-rotation mechanisms became optional. Market demand and crop prices became the main driving force behind the selection of crop types and crop rotations. As a result, over the past 20 years in many regions of Russia, the proportion of sunflower and corn crops have increased more than two-fold in Russia and more dramatically in some regions, whereas the proportion of winter and early spring crops has decreased. Sunflower and corn require a greater amount of moisture during the growing season and the expansion of these crop types has resulted in an overall decrease in soil moisture storage across cultivated lands. Consequently, even a relatively minor decrease in precipitation can provoke a sharp decline in crop yield.

Satellite-data analysis of crop conditions during the drought of 2010 in selected regions of Russia affirmed that crop rotation, with a preference for hydrophilic crop types, contributes to the magnitude of the drought impact on crops in European Russia. Investigations were carried out for 256 agricultural plots that were randomly selected in Chuvashia (Volga region of Russia), where crops were strongly impacted by the 2010 drought. The time profiles of MODIS-based NDVI between 2001 and 2010 were obtained for each selected field through the Satellite Service of the Analysis of Vegetation "VEGA" system (Lupian et al. 2011). The analysis of the NDVI trajectories does not allow for accurate identification of crop type; however, it is possible to sufficiently ascertain whether crops belong to the winter (winter wheat, winter barley, winter rye), early spring (spring barley, spring wheat) or late spring categories (potatoes, maize, sunflower, millet, sugar beets) based on the specific character of the curve shape at the beginning of the season, the date of the beginning of the vegetative season and the date of the seasonal NDVI maximum. In the dry years, crops frequently do not reach their typical seasonal maximum value. However, NDVI trajectories leading up to the onset of drought allow for the identification of crop groups (winter, early spring, or late spring) based on empirical relationships established from multi-year observations (Medvedeva et al. 2012). Within our study, the crop groups were identified annually for each of the 256 experimental plots during the 2001–2010 period.

MODIS data were also used to analyze the drought impact on croplands (Medvedeva et al. 2012). With this method, smoothed MODIS NDVI time series were used to determine the seasonal maximum value (peak NDVI) and its peak date for each year between 2001 and 2010 per field. Furthermore, the maximum and minimum NDVI values for each week were defined. Using the combined metrics, an "analogue" year was identified during the 2001–2009 period compared to 2010 per field, which was the year with the minimum difference in NDVI values during the available record. Damage to the crops by the 2010 drought began as early as the first half of the growth period, when, not having reached the normal date of the peak NDVI, the crops began to yellow from drought rather than crop ripening. This allowed for identification of damaged crops as those that reached their peak NDVI at least 3 weeks earlier than in the "analogue" year.

The results of this analysis indicated that crops were strongly impacted by the 2010 drought; however, the magnitude of the impacts was spatially highly variable where drought-impacted fields were frequently immediately adjacent to the fields that were largely unaffected by drought. Our findings show that late crops were cultivated on average 7.8 of the 10 years (1.03σ) on drought-impacted fields compared to 6.45 years (0.83σ) on unaffected fields. A more detailed assessment revealed that only 2 % of the fields with late spring crop cultivation for 5 of the 10 years were affected by drought, compared to 9, 24, and 40 % for 6, 7 and 8 years of late spring crop cultivation, respectively. Finally, over 63 % of fields where late spring crops were cultivated more than 8 of the 10 years were strongly impacted by drought. This analysis further showed that more than 60 % of the fields where early spring crops were cultivated for less than 3 of the 10 years were impacted by the 2010 drought. However, if early season crops were cultivated more frequently than 4 of the 10 years, less than 10 % of those fields were strongly impacted by the 2010 drought. Overall, there appears to be a strong relationship between the frequency of late spring crop planting and the degree of impact from extreme drought of 2010. This came out as a leading factor explaining the spatial heterogeneity of the 2010 drought impact on croplands.

Although the magnitude of the drought impact was influenced by crop rotation practices, the 2010 event had subsequent impacts on crop rotations in the following season. After the severe summer drought of 2010, there was a reduction in winter crop planting across many areas in Russia due to the excessive over-drying of the soil and insufficient soil moisture levels to support winter crop development in the fall of 2010. The reduction in the fall extent of winter crops was clearly detected by satellite imagery available through VEGA (Figs. 10 and 11). In turn, the reduction in the autumnal planting of winter crops led to an increase in early and late spring crop planting on the fields where winter crops would have been planted, subsequently further reducing drought tolerance of the crops in the immediate future.

Fig. 10 Landsat-based assessment of winter crop planting in the fall of 2009 (*left*) and 2010 (*right*) over one of the major grain producing regions of Russia

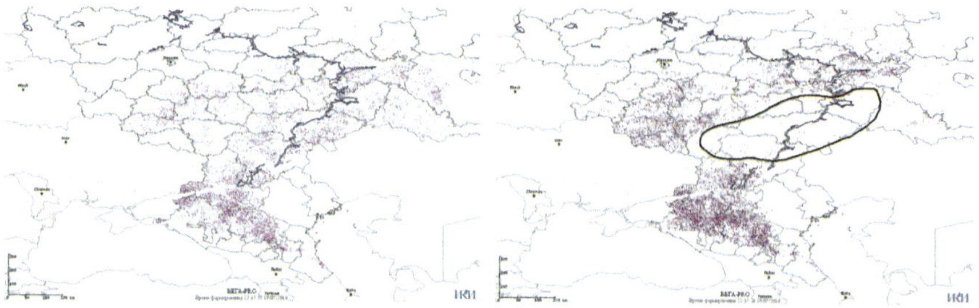

Fig. 11 Winter crop distribution in European Russia during the fall of 2009 (*left*) and in the fall of 2010 (*right*) from VEGA satellite-data service. *Black oval* outlines a region where winter crop planting was considerably reduced following the 2010 drought

In summary, market-driven changes in crop rotation practices over the main agricultural zone of Russia toward late spring crops has resulted in a widespread decline in drought resistance in cultivated crops. It is likely that due to excessive moisture removal by late spring crops from the already moisture limited soils, these crops will be susceptible to damage not only during extreme droughts like the mega-heatwave of 2010 but also during more minor fluctuations in precipitation. Additionally, extreme drought events are likely to propagate the undesirable crop rotation practices by preventing the planting of winter and early spring crops due to insufficient soil moisture content following the extreme events.

7 Conclusions

Recent studies have shown that the probability distribution of daily maximum and daily minimum temperatures has shifted globally toward a considerably higher proportion of high temperature events at the beginning of the 21st century (Dole et al. 2011; Gill et al. 2013). Although it is not clear whether the extreme drought of 2010 was related to the internal variability of the climate system or due to global warming (Rahmstorf and Coumou 2011; Dole et al. 2014), the overall trend in rising magnitude, extent, and frequency of high temperature anomalies in the northern extratropics has been repeatedly demonstrated. It is therefore likely that extreme droughts will become more frequent and possibly more severe with major consequences for natural and socio-economic systems regionally and potentially globally.

In this chapter, we focused on the impacts of land use decision-making on the consequences of the mega-heatwave of 2010 for forests and croplands of European Russia. Our findings indicate that the magnitude of these impacts was closely related in many (although not all) cases to forestry and crop rotation practices. Although planting closed pine forests as the preferred forestry species did not cause the 2010 fires, this forestry practice has supplied an excellent fuel source to sustain more intense large burns that would not have been possible within mixed forests. The magnitude of the 2010 drought impact on Russian forests has driven major changes in the forest management structure. Specifically, after the 2010 fire season, Russia's Federal Forestry Agency was extracted from under the oversight of the Ministry of Agriculture and is now reporting directly to the Russian Parliament (US Forest Service 2013).

Market-driven changes in crop rotation practices, where farmers chose the crop type based on its profitability rather than regional agronomy-driven rotations aimed at soil moisture conservation, led to crop losses above what would have been expected under a different management structure. Considering the importance of this region as a major grain producer in the world, drought-induced crop failure is a matter of global concern.

Finally, the studies described in this chapter highlight the benefits from satellite monitoring of vegetation and surface changes resulting from drought. Multiple coarse and moderate resolution satellite data sources have been used by the scientific community and operational agencies to monitor and study the impacts of mega-heatwaves on land surfaces and human safety. Much knowledge has been gained toward understanding the magnitude from the accumulation and circulation of pyrogenic emissions impacts on the death toll and severe health outcomes for populations as well as knowledge of the optimization of land use practices to adapt to changing environmental conditions. These findings now need to be translated into land management decisions with a major focus on adaption to a world with more frequent extreme heat events in the future and the mitigation of undesirable outcomes.

Acknowledgments This chapter reports results produced by projects supported by the Russian Science Foundation (grant # 15-16-30007) and the Ministry of Education and Science of the Russian Federation (research grant # 2394).

References

Barriopedro D, Fischer EM, Luterbacher J et al (2011) The hot summer of 2010: redrawing the temperature record map of Europe. Science 332:220–224

Bondur VG (2011) Satellite monitoring of wildfires during the anomalous heat wave of 2010 in Russia. Izv Atmos Oceanic Phys 47:1039–1048

Della-Marta P, Luterbacher J, von Weissenfluh H et al (2007) Summer heat waves over western Europe 1880–2003, their relationship to large-scale forcings and predictability. Clim Dyn 29:251–275

Department of Environmental Protection and Ecological Safety, The Republic of Mari El (2013) Report on the state of ecology in the Republic of Mari El for the year 2012, pp 207. Yoshkar-Ola, Russia (in Russian)

Dole R, Hoerling M, Perlwitz J et al (2011) Was there a basis for anticipating the 2010 Russian heat wave? Geophys Res Lett, vol 38

Dole R, Hoerling M, Kumar A, Eischeid J, Perlwitz J, Quan X.-W, et al (2014) The making of an extreme event: putting the pieces together. Bull Am Meteorol Soc 95(3):427–440

Elguindi N, Rauscher SA, Giorgi F (2013) Historical and future changes in maximum and minimum temperature records over Europe. Clim Change 117:415–431

FAO (1986) Irrigation water management: irrigation water needs. Training manual no. 3. Available via http://www.fao.org/docrep/s2022e/s2022e00.htm#Contents

FAS/USDA (2007) World agricultural production. Circular series. (August 2007). Available via http://www.pecad.fas.usda.gov/

FAS/USDA (2008) World agricultural production. Circular series. (August 2008). Available via http://www.pecad.fas.usda.gov/

FAS/USDA (2010a) World agricultural production. Circular series. (August 2010). Available via http://www.pecad.fas.usda.gov/

FAS/USDA (2010b) World agricultural production. Circular series. (September 2010). Available via http://www.pecad.fas.usda.gov/

FAS/USDA (2011) World agricultural production. Circular series. (August 2011). Available via http://www.pecad.fas.usda.gov/

Filleul L, Cassadou S, Medina S et al (2006) The relation between temperature, ozone, and mortality in nine french cities during the heat wave of 2003. Environ Health Perspect 114:1344–1347

Friedl MA, McIver DK, Hodges JCF et al (2002) Global land cover mapping from MODIS: algorithms and early results. Remote Sens Environ 83:287–302

Gao BC (1996) NDWI—a normalized difference water index for remote sensing of vegetation liquid water from space. Remote Sens Environ 58:257–266

Giglio L, Descloitres J, Justice CO et al (2003) An enhanced contextual fire detection algorithm for MODIS. Remote Sens Environ 87:273–282

Gill EC, Chase TN Sr, Pielke RA Sr et al (2013) Northern Hemisphere summer temperature and specific humidity anomalies from two reanalyses. J Geophys Res Atmos 118:8297–8305

Goskomstat (2010) All Russian 2010 census results. Available via http://www.perepis-2010.ru/results_of_the_census/. Accessed 5 Nov 2013

Grumm RH (2011) The Central European and Russian heat event of July–August 2010. Bull Am Meteorol Soc 92:1285–1296

Guha-Sapir D (2010) Disasters in numbers 2010—presentation. EM-DAT: the international disaster database centre for research on the epidemiology of disasters—CRED. Available via http://cred.be/sites/default/files/Disaster_numbers_presentation_2010.pdf. Accessed 23 Oct 2012

Jones TS, Liang AP, Kilbourne EM et al (1982) Morbidity and mortality associated with the July 1980 heat-wave in St-Louis and Kansas-City, Mo. Jama J Am Med Assoc 247:3327–3331

Kalnay E, Kanamitsu M, Kistler R et al (1996) The NCEP/NCAR 40-year reanalysis project. Bull Am Meteorol Soc 77:437–471

Kaufman YJ, Remer L, Ottmar R, et al (1996) Global biomass burning: relationship between remotely sensed fire intensity and rate of emission of smoke: SCAR-C experiment. MIT Press, Cambridge, pp 685–696

Konovalov IB, Beekmann M, Kuznetsova IN, Yurova A, and Zvyagintsev AM (2011) Atmospheric impacts of the 2010 Russian wildfires: integrating modelling and measurements of an extreme air pollution episode in the Moscow region. Atmos Chem Phys 11:10031–10056, doi:10.5194/acp-11-10031-2011

Kramer AE (2010) Drought in Russia ripples beyond the wheat fields. The New York Times: Global Business. Available via http://www.nytimes.com/2010/08/28/business/global/28wheat.html?_r=0. Accessed 27 Aug 2010

Krankina ON, Dixon RK (1992) Forest Management in Russia—challenges and opportunities in the era of Perestroika. J For 90:29–34

Krankina ON, Dixon RK, Kirilenko AP et al (1997) Global climate change adaptation: examples from Russian boreal forests. Clim Change 36:197–215

Loboda T, O'Neal KJ, Csiszar I (2007) Regionally adaptable dNBR-based algorithm for burned area mapping from MODIS data. Remote Sens Environ 109:429–442

Loupian EA, Savin IY, Bartalev CA et al (2011) Satellite monitoring of vegetative cover ("Vega"). Proc All Russia Conf Cur Aspects Remote Sens Earth Space 8:190–198 (in Russian)

Medvedeva MA, Savin IY, Isaev VA (2012) Determination of area of drought-affected crops based on satellite data (exemplified by crops in Chuvashia in 2010). Russ Agric Sci 38:121–124

Meehl GA, Tebaldi C (2004) More intense, more frequent, and longer lasting heat waves in the 21st century. Science 305:994–997

Moscow City Government (2015) Moscow city government official web portal: Moscow history (http://www.mos.ru/about/history)

Rahmstorf S, Coumou D (2011) Increase of extreme events in a warming world. Proc Natl Acad Sci USA 108:17905–17909

Romanov EM, Eremin NV, Nureeva TV (2009) Situation with reforestation in the Middle Povolgie. In: Proceedings of the international conference on international cooperation in forest sector: balancing education, science and industry, Yoshkar-Ola, Russia, pp 45–50

Sannikov SN, Goldammer JG (1996) Fire ecology of pine forests of northern Eurasia. In: Anonymous. Springer, pp 151–167

Schaaf CB, Gao F, Strahler AH et al (2002) First operational BRDF, albedo nadir reflectance products from MODIS. Remote Sens Environ 83:135–148

Schneidereit A, Schubert S, Vargin P et al (2012) Large-scale flow and the long-lasting blocking high over Russia: summer 2010. Mon Weather Rev 140:2967–2981

Semenza JC, McCullough JE, Flanders WD et al (1999) Excess hospital admissions during the July 1995 heat wave in Chicago. Am J Prev Med 16:269–277

Stolbovoi V, McCallum I (2002) Land resources of Russia. Int Inst Appl Syst Anal. Available via www.iiasa.ac.at/Research/FOR/russia_cd/download.htm

Tanskanen H, Venalainen A, Puttonen P et al (2005) Impact of stand structure on surface fire ignition potential in *Picea abies* and *Pinus sylvestris* forests in southern Finland. Can J For Res Revue Canadienne De Recherche Forestiere 35:410–420

Tucker CJ, Townshend JRG, Goff TE (1985) African land-cover classification using satellite data. Science 227:369–375

Twardosz R, Kossowska-Cezak U (2013) Exceptionally hot summers in central and eastern Europe (1951–2010). Theoret Appl Climatol 112:617–628

US Forest Service (2013) US forest service international programs: Russia, Europe and near Asia. Availble via http://www.fs.fed.us/global/globe/europe/russia.htm. Accessed 5 Nov 2013

Vorobyov ON, Kurbanov EA, Gubaev AV et al (2012) Remote sensing monitoring of burns in Mari El section of the Volga region. Vestnik of the Povolzhye State Technological University 1:12–22

Witte JC, Douglass AR, da Silva A, Torres O, Levy R, and Duncan BN (2011) NASA A-Train and Terra observations of the 2010 Russian wildfires. Atmos Chem Phys 11:9287–9301, doi:10.5194/acp-11-9287-2011

Zvyagintsev AM, Blum OB, Glazkova AA et al (2011) Air pollution over European Russia and Ukraine under the hot summer conditions of 2010. Izv Atmos Oceanic Phys 47:699–707

Agricultural Fires in European Russia, Belarus, and Lithuania and Their Impact on Air Quality, 2002–2012

Jessica L. McCarty, Alexander Krylov, Alexander V. Prishchepov, David M. Banach, Alexandra Tyukavina, Peter Potapov and Svetlana Turubanova

Abstract This chapter describes the first research to quantify air pollution emissions at a moderate to coarse scale from agricultural burning in Belarus, Lithuania, and European Russia using MODIS and Landsat-based estimates of fire, land-cover and land-use. Agricultural burning in Belarus, Lithuania, and European Russia showed a strong and consistent seasonal geographic pattern from 2002 to 2012, with the majority of fires occurring from March to June and a smaller peak in July and August. Over this 11-year period, there was a decrease in both cropland and pasture burning throughout the region. For Smolensk Oblast, a Russian administrative region with comparable agro-environmental conditions to Belarus and Lithuania, a detailed analysis of Landsat-based burned area estimations for croplands, pastures and field data collected in summer 2014 showed that the agricultural burning area can be up to 10 times larger than the 1 km MODIS active fire estimates. Using the annual MODIS and Landsat-based burned area estimations, we identified 25 carbon, particulate matter, volatile organic carbon (VOCs), and harmful air pollutants (HAPs) emissions for all agricultural burning, including both croplands and pastures. In general, European Russia is the main source of agricultural burning emissions. Lithuania and Belarus have relatively minor contributions. Indeed, emissions from certain agricultural burning air pollutants in European

J.L. McCarty (✉) · D.M. Banach
Michigan Tech Research Institute, Michigan Technological University, Ann Arbor, MI 48105, USA
e-mail: jmccarty@mtu.edu

A. Krylov · A. Tyukavina · P. Potapov · S. Turubanova
Department of Geographical Sciences, University of Maryland, College Park, MD 20742, USA
e-mail: amkrylov@umd.edu

A.V. Prishchepov
Department of Geosciences and Natural Resources Management, University of Copenhagen, Øster Voldgade 10, 1350 København K, Denmark
e-mail: alpr@ign.ku.dk

A.V. Prishchepov
Leibniz Institute of Agricultural Development in Transition Economies (IAMO), Theodor-Lieser-Strasse 2, 06120 Halle, Saale, Germany

G. Gutman and V. Radeloff (eds.), *Land-Cover and Land-Use Changes in Eastern Europe after the Collapse of the Soviet Union in 1991*,
DOI 10.1007/978-3-319-42638-9_9

Russia are so large that they are equivalent to 5 % of emissions from all sectors (industry, energy, transportation, all sources of fire) in Lithuania and likely in other neighboring Eastern European countries.

1 Introduction

Cropland and pasture burning are common agricultural management practices used across the globe to reduce crop residues that impede planting (Yevich and Logan 2003; Ortiz de Zarate et al. 2005; Badarinth et al. 2006; McCarty et al. 2007), to reduce the risk of pests and plant disease (Mazzola et al. 1997; Bescansa et al. 2006; Brye et al. 2006) and to rejuvenate grasses while reducing woody encroachment (Briggs et al. 2002; Chen et al. 2005; Venkataraman et al. 2006). Cropland burning is a source of greenhouse gases and carbonaceous emissions, including emissions that negatively affect air quality (Streets et al. 2001; Dhammapala et al. 2006; Yang et al. 2008; McCarty 2011) and short-lived climate pollutants like black carbon (McCarty et al. 2012) that can be deposited as far from agricultural areas as the Arctic or glaciers in the Himalayas and Andes (Bond et al. 2013).

Within Russia, the burning of agricultural areas, including both croplands and pastures, has been a well-documented phenomenon (Soja et al. 2004; Dubinin et al. 2011). Agricultural burning became a subject of governmental and scientific interest as early as the 1980s (Derevyagin 1987), and interest has continued since. For example, in the spring of 2006, the Baltic Times reported that out-of-control agricultural fires, set by farmers in western Russia, Belarus, and Ukraine, ignited nearby forests and resulted in the air pollution deaths of five people in Latvia due to decreased air quality (Stohl et al. 2007). Romanenkov et al. (2014) noted that a peak of satellite fire detections occurred in cropland areas in Russia, Baltic countries, Belarus, Ukraine, and Kazakhstan directly after the snow melt in the spring (indicating field preparation) and after agricultural harvests in the fall. However, prescribed fires in forests, grasslands, and croplands may be illegal or may go unreported by national agencies in Lithuania, Belarus, and Russia (Narayan et al. 2007), making it difficult to ascertain whether the number of agricultural fires has increased in recent decades.

One of the hypotheses is that the increased spread of agricultural fires may be a result of the collapse of the Soviet Union and the cessation of state subsidies for the cultivation of less productive agricultural lands, particularly in the livestock fodder crop production sector (Alcantara et al. 2012, 2013; Prishchepov et al. 2013). Official statistics indicate that approximately 42 Mha of croplands have been abandoned from 1985 to 2010 in Russia alone (ROSSTAT 2014), of which 5 Mha have returned to forests (Potapov et al. 2014). Recent studies suggest that afforestation of abandoned croplands is likely to have major impacts on the carbon budget of the region (Schierhorn et al. 2013; Kuemmerle et al. 2015). However, afforestation of abandoned croplands is currently not included in the official forestry reports (Potapov et al. 2012) and national carbon budgets, and, often, the legal status of these lands remains uncertain (Lerman and Shagaida 2007). Abandoned agricultural lands can accumulate old grass, which can help to spread fires (Dubinin et al. 2011). This means that such abandoned lands can be prone to fires, particularly due to lack of fire suppression management on socio-economically and agro-environmentally marginal lands with unclear legal status.

Belarus, Lithuania, and Russia represent for us an interesting case study because they share a similar starting point in transitioning from state-command to market-driven economies after the breakup of the Soviet Union, but they have used different approaches during the transition regarding governance of agricultural land use and land tenure (Lerman et al. 2004). For instance, in Belarus, the government has continued to provide strong support for agriculture, and most agricultural lands remain largely under governmental control (Sakovich 2008). In Lithuania, the government stopped subsidizing farming and returned agricultural lands to the previous land owners and their heirs (Lerman et al. 2004). Similarly to Lithuania, the Russian government cut agricultural support by 90 % and privatized the lands, but the land market did not function until the adoption of amendments to the Land Code in 2007, which allowed the buying and selling of agricultural lands (Lerman and Shagaida 2007). This all resulted in massive agricultural land abandonment across the study area, but with drastic differences in agricultural land abandonment rates across post-Soviet Eastern Europe (Prishchepov et al. 2012). In areas with similar agro-environmental conditions, higher abandoned rates occurred in Russia, followed by Lithuania, with lower rates in Belarus (Prishchepov et al. 2012).

The governance of agriculture and the environment in post-Soviet countries most likely reflects the strength of fire suppression as well. In Europe, most countries prohibit agricultural burning, with few exceptions (Saarikoski and Hillamo 2013). However, fires are common in agricultural landscapes in Russia, suggesting a lack of well-functioning institutional mechanisms to control and suppress illegal agricultural burning. Comparisons of fire dynamics across Belarus, Lithuania and Russia would allow us to analyze the strengths of the institutional frameworks that would allow them to suppress burning on agricultural lands.

Air pollution trend estimates for Russia and Belarus related to the collapse of the Soviet Union in the early 1990s have shown improvements in air quality due to decreases in industrial activity, pollution, and automobile emissions (Genikhovich et al. 2009). Similarly, Lithuania experienced a reduction from all-time highs of NO_x, SO_2, volatile organic compounds (VOCs), CO, and CH_4 emissions in 1990 (EEA 2014). However, these estimates do not include agricultural burning, which can be a major source of air pollution (Saarikoski and Hillamo 2013). Because it is a major source, it is important to understand its patterns and trends.

Our goal was to provide an estimate of carbon and air pollution emissions from anthropogenic fires in agricultural landscapes of European Russian, Belarus, and Lithuania. We focused on recent land-cover/land-use change (LCLUC) related to abandonment or transitioning of agricultural landscapes in the region (Alcantra et al. 2013; Potapov et al. 2014). Air pollution from anthropogenic fires in agroecosystems is currently overlooked in the systemic air quality monitoring for these countries as well as in the UN Framework Convention on Climate Change

(UN FCCC), which is part of the report of the Intergovernmental Panel on Climate Change (IPCC). Thus, this chapter is the first effort to develop an emissions inventory of agricultural burning in an area that has experienced a large amount of LCLUC related to agricultural abandonment.

The specific objectives of this chapter are as follows:

1. Characterize country- and region-scale differences in agricultural fire regimes using a Moderate Resolution Imaging Spectroradiometer (MODIS) and Landsat data, including the impact of abandoned lands.
2. Calculate emissions for air pollutants at the region- and country-level scales.
3. Assess the degree to which differences in institutional settings affect agricultural burning across Belarus, Lithuania and European Russia.

2 Data and Methods

Anthropogenic fires and related emissions from agricultural ecosystems (i.e., croplands and pastures) in European Russia, Belarus, and Lithuania were measured by using remote sensing-based products for fire and land cover. This analysis used coarse resolution data (1 km and 500 m) from MODIS to assess changes in observed cropland burning across national and regional scales, including the 1 km MODIS Active Fire Product (Giglio et al. 2003, 2006; MCD14ML collection 5) available from the NASA FIRMS fire mapper archive (https://firms.modaps.eosdis.nasa.gov/firemap/) and the 500 m MODIS Burned Area Product (Roy et al. 2008; MCD45A1 collection 5.1) to analyze annual and seasonal fire dynamics of anthropogenic fires for 2002 through 2012. MODIS MCD14ML and MCD45A1 have been shown to underestimate small fires and low intensity fires (Hawbaker et al. 2008; Hall et al. 2015). However, for the purposes of quantifying different fire regimes and establishing emissions baselines at the regional and country levels, MODIS data and products are appropriate (McCarty et al. 2012). We characterized fire rates using MODIS active fire per MODIS cropland and natural vegetation mosaic land cover maps and MODIS active fire per MODIS grassland land cover maps. Additionally, we characterized fire rates using MODIS active fire detection within non-forest areas mapped at moderate resolution (30 m) from the Landsat data (Potapov et al. 2014). We also used Landsat 7 and Landsat 8 burn scar visual interpretation within the probability-based samples (Krylov et al. 2014a, b) to calculate local emissions for Smolensk Oblast for 2014. Finally, an intensive field campaign completed in Smolensk Oblast of Russia in May and June 2014 validated the MODIS active fire detections over croplands, pastures, and forests and provided context for the Landsat burn scar analysis. It also improved the accuracy of the MODIS MCD14ML and MCD45A1 products.

Cropland extent was determined by using three separate land cover products for European Russia, Lithuania, and Belarus, and for the targeted analysis of Smolensk Oblast. The annual 500 m MODIS Land Cover Type product (MCD12Q1)

Collection 5 was used to analyze fire and emissions dynamics from anthropogenic sources, namely the land cover classes of cropland and cropland/natural vegetation mosaic. MCD12Q1 is derived from 8-day MODIS observations from both Terra and Aqua (Friedl 2013). We selected the Intergovernmental Geosphere-Biosphere Programme (IGBP) land cover schema, and we defined anthropogenic sources of burning as fires occurring in the land cover classes of croplands, croplands/natural vegetation mosaic, grasslands (i.e., potential pasture sources), and permanent wetlands (Table 1). Permanent wetlands were included because reed and peatland burning is common throughout Eastern Europe, as has been noted in local media sources (Evseev 2013; UNDP 2014). All other land cover types, including evergreen and deciduous forests, shrublands, barren lands, and urban lands, were grouped into the category 'all other land cover classes' because fires in these land cover types are likely to be wildland forest fires or industrial fires.

We used the IIASA-IFPRI Global Cropland Map to determine cropland burning patterns and rates using the 1 km MODIS Active Fire product. We examined whether the patterns changed for year 2012 when compared to the MOD12Q1 land cover product. The International Institute for Applied Systems Analysis (IIASA) and International Food Policy Research Institute (IFPRI) Global Cropland Map harmonized cropland maps from GlobCover 2005, AFRICOVER, and other national-level maps and estimated the global cropland percentages (0–100 %) for each 1 km pixel for circa 2005. When defining croplands, developers based final decisions on the level of agreement between input datasets (IIASA 2015). Fritz et al. (2015) reports an overall accuracy of 82.4 %.

We used the IIASA Global Field Size Map to determine rates and patterns of cropland burning in 2012 as detected by MODIS 1 km active fire detections. The 1 km Global Field Size Map used crowdsourced data obtained through the Geo-Wiki project (www.geo-wiki.org), a public high-resolution image interpretation platform (IIASA 2015). Additionally, we conducted a secondary analysis to

Table 1 MODIS MOD12Q1 land cover classification schema and corresponding descriptions

Land cover classification scheme	Class	Description
IGBP	Croplands	Temporary crops followed by harvest and a bare soil period (single and multiple cropping systems). Excludes perennial woody crops
	Croplands/natural vegetation mosaic	Mosaic of croplands, forest, shrublands, and grasslands; no one component comprises >60 % of the pixel
	Grasslands	Lands with herbaceous types of cover. Tree and shrub cover is less than 10 %
	Permanent wetlands	Lands with permanent mixture of water and herbaceous or woody vegetation that cover extensive areas. The vegetation can be present in either salt, brackish, or fresh water

Descriptions taken from FAO (2000)

determine whether field size is related to burning patterns in cropland landscapes in Belarus, Lithuania, European Russia, and the Smolensk Oblast of Russia.

We used the bottom-up approach develop by Seiler and Crutzen (1980) to calculate the atmospheric emissions. The equation for the bottom-up approach is:

$$\text{Emissions} = A * B * \text{CE} * e_i \tag{1}$$

where A is cropland burned area, B is the fuel load variable (mass of biomass per area), CE is combustion efficiency (fraction of biomass consumed by fire), and e_i is the emission factor for atmospheric species (mass of atmospheric species per mass of biomass burned). For this analysis, all agricultural burning in European Russia, Belarus, and Lithuania was attributed to grain production (wheat, barley, rye, rapeseed/canola), which accounts for the majority of the area of planted croplands in these countries (USDA FAS 2013, 2014; Republic of Belarus 2015). As previously mentioned, burned area (A) was calculated from the MODIS burned area (MCD45A1) or the MODIS active fire data (MCD14ML) product, and cropland extent was derived from the annual MODIS MOD12Q1 land cover products and the IIASA 1 km Global Cropland Map.

We focused our emissions analysis on carbon, ozone precursors, particulate matter, volatile organic carbon (VOC), and harmful air pollutants (HAPs). Table 2 lists the 25 atmospheric species and pollutant species (e_i) used to calculate the impact of cropland burning on air quality. These atmospheric species are the pollutant species targeted by the U.S. Environmental Protection Agency (EPA) National Emissions Inventory and the World Health Organization (2000) air quality monitoring program in Europe. Akagi et al. (2011) was the source for the pasture burning e_i values (Table 3).

Table 2 Emission factors used to estimate emissions from cropland burning

Pollutant species	Emission factor (g/kg)	Pollutant species	Emission factor (g/kg)
CO_2	1671	Benzo(a)pyrene	0.001
CH_4	2.04	Benzo(e)pyrene	0.001
CO	55.1	Benzo(ghi)perylene	0.002
NO_x	2.36	Benzo(k)fluoranthene	0.001
SO2	0.44	Chrysene	0.002
PM2.5	4.03	Fluoranthene	0.004
PM10	7.05	Formaldehyde	1.69
VOC	3.80	Indeno(1,2,3-cd)pyrene	0.001
1,3-butadiene	0.177	Perylene	0.001
Acetaldehyde	0.722	Phenanthrene	0.005
Anthracene	0.002	Pyrene	0.004
Benz(a)anthracene	0.002	Toluene	0.235
Benzene	0.357		

Emission factors are derived from the 2011 U.S. Environmental Protection Agency National Emissions Inventory for agricultural burning activity data (http://www.epa.gov/ttnchie1/net/2011inventory.html)

Table 3 Emission factors (g/kg) used to estimate emissions for pasture burning in Smolensk Oblast

Pollutant species	Emission factor (g/kg)
CO_2	1548
CH_4	8.71
CO	135
NO_x	0.75
SO_2	0.32
$PM_{2.5}$	14.8
PM_{10}	28.9

Emission factors from Akagi et al. (2011)

Fuel loading values were obtained from the average annual grain yield statistics collected from the Federal State Statistics Service database for each grain-producing oblast in Russia in years 2002 through 2010 (McCarty et al. 2012). For this analysis, the fuel load value (*B*) applied for all three countries was 1983 kg/ha. Without country-specific values, the combustion efficiency (CE) value was assumed to be 0.80, which is slightly less than the value used by McCarty (2011) for wheat residue burning in the contiguous U.S. The value was then used for Russia-specific cropland burning emissions estimates in McCarty et al. (2012), but it matches the values for European cereal crop burning reported by van Leeuwen et al. (2014).

This analysis represents a convergence of satellite-based products with a bottom-up approach for calculating emissions. We found it difficult to accurately assess the uncertainty and error associated with the calculated agricultural burning emissions. First, the emission factors, fuel loading, and combustion completeness used in this analysis represent a limited sample even if they are taken from state-of-the-art scientific sources. Second, where appropriate, we indicated the performances of the fire and land-cover/land-use products in detecting and classifying agricultural burning and conducting direct comparisons with limited in situ data. Therefore, a method to assess the uncertainty is not included in this analysis, but it may be an aim in future studies involving emissions calculations for agricultural burning in this region.

3 Results

3.1 Anthropogenic Fire Patterns and LCLUC

Most of the agricultural fires detected by the MODIS MCD14ML active fire product and the MODIS MOD12Q1 land cover product in Belarus and Lithuania occurred during spring (March through May), with a smaller peak in late summer/early fall (July through September) (Fig. 1). On average, European Russia experienced a bimodal agricultural fire season, with the largest peak in late summer/early fall (July through September). Between 2002 and 2012, fires were detected in grasslands, and all other land cover types (including forest fires)

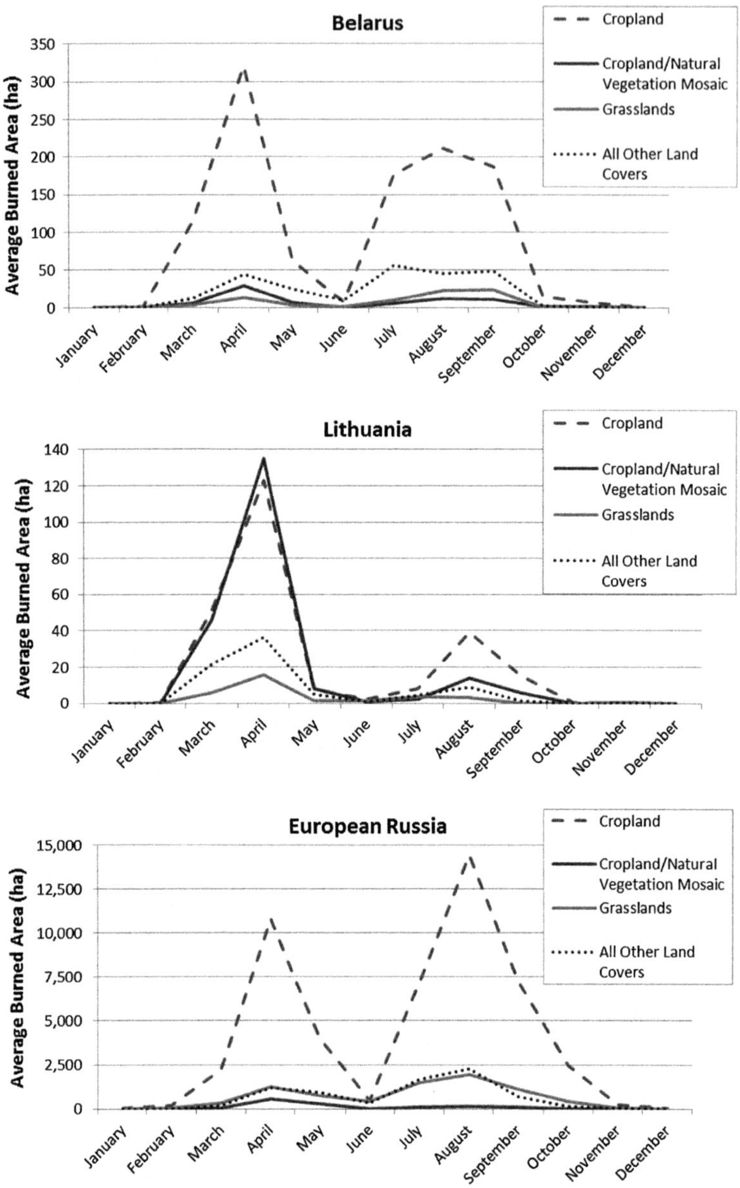

Fig. 1 Average seasonal variability of estimated annual burned area (ha) in Belarus, Lithuania, and European Russia based on MODIS 1 km Active Fire Data for the IGBP land cover classes of cropland, cropland and natural vegetation mosaic, grasslands, and all other land covers, 2002–2012; annual burned area was estimated by classifying all active fire detections by land cover and multiplying each event by 25 ha to estimate burned area. This was repeated for each year in the period of 2002–2102, and the results were averaged

experienced the same bimodal peaks in the spring and late summer/early fall, but the magnitude of the agricultural burning (indicated by the cropland and cropland and natural vegetation mosaic land cover classes) was much greater. On average, this analysis found that agricultural burning accounts for 78 % of all burning and 81 % of all burned area across all land cover types (including forest fires) in Belarus, 81 % of all fire detections and 82 % of all burned area in Lithuania, and 77 % of all fire detections and 76 % of all burned area in European Russia.

Agricultural burning in Belarus, Lithuania, and European Russia showed a strong and consistent seasonal geographic pattern from 2002 to 2012 (Fig. 2). In early spring, fires started in southern Belarus and transitioned to burning along the border with European Russia and throughout much of the country during spring (March to June) and summer (July to September). By fall (October to December), burning in cropland landscapes was more scattered, and it partially retreated back to southern Belarus. During the late winter and spring (March to June), fires in Lithuania started along the coast of the Baltic Sea and also along the border with Belarus and Latvia. This transitioned to a distinct summer burning pattern (July to September) in central Lithuania. However, the summer burning in central Lithuania began declining in 2008 and disappeared by 2012. Cropland burning in European Russia showed a strong seasonal geographic pattern from 2002 to 2012. During late winter/early spring, cropland burning occurred mainly in southern Russia and along the border with Ukraine and Belarus, specifically in the months of February and March. From 2005 to 2008 and in 2011, significant cropland burning occurred in January in Krasnodar Krai along the Black Sea. By April, burning across southern Russia increased and spread north into the vicinities of Moscow and St. Petersburg. In May and June, much of southern Russia experienced cropland burning, from the Black Sea to the Volga River Valley. In contrast, summer burning (July, August, and September) encompassed much of temperate European Russia. By October, fires were generally concentrated along the border with Kazakhstan and Ukraine, with more burning near the Caucasus in November and December. Even Kaliningrad Oblast, which is located next to Poland and Lithuania and is not contiguous with the rest of Russia, experienced agricultural burning between March and June (Fig. 2). This seasonality and the occurrence of agricultural burning in Kaliningrad Oblast indicates that institutions and policies to prevent and/or to control agricultural burning are not properly working across Russia, disregarding geographic location.

Over the time period that we analyzed, the MCD14ML MODIS Active Fire product showed a decrease in non-forest fire activity in Belarus, Lithuania, and European Russia and a negative but not statistically significant linear trend in cropland burning specifically (Fig. 3). Belarus experienced 6859 fewer fires (i.e., MODIS pixels detected as thermal anomalies) in 2012 than in 2002 across all land cover types. Small peaks of burning in croplands and cropland and natural vegetation mosaic land cover types occurred in both 2006 and 2009. Lithuania

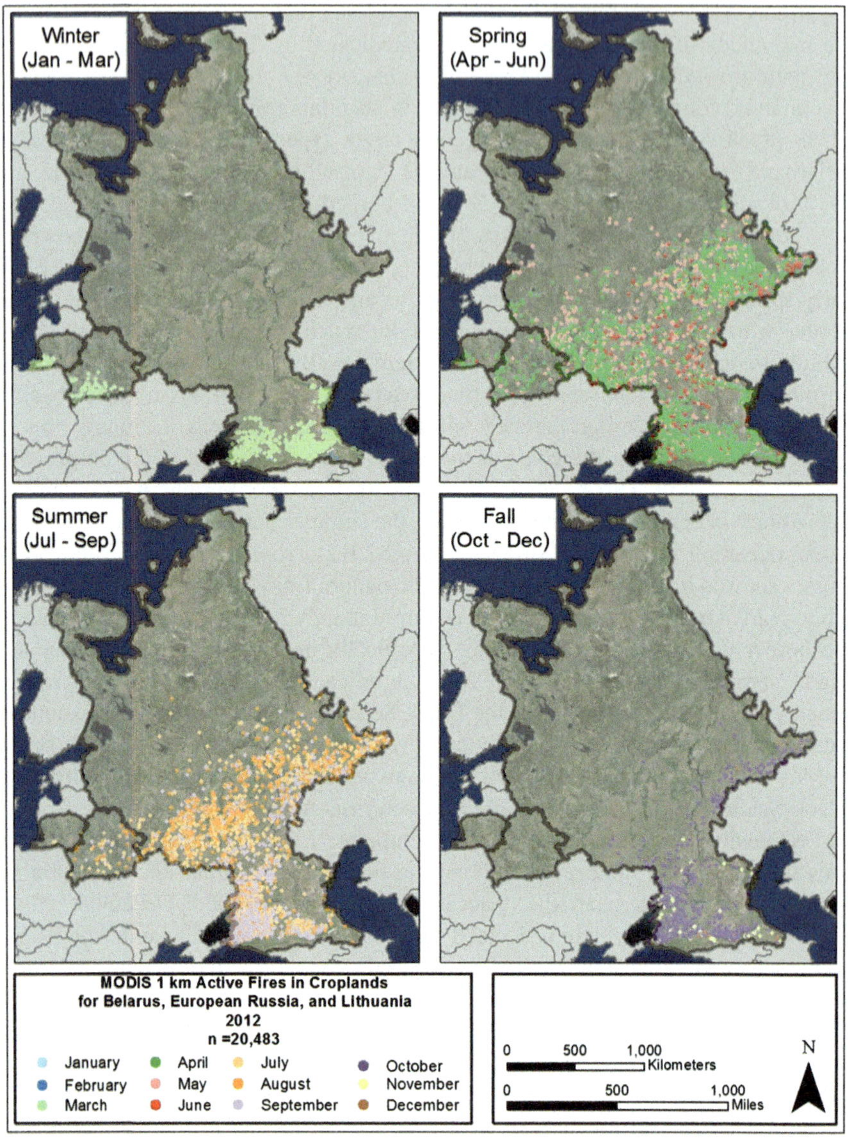

Fig. 2 Seasonal and monthly cropland fire observations for the study region in 2012; fire observations from the 1 km MODIS active fire product and cropland extent from the MOD12Q1 MODIS Land Cover Product (land cover class = 12)

experienced the same decrease in fire activity from 2002 to 2012 in all non-forest land cover types, with 597 fewer fires in 2012 than in 2002 and small peaks of burning in the cropland and cropland and natural vegetation mosaic land cover types in 2005, 2006, and 2009. The MODIS Active Fire product showed an

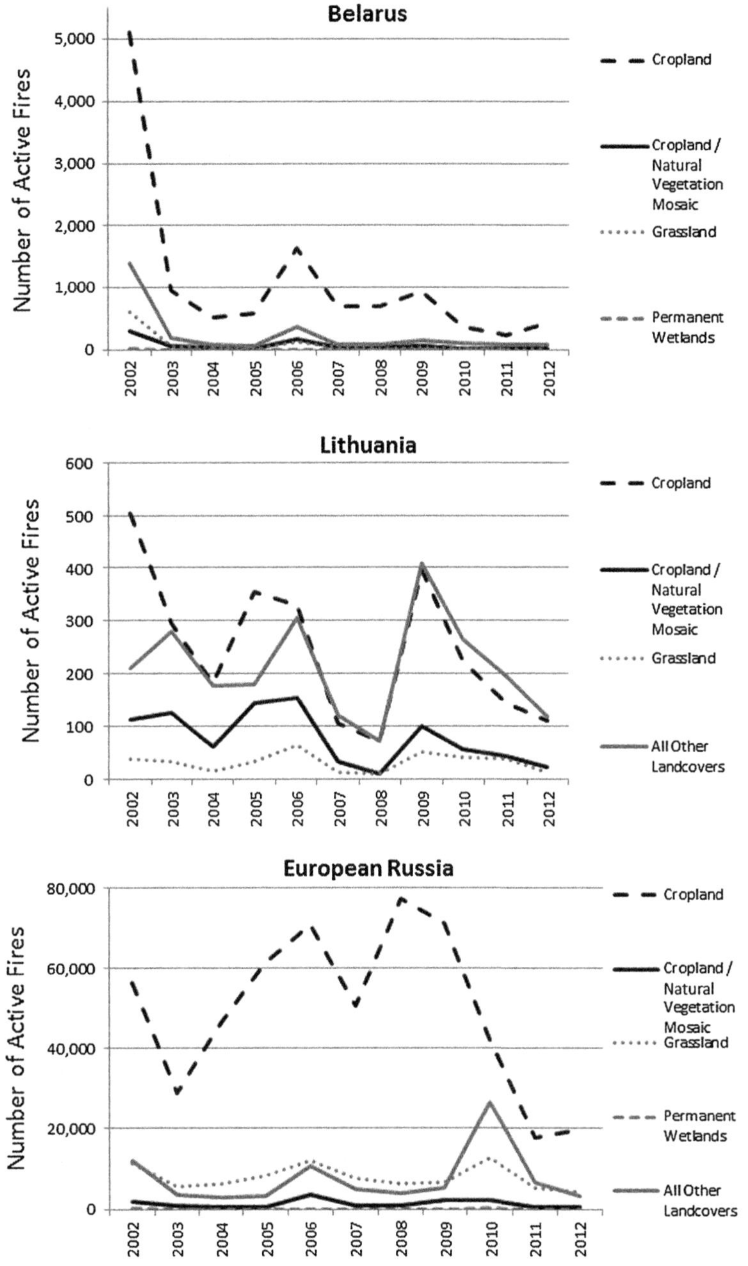

Fig. 3 Interannual active fire change in Belarus, Lithuania, and European Russia based on MODIS 1 km active fire data (MCD14ML) for the IGBP land cover classes of cropland, cropland and natural vegetation mosaic, grassland, permanent wetland, and all other land covers, 2002–2012. Note that permanent wetland burning was not detected in Lithuania

increase in cropland fire activity in European Russia from 2002 to 2008 followed by a decline, with 53,860 fewer fire detections in 2012 than in 2002. Peaks of burning in the cropland and cropland and natural vegetation mosaic land cover types occurred in 2006, 2008, and 2009, with even more agricultural burning than in 2002. Bowman et al. (2011) noted a peak in grassland, permanent wetland, and cropland fires in 2010, which was an extreme fire year in European Russia. However, we did not observe a large increase in cropland fire detections in 2010, but rather a decreasing and not statistically linear significant trend in all cropland fires in European Russia. In particular, there was a sharp drop in all types of agricultural burning in European Russia in 2011 (Fig. 3), which was also discovered by Loboda et al. (2017; chapter in this volume).

For Belarus, both the MODIS active fire and burned area datasets recorded maximum fire activity in 2002 with a large drop in subsequent years (Fig. 4). The MCD45A1 MODIS burned area product showed little to no agricultural fire detections in Belarus and Lithuania. The MCD14ML active fire product indicated highly variable agricultural burning in Lithuania while the burned area product did not identify any agricultural fires. For European Russia, the MODIS Burned Area Product showed peaks of agricultural burning in 2002, 2005–2010, and 2012. In general, the MODIS Active Fire Product showed the highest burning in 2002 and from 2006 to 2009 (though with a small decreases in 2007 relative to the 2002 fire levels). The comparison of these two fire products helps us to understand how two satellite-based global products related to fires can produce very different characterizations of agricultural burning at the country and region scales. In general, the MCD45A1 MODIS burned area product was not designed to monitor the smaller burned area sizes of agricultural burning but rather to detect large-scale wildfires in grasslands and forests (Roy et al. 2008). In contrast, the MCD14ML active fire product detects fires as small as 100 m^2 across all land types (Giglio et al. 2003).

We compared all of the MODIS active fire detections for 2012 in Belarus, Lithuania, and European Russia with the estimated cropland field size from the IIASA-IFPR Field Size product (Fritz et al. 2015). The detected cropland fires in Belarus (where the crop percentage is equal to or greater than 70 % per pixel) occurred only in medium (25–32 ha) and large (33–40 ha) fields (Fig. 5). Approximately 42 % of the MODIS active fire detections occurred in medium-sized fields, and 58 % occurred in large fields. No fires occurred in the very small (10–17 ha) or small (18–24 ha) fields. Agricultural burning in Lithuania occurred mainly in medium and large fields. Approximately 79 % of the MODIS active fires were detected in medium-sized fields, and 15 % were detected in large fields. Approximately 5 % were detected in small fields. Like Belarus, no fires occurred in the very small fields. In European Russia, 83 % of the fire detections occurred in large fields. Additionally, 16 % of fire detections occurred in medium fields. Although there were a few fires in small and very small fields, the numbers were negligible compared to the other two categories. The MODIS active fire product can detect fires as small as 100 m^2 (or 0.01 ha) (Giglio et al. 2003), meaning that even the smallest field size category could be actively burning and detected. However, the results from our analysis using a map of field sizes seems to suggest otherwise.

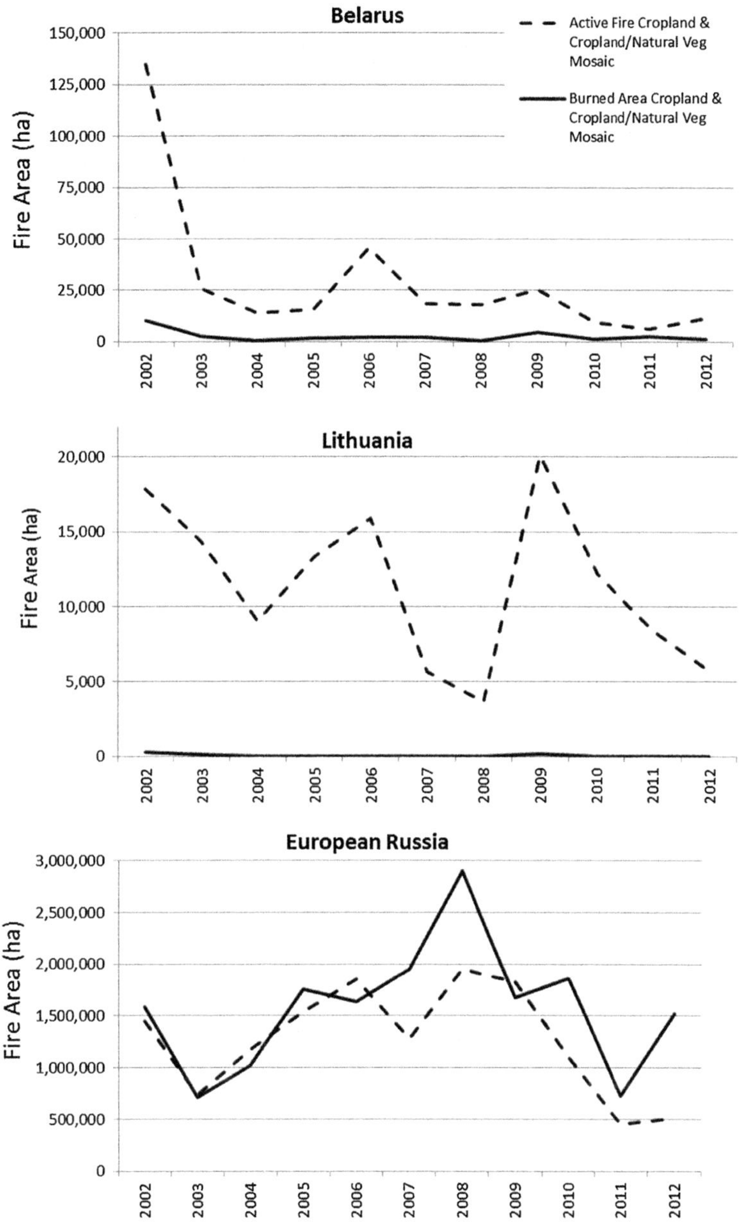

Fig. 4 Comparison of calculated burned area for cropland and natural vegetation mosaic based on the MODIS active fire product (MCD14ML) and the MODIS burned area product (MCD45A1) for Belarus, 2002–2012

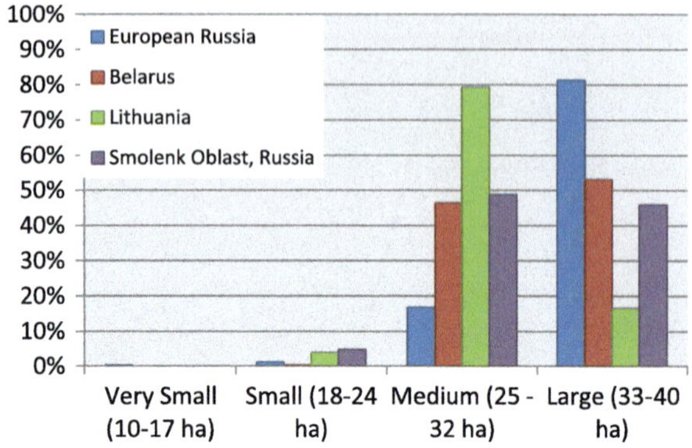

Fig. 5 Percentages of agricultural field sizes for European Russia, Belarus, Lithuania, and Smolensk Oblast, Russia, according to the IIASA-IFPRI Field Size dataset

Whether this is because there were no fires or because the fires were undetected is unclear, but we deem it likely that the MODIS active fire product does not detect fires at the same rate in small and very small fields.

The 2012 active cropland fire detections in European Russia based on the IIASA-IFPRI (where crop percentage is equal to or greater than 70 %) dataset were much lower than the number of cropland fires based on the MODIS land cover product. This suggests that the cropland extent as determined by the global products was very important for determining the actual cropland burning rates. The IIASA-IFPRI data indicated approximately 7700 cropland fire detections, whereas the MODIS MOD12Q1 product indicated 19,939 fire detections in the croplands of all land cover classes, a −61 % difference. In Lithuania, only 18 cropland fires were detected in 2012 using the IIASA-IFPRI cropland extent, and 111 cropland fires were detected using the MOD12Q1 cropland extent, a −84 % difference. However, the seasonal patterns of active fire detection assigned to the land-cover/land-use type of croplands based on the two land cover products was the same, with most cropland fire detections occurring in March (62 %) and April (35 %) in Lithuania. For Belarus, the seasonality of the cropland burning did not change; the majority of fires occurred in March or April. Only 60 cropland fires were detected in Belarus in 2012 using the IIASA-IFPRI 70 % cropland extent. In contrast, 433 cropland fires were detected by using the MOD12Q1 cropland extent, a difference of −86 %.

3.1.1 Fine-Scale Analysis: Smolensk Oblast, Russian Federation

Fine-scale quantification of the agricultural burning in Smolensk Oblast in Russia was completed to provide a preliminary sensitivity analysis of the global MODIS fire products, MCD14ML and MCDA45A1. This analysis can be used for characterizing cropland and grassland/pasture burning. An intensive field campaign was completed in the Smolensk Oblast in May and June 2014 to estimate the current fire regime in croplands, pastures, and forests. We used the Potapov et al. (2014) forest cover data and forest cover change map covering the years 1985–2012 to calculate stable forest and non-forest area and to determine the areas of re- and afforestation since 1985. To disaggregate the non-forested land into land cover classes, we used a sample-based approach. 500 random points allocated within areas mapped as 1985–2012 stable non-forest to complete the field validation. For each sample point, we identified the uniform land cover patch (e.g., field). Each land cover patch was categorized as wetland, urban/road, or agricultural land (active or abandoned) using the Landsat time-series and Google Earth high-resolution data. In the field portion of the study, we targeted agricultural lands to estimate the areas of active and abandoned agriculture. We sampled 106 points within this class and identified the corresponding fields on the ground. Each sample field was attributed as actively managed cropland, hay field/pasture, or abandoned land.

Landsat 8 OLI and Landsat 7 ETM+ data for 2014 were visually analyzed within the same 500 sample fields to estimate the rate of burning (% of field burnt). Based on Landsat-derived burned area estimations (Fig. 6), we estimate that 0.65 Mha of pasture, hay and abandoned lands was burned in Smolensk in the spring of 2015, but only 0.06 Mha was burned in the fall (October and November) of 2014, meaning that spring fire activity is 10 times greater than fall.

The MODIS MCD14ML Active Fire data were clipped to the MODIS MOD12Q1 land cover product for two classes: Cropland/Natural Vegetation Mosaic (IGBP class = 14) and Grasslands/Pasture (IGBP class = 10) and then burned area estimations assumed a 25 ha fire for each active fire pixel. For 2014, 24,300 ha of croplands burned in Smolensk, with 91.7 % of the fires occurring in spring (March–May), 0.8 % in summer (June–July), and 7.5 % in fall (August–November) (Table 4). Only 5375 ha of grassland/pasture burned in 2014, with

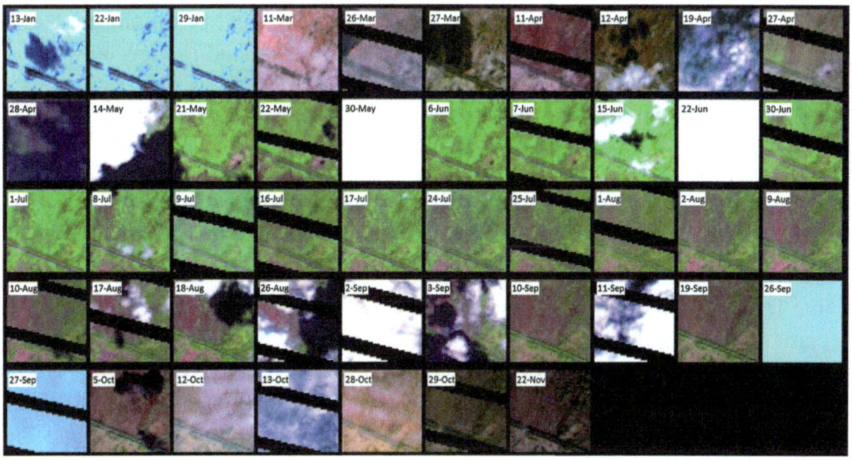

Fig. 6 Time series of Landsat 7 and Landsat 8 true color images in Smolensk Oblast from 13 January 2014 to 22 November 2014; note the presence of an active fire line in 26-Mar and 27-Mar with a distinctive green-up during the growing season in this landscape that is mainly composed of pasture

Table 4 Burned areas in Smolensk Oblast from 2014 cropland and grassland burning detected by MODIS active fire product and pasture burning derived from Landsat and field validated burn scars

Season (months)	Area (ha)
MODIS active fire cropland burning	
Spring (March–May)	22,275
Summer (June–July)	200
Fall (August–November)	1825
Total	24,300
MODIS active fire grassland/pasture burning	
Spring (March–May)	4950
Fall (August–November)	325
Total	5275
Landsat/field validated pasture burn scars	
Spring (March–May)	649,108
Fall (August–November)	59,917
Total	709,025

94 % of the fires occurring in spring and 6 % occurring in fall. The Landsat pasture burn scars combined with the field validated pasture burned area resulted in an estimate of 673,400 ha burned in spring and 60,000 ha burned in fall, for a total of 733,400 ha of grassland/pasture burning in Smolensk in 2014. Similar to the cropland and grassland fires detected by the MODIS active fire product, the Landsat/Field validated burn scars also indicated that the vast majority of burning occurs in spring (~92 %). However, the MODIS active fire-based estimate of cropland burned area was 97 % smaller than the Landsat/field-based pasture burned area, and the MODIS active fire-based estimate of grassland/pasture burned area was 99 % smaller than the Landsat/field-based pasture burned area. In summary, our detailed field- and Landsat-based assessment of fire activity and land cover for Smolensk Oblast showed major underestimation of anthropogenic fire by the MODIS active fire data. We also found inaccuracies in land cover mapping. Specifically, many areas mapped as croplands in the global land cover products were likely pastures on the ground.

According to the IIASA-IFPRI cropland map, 2014 active cropland fires (i.e., active fire pixels for which the crop percentage was ≥70 %) for Smolensk Oblast occurred mostly in April (17 fires) or May (11 fires), with only one fire detected in July. Active cropland fire detections determined from the MODIS MOD12Q1 land cover product followed this seasonal trend, with 176 fires detected in April, 180 in May, and 7 in July, but the differences in the absolute numbers of cropland fires between the two datasets were stark.

The 2012 active cropland fire detections in Smolensk Oblast based on the IIASA-IFPRI dataset were also much lower than the number of cropland fires based on the MOD12Q1 product. The IIASA-IFPRI cropland extent data indicated 29 fire detections in Smolensk, whereas the MODIS MOD12Q1 product detected 363 cropland fire detections, a −170 % difference. Global land cover products at the regional level for Smolensk Oblast did not show much agreement.

The 29 detected cropland fires that were detected according to IIASA-IFPRI mostly occurred (62 %) in large fields (33–44 ha). 38 % of the fires occurred in medium fields (25–32 ha), and no fires were detected in small and very small fields. Our randomly sampled field observations showed a large number of pastures burning in Smolensk while the IIASA-IFPRI Field Size dataset indicates that some cropland and crop residue is burning in Smolensk, but mainly in medium and large fields. Similar to European Russia, Belarus, Lithuania, few agricultural fires in Smolensk Oblast occurred in very small or small fields (Fig. 6) according to the IIASA-IFPRI Field Size dataset.

3.1.2 LCLUC and Agricultural Burning

Agricultural burning includes fires in croplands and fires in managed or unmanaged grasslands, with unmanaged grasslands being referred to in this chapter as abandoned lands and managed grasslands being referred to as pastures. In cropland agroecosystems, farmers use fire to remove crop residues after harvesting. Grains are the most common crops that are burned in Belarus, Lithuania, and European Russia. If the farmers are burning crop residues directly after harvest, then fires occur in late summer and early fall. On managed grasslands, like hay or pasture, fire prevents shrub encroachment and stimulates the greening of pastures in early spring (Mierauskas 2012). On abandoned lands, farmers also use fire to prevent shrub encroachment, but it is also done to avoid potential government penalties for not properly maintaining agricultural landscapes (http://www.rg.ru/2014/02/04/zemli-site-dok.html). Farmers may also convert land back to hay or pasture. This means that fire regimes in agricultural areas depend on land use practices (Tulbure et al. 2010). In our study area, there are strong geographic differences in land-use and land-cover, and the differences generally follow latitudinal gradients. The largest percentage of cropland is found in southern Russia.

In contrast, non-chernozem soil regions of Russia have a low percentage of croplands and high percentages of hay, pasture, and abandoned agricultural lands. Approximately 14 Mha or 49 % of abandoned croplands in European Russia occurs in non-chernozem zones of boreal and temperate European Russia (Schierhorn et al. 2013; ROSSTAT 2010). According to official statistics, the sown agricultural area decreased by 30 % from 1990 to 2000 in Lithuania, but by 2012, the agricultural areas were cultivated at nearly the same acreage levels as in 1990 (LITHSTAT 2014). The recovery of abandoned croplands in Lithuania after 2000 was likely due to Lithuania becoming a member of the European Union and the impact of the EU Common Agricultural Policy and its second pillar, which financially supports maintenance of less favorable ("marginal") agricultural areas. According to official statistics in Belarus, sown agricultural areas increased by 1 % from 1990 to 2000 despite reports of ongoing agricultural abandonment

(Prishchepov et al. 2012). However, the actual abandonment occurred in the statistics reported only after 2000, so that by 2012, approximately 13 % of croplands cultivated in 1990 were abandoned (BELSTAT 2011). Most likely, strong governmental support of agriculture in Belarus postponed the agricultural abandonment, which was a common phenomenon across post-Soviet countries from 1990 to 2000. Compared to the massive agricultural abandonment in Russia, the cases of Belarus and Lithuania suggest relatively stable agricultural production and land governance, which might also reflect differences in the spread of agricultural fires across the three selected countries.

In southern European Russia, we see a strong fall peak in the incidence of fires, whereas in Lithuania, Belarus, and non-chernozem soil regions of European Russia, spring is the major season for fires. Magi et al. (2012) reported the same fire seasonality for agricultural land use types in this region. Using the MOD12Q1 MODIS Land Cover product and the MCD14ML MODIS Active Fire Product, we calculated that 73 % of total fire in Belarus, 43 % in Lithuania and 75 % in European Russia is cropland fire. Fires occurring in cropland and natural vegetation mosaics and grasslands account for 2 and 12 % of all fires in European Russia, 37 and 5 % of all fires in Lithuania, and 5 and 6 % of all fires in Belarus, respectively. Forest fires account for 16 % of all fires in Belarus, 14 % in Lithuania, and 11 % in European Russia.

The negative trend in fire activity in Belarus, Lithuania, and European Russia over the time period that we analyzed was not statistically significant because of high annual variation. The relative MODIS active fire detection per arable land area value is dramatically lower in Lithuania and Belarus than in European Russia (Fig. 7), likely because the southern region of European Russia has a high agricultural burning rate. When we compared Belarus to the neighboring Russian

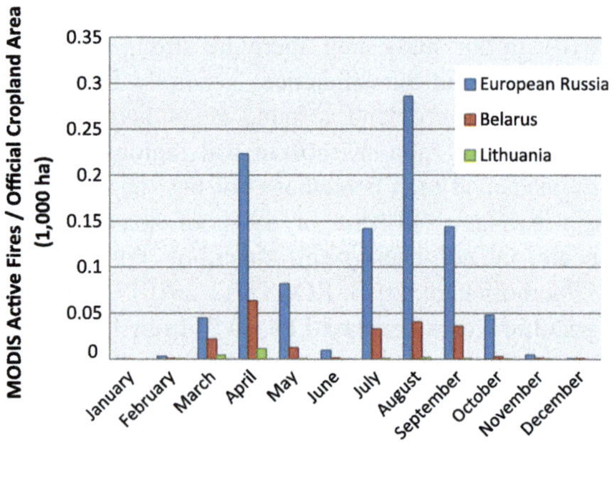

Fig. 7 Average seasonal distribution of cropland burning for 2002–2012 per 1000 ha of arable land; MODIS active fires detected in cropland and cropland/natural vegetation mosaic land cover classes normalized by official arable land statistics (1000 ha) for 2012; Belarus and Lithuania arable land estimates come from FAOSTAT (2012), and the European Russia arable land estimates come from ROSSTAT (2012)

region of Smolensk Oblast, we found that Smolensk has 166 % more active fires detected in croplands and grasslands than Belarus. There are likely two explanations for this difference. Smolensk has experienced much more land abandonment than Belarus (Prishchepov et al. 2012) (Fig. 7). The fire rate on abandoned land is twice as high as the fire rate for hay and pastures (Krylov et al. 2014a, b). Second, agricultural burning is more regulated in Belarus than in Russia, and it was declared illegal in Lithuania (with a few exceptions) after its entrance into the European Union, though some illegal field burning has been documented (Mierauskas 2012). In Russia, agricultural burning is officially illegal, and the Ministry of Emergency passed an agricultural burning regulation in 2014 (http://www.rg.ru/2014/02/04/zemli-site-dok.html). In practice, however, this rule has not been enforced in most administrative areas of Russia (Bellona and Yabloko 2010; Kobets et al. 2011).

3.2 Emissions from Agricultural Burning: Croplands Versus Pasture

Burned area estimates from different satellite-based products resulted in highly variable ranges of fire activity and fire area. The MCD45A1 MODIS Burned Area product greatly underestimates the burned area. The MCD14ML Active Fire product performed better in terms of its ability to capture agricultural burning, but it still missed some fires due to cloud cover, and when the fires were small or burned with low intensity, they were more likely to be missed. For Russia, MCD14ML had been assumed to be equal to 1 km^2 of burned area per detection (McCarty et al. 2012) for agricultural burning, but we found this estimate to be too high when we compared it to high-resolution field boundaries (McCarty et al. 2012; Romanenkov et al. 2014). For our analysis, we normalized each active fire by an average field area of 25 ha, resulting in a more conservative estimate for agricultural burning. However, some published (Krylov et al. 2014a, b) and preliminary results (Hall et al. 2015) indicate that this approach may lead underestimation of the burned area in Russia. Thus, we provide somewhat conservative emissions estimates.

We calculated emissions to four significant digits to match the significant digits of the emission factor input variables, with minor emissions from the HAPs atmospheric species represented in scientific notation. Table 5 reports the minimum, maximum, annual average, and total carbonaceous matter, particulate matter, VOCs, and HAPs emissions from cropland burning in Belarus, Lithuania, and European Russia for 2002 through 2012. 2008 and 2002 were the minimum and maximum years for cropland burning emissions for both Lithuania and Belarus. For Belarus, 2010 and 2012 had cropland burning emissions 60 and 40 % larger, respectively, than the minimum year of 2008. 2002 was much higher in terms of emissions than any other year in Belarus. For example, the second highest year in terms of emissions, 2003, was 81 % less than 2002. In Lithuania, 2007 and 2012 were the next lowest years in terms of emissions, but they were still approximately

Table 5 Carbon, NO_x, SO_2, particulate matter, VOC, and HAPs emissions from cropland burning (LC = 12 and 14 in IGBP classification schema) as quantified from 1 km MODIS active fire data for Belarus, Lithuania, and European Russia

Atmospheric species	Belarus emissions: average (*Range*)	Lithuania emissions: average (*Range*)	European Russia emissions: average (*Range*)
CO_2	51.3 (*5.00–339*)	16.37 (*4.84–33.3*)	1608 (*187–4057*)
CH_4	0.062 (*0.006–0.413*)	0.020 (*0.006–0.041*)	1.96 (*0.23–4.95*)
CO	1.68 (*0.17–11.16*)	0.538 (*0.157–1.099*)	53.0 (*6.2–133.8*)
NO_x	0.072 (*0.007–0.478*)	0.023 (*0.007–0.047*)	2.27 (*0.26–5.73*)
SO_2	0.013 (*0.001–0.089*)	0.004 (*0.001–0.009*)	0.42 (*0.05–1.07*)
$PM_{2.5}$	0.123 (*0.012–0.816*)	0.039 (*0.012–0.08*)	3.88 (*0.45–9.79*)
PM_{10}	0.216 (*0.021–1.428*)	0.094 (*0.02–0.141*)	6.78 (*0.79–17.12*)
VOC	0.116 (*0.011–0.770*)	0.036 (*0.011–0.076*)	3.66 (*0.43–9.23*)
1,3-butadiene	0.005 (*0.001–0.036*)	0.002 (*0.001–0.004*)	0.170 (*0.020–0.430*)
Acetaldehyde	0.022 (*0.002–0.146*)	0.007 (*0.002–0.014*)	0.695 (*0.081–1.753*)
Benzene	0.011 (*0.001–0.072*)	3.49E−03 (*1.02E−03-7.12E−03*)	0.344 (*4.00E−02-8.67E−01*)
Formaldehyde	0.052 (*0.005–0.342*)	1.65E−02 (*4.83E−03-3.37E−02*)	1.627 (*0.19–4.103*)
Toluene	0.007 (*0.001–0.048*)	2.11E−03 (*6.71E−04-4.69E−03*)	0.226 (*0.026–0.571*)

Average emissions calculated for 2002–2012 with range derived from the min and max year of cropland burning emissions; 2008 and 2002 were min and max years for Lithuania and Belarus; 2004 and 2004 were minimum and maximum years for European Russia; HAPs emissions from anthracene, benz(a)anthracene, benzo(a)pyrene, benzo(e)pyrene, benzo(ghi)perylene, benzo(k) fluoranthene, chrysene, fluoranthene, indeno(1,2,3-cd)pyrene, perylene, phenanthrene, and pyrene are not included in the table but were calculated at consistently less than 7.00E−04 for Belarus and Lithuania and 5.00E−03 for European Russia; all emissions in Gg

44 and 52 % larger, respectively, than the minimum emissions estimated for 2008. 2009 was the next highest year in terms of cropland burning emissions in Lithuania, but it was still 21 % less than the maximum year, 2002. The minimum and maximum years of cropland burning emissions in European Russia were 2004 and 2005, respectively. The next lowest years were 2008 (42 % higher than 2004) and 2012 (16 % higher than 2004). 2002 and 2007 had emissions 8 % and 18 % less than the maximum emissions year of 2005.

Agricultural burning in European Russia was the main source of fire-related emissions in the study area, with PM_{10} emissions approximately 99 % greater than those of Belarus and Lithuania. To put this into perspective, the NO_x emissions from agricultural burning in European Russia, as estimated from the MCD14ML Active Fire product, are equivalent to 5 % of all NO_x emissions from all sectors (industry, energy, transportation, etc.) in Lithuania (EEA 2014). On average, agricultural burning in Belarus emits 1.68 Gg of CO, 0.54 Gg of CO in Lithuania, and 53.0 Gg of CO in European Russia, much less than agricultural burning in the contiguous U.S., which averages 278 Gg CO from the burning of rice, sugarcane, wheat, and grass seeds (McCarty 2011). Similarly, the contiguous U.S. on average emits 23.6 Gg of $PM_{2.5}$ from crop residue burning, whereas agricultural burning in Belarus emitted 0.12 Gg of $PM_{2.5}$, 0.04 Gg of $PM_{2.5}$ in Lithuania, and 3.88 Gg of $PM_{2.5}$ in European Russia. Average formaldehyde emissions from agricultural burning in European Russia (1.627 Gg) are 6.4 % of the 2011 estimates of the contiguous U.S., and the maximum formaldehyde emissions in European Russia (4.103 Gg) are 16.2 % of the estimated 25.32 Gg reported under agricultural field burning for the 2011 U.S. National Emissions Inventory (U.S. EPA 2015).

Several federal administrative regions were the main sources of agricultural burning in European Russia (Table 6). These 27 Russian regions accounted for approximately 87 % of all emissions from 2002 to 2012, with the exclusion of Belarus and Lithuania. Geographically, these federal administrative units are located in southern Russia. The regions with less agricultural burning than

Table 6 Source regions in European Russia with larger than 1 % of the total agricultural burning emissions for the entire study region, 2002–2012

Region	Percent of total agricultural burning emissions, 2002–2012 (%)	Region	Percent of total agricultural burning emissions, 2002–2012 (%)
Orenburg	7.36	Penza	2.41
Rostov	7.20	Oryol	2.13
Krasnodar	7.09	*Belarus*	*2.09*
Volgograd	6.56	Bashkortostan	2.03
Stavropol	6.36	Smolensk	2.00
Saratov	5.95	Tatarstan	1.78
Tambov	3.99	Nizhny Novgorod	1.71
Voronezh	3.81	Ulyanovsk	1.67
Samara	3.60	Kalmykia	1.66
Astrakhan	3.28	Mordovia	1.47
Lipetsk	2.98	Moskva	1.44
Ryazan	2.83	Bryansk	1.39
Kursk	2.78	Belgorod	1.14
Tula	2.41	*Lithuania*	*0.21*

Belarus and Lithuania were considered as one geographical unit and are italicized

Lithuania, i.e., less than 0.20 % of the total agricultural burning emissions, were those with large cities, like Moscow and St. Petersburg, mountainous areas of the Caucasus Mountains, like the Republic of Ingushetia, Russian federal subjects in the northern boreal and tundra zones, like the Komi Republic, Arkhangelsk Oblast, Republic of Karelia, Murmansk Oblast, and Nenets Autonomous Okrug, or areas dominated by forests, like the Urdmurt Republic and Perm Krai. The MODIS active fire algorithm (Giglio et al. 2003) likely misses fires occurring in small fields, so administrative regions with agricultural lands made up of mostly small fields also account for less than 0.20 % of total agricultural burning emissions.

3.2.1 Pasture Versus Cropland Emissions: Smolensk Oblast

In Smolensk Oblast, we used the MODIS MCD14ML Active Fire calculated area for a minimum field area of 25 ha and a maximum field area of 44 ha for cropland and grassland fire emissions, and we used Landsat-derived burn scars to estimate the burned area of pastures. The Landsat-based burned area estimate was more than 20 times greater than the combined cropland and grassland MCD14ML burned area estimate, assuming a minimum field area of 25 ha (Table 7). However, we cannot extrapolate this estimate of pasture burned area to the entire study area. We calculated emissions for the study region using a conservative MODIS active fire base estimate and, as an example, we calculated the local emissions for the Smolensk region based on Landsat data. Table 7 shows the selected air quality emissions for Smolensk Oblast. On average, all cropland burning in European Russia detected by the MCD45A1 MODIS Burned Area Product emitted 17.66 Gg of PM_{10}. Annual burning of pastures in Smolensk Oblastemitted 494.85 Gg of PM_{10}. Additionally, compared to $PM_{2.5}$ emission estimates for agricultural burning in the contiguous U.S. from sugar cane, rice, wheat, and grass seeds, Smolensk Oblast emitted a little over a 1000 % more $PM_{2.5}$ from agricultural burning based on the Landsat analysis than the contiguous U.S. based on a regionally-adapted MODIS product (McCarty 2011). This highlights that quantifying the contribution of pasture burning and burned area versus cropland burning in agricultural regions is important for accurately calculating carbonaceous emissions and emissions that negatively impact air quality.

4 Discussion and Conclusions

The majority of fires detected by satellite fire products for Belarus, Lithuania, and European Russia occurred in croplands, cropland and natural vegetation mosaic, and unmanaged and managed grasslands but not in forests. Despite the common belief about widespread wildfires occurring in forests, forest fires are rare events in this region (Krylov et al. 2014a, b). Fires occurred in a few agricultural land types, including abandoned lands, actively managed croplands, and grasslands, pastures,

Table 7 Selected emissions for Smolensk Oblast differentiating fire activity by land-use types of croplands and pasture; all emissions in Gg

Season (months)	Area (ha)	CO_2	CH_4	CO	NO_x	SO_2	$PM_{2.5}$	PM_{10}
MODIS active fire cropland burning								
Minimum field size of 25 ha								
Spring (March–May)	22,275	59.05	0.07	1.95	0.08	0.02	0.14	0.25
Summer (June–July)	200	0.53	6.47E−04	0.02	7.49E−04	1.40E−04	1.28E−03	0.00
Fall (August–November)	1825	4.84	0.01	0.16	0.01	1.27E−03	0.01	0.02
Total	24,300	64.42	0.08	2.12	0.09	0.02	0.16	0.27
MODIS active fire grassland/pasture burning								
Minimum field size of 25 ha								
Spring (March–May)	4950	185.05	1.04	16.14	0.09	0.04	1.77	3.45
Fall (August–November)	325	12.15	0.07	1.06	0.01	2.51E−03	0.12	0.23
Total	5275	197.2	1.11	17.20	0.10	0.04	1.89	3.68
MODIS active fire cropland burning								
Maximum field size of 44 ha								
Spring (March–May)	39,204	103.92	0.13	3.43	0.15	0.03	0.25	0.44
Summer (June–July)	352	0.93	1.14E−03	0.03	1.32E−03	2.46E−04	2.25E−03	3.94E−03
Fall (August–November)	3212	8.51	0.01	0.28	0.01	2.24E−03	0.02	0.04
Total	42,768	113.4	0.14	3.74	0.16	0.03	0.27	0.48
MODIS active fire grassland/pasture burning								
Minimum field size of 44 ha								
Spring (March–May)	8712	351.6	0.43	11.59	0.50	0.09	0.85	1.48
Fall (August–November)	572	23.08	0.03	0.76	0.03	0.01	0.06	0.10
Total	9284	374.7	0.46	12.35	0.53	0.10	0.90	1.58
Landsat/field validated pasture burn scars								
Spring (March–May)	649,108	24,266	136.5	2116	11.76	5.02	232.0	453.0
Fall (August–November)	59,917	2240	12.60	195.3	1.09	0.46	21.42	41.82
Total	709,025	26,506	149.1	2312	12.84	5.48	253.4	494.9

and hayfields. We found a strong geographical concentration of these fires with land use type. Grasslands, pastures, hay, and abandoned land were burned in the northern transition region, and actively managed croplands were burned in the southern regions of European Russia. The agricultural fires in Belarus, Lithuania, and European Russia had a distinct seasonality associated with distinct land use types. Spring fires were the most common, and abandoned land and pastures fires were widespread. Presumably, the burning was meant to suppress shrubs. Autumn fires were more common in actively managed croplands.

A comparison of agricultural burning by field size seems to indicate that though the MODIS active fire product can detect fires as small as 100 m^2 (or 0.01 ha) (Giglio et al. 2003), it did not perform as such in our study area. Whether this is because there were no fires in small areas or because the fires were undetected is unclear, but we deem it likely that the MODIS active fire product does not detect fires at the same rate in small and very small fields. Secondly, we found that the majority of fires in Lithuania take place in medium and large fields, suggesting that agricultural enterprises, not private small-scale farmers, are involved in burning crop residues and grassland/pastures. In European Russia, as illustrated by the fine scale analysis of Smolensk, agricultural enterprises likely set the fires in the large fields, or the fires may represent pasture fires rather than cropland fires. Future research comparing land tenure, LCLUC, and agricultural burning (pasture versus cropland) would reveal the role of agricultural enterprise size in anthropogenic burning.

Policy has an impact on agricultural burning. For instance, there is a large difference in fire regimes between Belarus and Smolensk Oblast, which are next to each other and share the same ecoregions and vegetation. However, enforcement of agricultural burning bans differs, and this is likely why more fires occurred in Smolensk Oblast. The Russian Federal Assembly passed legislation in 2014 to fine land owners in Russia for not managing abandoned lands, including lands already transitioning to shrublands and forests (http://www.rg.ru/2014/02/04/zemli-site-dok.html). This regulation can increase burning and burned area in European Russia because land owners may use fires to prevent shrub and tree regrowth. In 2010, a Normative Act of the Ministry of Emergency Response banned crop stubble burning and the building of fires in harvested fields under paragraph 327 (Kobets et al. 2010), which could potentially decrease burning and burned area. However, enforcement of these regulations appears to be weak. Lithuania and Belarus both have functioning bans on agricultural burning (Mierauskas 2012), and they have lower rates of burning than European Russia.

Several satellite products were used to characterize agricultural burning and emissions, and we focused on the MCD14ML MODIS Active Fire product to produce annual and average agricultural fire emissions, but we fully acknowledge the limitations of this approach. In Smolensk Oblast, Landsat-based burned area estimates for pasture area show that the agricultural burning area can be up to 10 times greater than the 1 km MODIS active fire estimates.

This chapter summarizes the first research to quantify air pollution at this level of detail due to agricultural burning in Belarus, Lithuania, and European Russia.

Others have noted the importance of this emissions source, but the atmospheric species and subsequent emissions that negatively impact air quality from anthropogenic burning in croplands and pastures had not previously been fully quantified (Kakareka and Kukharchyk 2003; Stohl et al. 2007; Witham and Manning 2007; Goldammer 2013; Targino et al. 2013). On average, agricultural burning in Lithuania is the lowest contributor to emissions followed by Belarus with the largest emissions from agricultural burning in European Russia. Indeed, certain air pollutants from agricultural burning in European Russia are so abundant that they are equivalent to 5 % of the emissions from all sectors in Lithuania and likely other Eastern European neighbors. Emissions from our study region were consistently much less than agricultural burning emissions estimates calculated for the contiguous U.S., with emissions from European Russia being approximately 400 % less than the contiguous U.S. However, the detailed Landsat analysis of pasture burning in Smolensk Oblast resulted in emissions estimates 91 % larger than the estimated emissions from crop residue burning for the entire contiguous U.S. and 98 % larger than the estimate for European Russia calculated from the MODIS active fire product. This shows that quantifying the contributions of pasture burning and burned area versus cropland burning and burned area in agricultural regions at finer resolutions—30 m rather than 1 km or 500 m—is important for accurately calculating carbonaceous emissions and emissions that negatively impact air quality.

References

Akagi SK, Yokelson RJ, Wiedinmyer C, Alvarado MJ, Reid JS, Karl T, Crounse JD, Wennberg PO (2011) Emission factors for open and domestic biomass burning for use in atmospheric models. Atmos Chem Phys 11:4039–4072

Alcantara CT et al (2013) Mapping the extent of abandoned farmland in Central and Eastern Europe using MODIS time series satellite data. Environ Res Lett 8

Alcantara C, Kuemmerle T, Prishchepov AV, Radeloff VC (2012) Mapping abandoned agriculture with multi-temporal MODIS satellite data. Remote Sens Environ 124:334–347

Badarinth KVS, Chand TRK, Prasad VK (2006) Agriculture crop residue burning in the Indo-Gangetic Plains—a study using IRS-P6 AWiFS satellite data. Curr Sci India 91:1085–1089

Bellona and Yabloko (2010) Conclusion of the public commission about investigations of the causes and consequences of nature fires in Russia in 2010. url:http://www.yabloko.ru/mneniya_i_publikatsii/2010/09/14. Accessed 13 July 2015

BELSTAT (2011) Regions of Belarus republic, 2002–2010, Statistical Compendium Volume 1 (Regioni Respubliki Belarus. Statisticheskii Sbornik). Minsk: BELSTAT

Bescansa P, Imaz MJ, Virto I, Enrique A, Hooqmoed WB (2006) Soil water retention as affected by tillage and residue management in semiarid Spain. Soil Till Res 87:19–27

Bond TC et al (2013) Bounding the role of black carbon in the climate system: a scientific assessment. J Geophys Res Atmos 118:5380–5552

Bowman DMJS et al (2011) The human dimension of fire regimes on Earth. J Biogeogr 38: 2223–2236

Briggs JM, Knapp AK, Brock BL (2002) Expansion of woody plants in tallgrass prairie: a fifteen-year study of fire and fire-grazing interactions. Am Midl Nat 147:287–294

Brye KR, Longer DE, Gbur EE (2006) Impact of tillage and residue burning on carbon dioxide flux in a wheat-soybean production system. Soil Sci 70:1145–1154

Chen Y, Tessier S, Cavers C, Xu X, Monero E (2005) A survey of crop residue burning practices in Manitoba. Appl Eng Agric 21:317–323

Derevyagin VA (1987) Practical guidance on straw utilization as fertilizer. All-Union Research and Development, USSR State Agro-Industrial Committee on design and technological institute for organic fertilizers and peat, Moscow, Russia, pp 1–11. http://quickr.mtri.org/croplandburning

Dhammapala R, Claiborn C, Corkill J, Gullett B (2006) Particulate emissions from wheat and Kentucky bluegrass stubble burning in eastern Washington and northern Idaho. Atmos Environ 40:1007–1015

Dubinin M, Potapov P, Lushchekina A, Radeloff VC (2011) Reconstructing long time series of burned areas in arid grasslands of southern Russia by satellite remote sensing. Remote Sens Environ 114:1638–1648

EEA (2014) Air pollution country fact sheets (2014): Lithuania. url:http://www.eea.europa.eu/themes/air/air-pollution-country-fact-sheets-2014. Accessed 2 May 2015

Evseev A (2013) Spring grass burning is troubling Russia. url:http://english.pravda.ru/russia/kremlin/11-04-2013/124256-grass_fires-0/. Accessed 2 May 2015

FAO (2000) The Forest Resource Programme 2000. url:http://www.fao.org/forestry/4031-0b6287f13b0c2adb3352c5ded18e491fd.pdf. Accessed 2 May 2015

FAOSTAT (2012) Arable land area. url:http://faostat3.fao.org/download/R/RL/E. Accessed 13 May 2015

Friedl M (2013) User guide for the MODIS land cover type product (MCD12Q1). url:lpdaac.usgs.gov/sites/default/files/public/modis/docs/MCD12Q1_UserGuide_07302013.pdf. Accessed 26 Feb 2015

Fritz S et al (2015) Mapping global cropland and field size. Glob Change Biol 21:1980–1992

Genikhovich E, Polischuk A, Pershina N (2009) Air pollution in eastern Europe. Regional aspects of climate-terrestrial-hydrologic interactions. In: Grosima PY, Ivanov SV (eds) Non-boreal eastern Europe. NATO Science for Peace and Security Series C: Environmental Security 2009, Spring Science+Business Media, New York

Giglio L, Descloitres J, Justice CO, Kaufman Y (2003) An enhanced contextual fire detection algorithm for MODIS. Remote Sens Environ 87:273–282. doi:10.1016/S0034-4257(03)00184-6

Giglio L, van der Werf GR, Randerson JT, Collatz GJ, Kasibhatla P (2006) Global estimation of burned area using MODIS active fire observations. Atmos Chem Phys 6:957–974. doi:10.5194/acp-6-957-2006

Goldammer JG (2013) White paper on use of prescribed fire in land management, natura conservation and forestry in temperate-boreal Eurasia. In: Goldammer JG (ed) Prescribed burning in Russia and neighbouring temperate-boreal Eurasia. Global Fire Monitoring Center (GFMC) and Kessel Publishing House, Remagen-Oberwinter, Germany

Hall J, Loboda T, McCarty G (2015) Mapping and monitoring cropland burning in Russia: a multi-sensor approach. Poster presented at the NASA Carbon Cycle and Ecosystems workshop, College Park, MD 20–24 Apr. url:http://cce.nasa.gov/cgi-bin/meeting_2015/mtg2015_ab_search.pl?action=3&ab_id=86&search_pres_type=Poster&limit_last=100&search_free_text=Hall%20Loboda&limit_first=1&asc=1&desc=0&order_by=

Hawbaker TJ, Radeloff VC, Syphard AD, Zhu Z, Stewart SI (2008) Detection rates of the MODIS active fire product in the United States. Remote Sens Environ 112:2656–2664

International Institute for Applied Systems Analysis (IIASA) (2015) Finding farmland: new maps offer a clearer view of global agriculture. url:www.iiasa.ac.at/web/home/about/news/150116-Cropland-Maps.html. Accessed 26 Feb 2015

Kakareka SV, Kukharchyk TI (2003) PAH emission from the open burning of agricultural debris. Sci Total Environ 308:257–261

Kobets EM, Romanenkov V, Rukhovitch D (2010) Agricultural burning and forest fires in Russia. Presented on Sept 2010. url:https://www.eionet.europa.eu/events/Arctic%20Council/Elena%20Kobets

Kobets E, Blokova E, Tsepilova O (2011) The impact of agricultural burning on atmospheric air pollution and arctic climate. url:http://bellona.ru/filearchive/fil_otcheteng.pdf

Krylov A, McCarty JL, Potapov P, Loboda T, Tyukavina A, Turubanova S, Hansen MC (2014a) Remote sensing estimates of stand-replacement fires in Russia, 2002–2011. Environ Res Lett 9:105007–105014

Krylov A, McCarty JL, Potapov P, Turubanova S, Prishchepov A, Manisha A, Romanenkov V, Rukhovitch D, Koroleva P, Hansen MC (2014b) Fire regime and land abandonment in European Russia: case study of Smolensk Oblast. Abstract GC33E-0576 presented at the 2014 Fall Meeting, American Geophysical Union (AGU), San Francisco, CA 15–19 Dec 2014

Kuemmerle T, Kaplan JO, Prishchepov AV, Rylsky I, Chaskovskyy O, Tikunov VS, Müller D (2015) Forest transitions in eastern Europe and their effects on carbon budgets. Glob Change Biol. doi:10.1111/gcb.12897

Lerman Z, Shagaida N (2007) Land policies and agricultural land markets in Russia. Land Use Policy 24:14–23

Lerman Z, Csaki C, Feder G (2004) Agriculture in transition: land policies and evolving farm structures in post-soviet countries. Lexington Books, Lanham, Boulder, New York, Toronto, Oxford

LITHSTAT (2014) Lithuanian statistics online dataset. url:http://www.stat.gov.lt/en/home. Accessed 13 July 2015

Loboda T, Krankina O, Savin I, Kurbanov E, Hall J (2017) Land management and impact of 2010 extreme drought event on agricultural and ecological systems of European Russia. In Radeloff V, Gutman G (eds) Land-cover and land-use changes in Eastern Europe after the Collapse of the Soviet Union in 1991. Springer (this volume)

Magi BI, Rabin S, Shevliakova E, Pacala S (2012) Separating agricultural and non-agricultural fire seasonality at regional scales. Biogeosciences 9:3003–3012

Mazzola M, Johnson TE, Cook RJ (1997) Influence of field burning and soil treatments on growth of wheat after Kentucky bluegrass, and effect of Rhizoctonia cerealis on bluegrass emergence and growth. Plant Pathol 46:708–715

McCarty JL (2011) Remote sensing-based estimates of annual and seasonal emissions from crop residue burning in the contiguous United States. JAPCA J Air Waste Manag Assoc 61:22–34

McCarty JL, Justice CO, Korontzi S (2007) Agricultural burning in the southeastern United States detected by MODIS. Remote Sens Environ 108:151–162

McCarty JL, Ellicott EA, Romanenkov V, Rukhovitch D, Koroleva P (2012) Multi-year black carbon emissions from cropland burning in the Russian Federation. Atmos Environ 63:223–238. doi:10.1016/j.atmosenv.2012.08.053

Mierauskas P (2012) Policy and legislative framework overview of fire management in Lithuanian protected areas. Flamma 3:1–5

Narayan C, Fernades PM, van Brusselen J, Schuck A (2007) Potential for CO_2 emissions mitigation in Europe through prescribed burning in the context of the Kyoto Protocol. For Ecol Manage 251:164–173

Ortiz de Zarate I, Ezcurra A, Lacaux JP, Van Dinh P, Diaz de Argandona J (2005) Pollution by cereal waste burning in Spain. Atmos Res 73:161–170

Potapov P, Turubanova S, Zhuravleva I, Hansen M, Yaroshenko A, Manisha A (2012) Forest cover change within the Russian European North after the Breakdown of Soviet Union (1990–2005). Int J For Res

Potapov PV, Turubanova SA, Tyukavina A, Krylov AM, McCarty JL, Radeloff VC, Hansen MC (2014) Eastern Europe's forest cover dynamics from 1985 to 2012 quantified from the full Landsat archive. Remote Sens Environ 159:28–43

Prishchepov AV, Radeloff VC, Baumann M, Kuemmerle T, Müller D (2012) Effects of institutional changes on land use: agricultural land abandonment during the transition from state command to market driven economics in post-soviet eastern Europe. Environ Res Lett 7:024021

Prishchepov AV, Muller D, Dubinin M, Baumann M, Radeloff VC (2013) Determinants of agricultural land abandonment in post-Soviet European Russia. Land Use Policy 30:873–884

Republic of Belarus (2015) Agriculture in Belarus. http://www.belarus.by/en/invest/key-sectors-for-investment/agriculture. Accessed 2 Apl 2015

Romanenkov V, Rukhovitch D, Koroleva P, McCarty JL (2014) Estimating black carbon emissions from agricultural burning. In: Mueller L, Lischeid G, Saparov A (eds) Novel measurement and assessment tools for monitoring and management of land and water resources in agricultural landscapes of central Asia. Springer, New York, pp 347–364

ROSSTAT (2010) Regions of Russia. url:www.gks.ru. Accessed 12 July 2015

ROSSTAT (2012) Socio-economic data for Russian Region. url:www.gks.ru/wps/wcm/connect/rosstat_main/rosstat/ru/statistics/publications/catalog/doc_1138623506156. Accessed 13 May 2015

ROSSTAT (2014) Russian Federation Federal State Statistics Service. url:www.gks.ru/wps/wcm/connect/rosstat_main/rosstat/en/main/. Accessed 13 July 2015

Roy DP, Boschetti L, Justice CO, Ju J (2008) The collection 5 MODIS burned area product—global evaluation by comparison with the MODIS active fire product. Remote Sens Environ 112:3690–3707. doi:10.1016/j.rse.2008.05.013

Saarikoski S, Hillamo R (2013) Wildfires as a source of aerosol particles transported to the northern European regions. In: Viana M (ed) Urban air quality in Europe. Springer, Heidelberg, pp 101–121

Sakovich VS (2008) Agriculture in Belarus between 1980–2007: tendency of the development (Selskoe Hozjaistvo v Respublike Belarus v 1980–2007 g.: tendencii razvitija). Minsk: Belorusskaja nauka

Schierhorn F, Muller D, Beringer T, Prishchepov AV, Kuemmerle T, Balmann A (2013) Post-Soviet cropland abandonment and carbon sequestration in European Russia, Ukraine and Belarus. Glob Biochem Cycles 27:1175–1185

Seiler W, Crutzen PJ (1980) Estimates of gross and net fluxes of carbon between the biosphere and atmosphere from biomass burning. Clim Change 2:207–247

Soja AJ, Cofer WR, Shugart HH, Sukhinin AI, Stackhouse PW Jr, McRae DJ, Conard SG (2004) Estimating fire emissions and disparities in boreal Siberia (1998–2002). J Geophys Res 109: D14S06. doi:10.1029/2004JD004570

Stohl A, Berg T, Burkhart JF, Fjæraa AM, Forster C, Herber A, Hov Ø, Lunder C, McMillan WW, Oltmans S, Shiobara M, Simpson D, Solberg S, Stebel K, Ström J, Tørseth K, Treffeisen R, Virkkunen K, Yttri KE (2007) Arctic smoke—record high air pollution levels in the European Arctic due to agricultural fires in Eastern Europe in spring 2006. Atmos Chem Phys 7:511–534. doi:10.5194/acp-7-511-2007

Streets DG, Gupta S, Waldhoff ST, Wang MQ, Bond TC, Yiyun B (2001) Black carbon emissions in China. Atmos Environ 35:4281–4296

Targino AC, Krecl P, Johansson C, Swietlicko E, Massling A, Coraiola GC, Lihavainen H (2013) Deterioration of air quality across Sweden due to transboundary agricultural burning emissions. Boreal Environ Res 18:19–36

Tulbure M, Wimberly MC, Roy DP, Henebry GM (2010) Spatial and temporal heterogeneity of agricultural fires in the central United States in relation to land cover and land use. Landsc Ecol 26:211–224

UNDP (2014) 2894 Belarus—catalyzing sustainability of the wetland protected area system in Belarusian Polesie through increased management efficiency and realigned land use practices. http://www.undp.org/content/undp/en/home/ourwork/environmentandenergy/focus_areas/ecosystems_and_biodiversity/projects/belarus_-_catalyzingsustainabilityofthewetlandprotectedareasyste.html

US EPA (2015) The 2011 national emissions inventory, Version 2. url:http://www.epa.gov/ttnchie1/net/2011inventory.html

USDA FAS (2013) Favorable crop prospects in Poland and Lithuania. http://www.pecad.fas.usda.gov/highlights/2013/07/PolandLithuania/. Accessed 2 Apr 2015

USDA FAS (2014) Grain and feed September 2014 update: Russian Federation. http://gain.fas.usda.gov/Recent%20GAIN%20Publications/Grain%20and%20Feed%20September%202014%20Update_Moscow_Russian%20Federation_8-29-2014.pdf. Accessed 2 Apr 2015

Van Leeuwen TT, Van der Werf GR, Hoffmann AA, Detmers RG, Rücker G, French NHF, Archibald S, Carvalho JA, Jr, Cook GD, De Groot WJ, Hély C, Kasischke ES, Kloster S, McCarty JL, Pettinari ML, Savadogo P, Alvarado EC, Boschetti L, Manuri S, Meyer CP, Siegert F, Trollope LA, Trollope WSW (2014) Biomass burning fuel consumption rates: a field measurement database. Biogeosciences 11:7305–7329

Venkataraman C, Habib G, Kadamba D, Shrivastava M, Leon J-F, Crouzille B, Boucher O, Streets DG (2006) Emissions from open biomass burning in India: integrating the inventory approach with high-resolution moderate resolution imaging spectroradiometer (MODIS) active-fire and land cover data. Global Biogeochem, Cy. 20

WHO (2000) World Health Organization: air quality guidelines for Europe, 2nd edn. WHO Regional Publication, European Series, No. 91. url:http://www.euro.who.int/__data/assets/pdf_file/0005/74732/E71922.pdf. Accessed 2 Apr 2015

Witham C, Manning A (2007) Impacts of Russian biomass burning on UK air quality. Atmos Environ 41:8075–8090

Yang S, He H, Lu S, Chen D, Zhu J (2008) Quantification of crop residue burning in the field and its influence on ambient air quality in Suqian, China. Atmos Environ 42:1961–1969

Yevich R, Logan JA (2003) An assessment of biofuel use and burning of agricultural waste in the developing world. Glob Biogeochem Cycles 17:1095

Land Change in European Russia: 1982–2011

Kirsten de Beurs, Grigory Ioffe, Tatyana Nefedova and Geoffrey Henebry

Abstract In this chapter, we use change analysis at three spatial resolutions (8 km, 500 m, and 30 m) to investigate land changes in European Russia occurring between 1982 and 2011. We first apply the nonparametric Seasonal Kendall trend test to the improved GIMMS 3g AVHRR NDVI dataset in three ten-year epochs: 1982–1991, 1991–2000, and 2000–2009. We investigate the changes in each individual period to determine the consistency of the change analysis. We then use Landsat and MODIS imagery to identify the arable lands in the grain belt of European Russia. We report on cultivation frequency, which is a key management decision that affects soil carbon stocks in croplands. We previously demonstrated for two MODIS tiles that the cultivation frequency strongly depends on location. Here we extend the analysis to a third MODIS tile. We conclude with a discussion of visible changes on the ground for four study regions: Kostroma, Chuvash Republic, Samara, and Stavropol.

K. de Beurs (✉)
Department of Geography and Environmental Sustainability,
The University of Oklahoma, 100 East Boyd Street, Sarkeys Energy Building,
566, Norman, OK 73019, USA
e-mail: kdebeurs@ou.edu

G. Ioffe
College of Science and Technology, Geospatial Science,
Radford University, Cook Hall, Room 134, Radford, VA 24142, USA
e-mail: gioffe@radford.edu

T. Nefedova
Institute of Geography, Russian Academy of Sciences, Moscow, Russia
e-mail: Trene12@yandex.ru

G. Henebry
Geospatial Sciences Center of Excellence (GSCE), South Dakota State University,
1021 Medary Avenue, Wecota Hall 506B, Brookings, SD 57007-3510, USA
e-mail: geoffrey.henebry@sdstate.edu

G. Gutman and V. Radeloff (eds.), *Land-Cover and Land-Use Changes in Eastern Europe after the Collapse of the Soviet Union in 1991*,
DOI 10.1007/978-3-319-42638-9_10

1 Introduction

This chapter builds on the work of four previously published papers. The first paper discussed a vegetation trend analysis at multiple scales in Northern Eurasia (de Beurs et al. 2009). In this chapter, we extend the analysis using updated satellite data. We analyze AVHRR data in ten-year epochs and compare significant positive and negative trends between the MODIS and AVHRR data. We also discuss these changes in light of the ongoing anthropogenic and climatic changes in the region. The second paper discussed the agricultural and arable land mapping and development in European Russia covered by two MODIS tiles (de Beurs and Ioffe 2013). In this chapter, we extend the mapping of agriculture and crop intensity towards a third tile located farther south and overlapping the intensely cultivated Stavropol kray (republics, krays, and oblasts are districts comparable to a US states or Canadian provinces, rayons are subdistricts comparable to US counties). The third and fourth papers discussed the socio-economic development on the ground in Kostroma oblast, Samara oblast, and Stavropol kray (Ioffe et al. 2012, 2014). Here we highlight the main findings for these regions and discuss them in light of the observed satellite changes that we noted.

1.1 Study Regions

We specifically focus our analyses on four areas we visited on the ground: Kostroma oblast, Chuvash Republic, Samara oblast, and Stavropol kray. We provide a short description of these four areas, from north to south.

Kostroma oblast (60,100 km^2) is located in the northeastern corner of the Russian heartland; it is humid and relatively cold for crop farming, resulting in very aggressive forest growth with a scattering of smaller pockets of arable lands. Peak agricultural colonization of this oblast occurred in the middle of the 20th century, and its farmland percentage has been declining since the 1950s. In 2000, money-losing collective farms were still abundant in this region, with more than 70 % of the farm units operating at a loss (Ioffe et al. 2006). The rural population density in 2006 was low, at 3.7 people/km^2, which is less than the population density of the US state of North Dakota and is just slightly more than the population density of the country of Australia. Land abandonment is relatively high: only 49 % of the areas cultivated in the 1990s were still being cultivated in 2006.

Chuvash Republic is located farther south than Kostroma and lies in the heart of the Volga region, with more favorable temperatures for agriculture. It is the most densely populated region among our study areas and has more fertile soils than Kostroma. The density of the rural population is higher, at 30 people/km^2 in 2006. Ethnic Russians are a minority group, while the Chuvash form the largest ethnic group (69 %). Land abandonment is moderate: only 71 % of the area cultivated in 1990 was still being cultivated in 2006.

Samara oblast is even farther south than Chuvash Republic but still lies in the central Volga River basin, with favorable temperatures for agriculture. The rural population density is 11.6 people/km^2. The ethnicity of Samara's population is principally Russian (83.6 %), while Chuvash (6 %) and Tatar (4.3 %) comprise the largest of several minorities. Non-Russians predominantly live in the northern areas of Samara. Land abandonment in this area is moderate: just 69 % of the area cultivated in 1990 was still cultivated in 2006.

Stavropol kray is located farthest south, with very fertile soils but a continuous threat of drought. It is one of the most agriculturally productive regions of Russia. The population density is highest among the four study areas, at 41 people/km^2 in 2006. Caucasian ethnic groups account for 11.6 % of the population. Almost 85 % of the region is cropland, and the share of wheat is growing.

2 Data

2.1 GIMMS3g AVHRR Data

We used third-generation Normalized Difference Vegetation Index (NDVI) data from the Global Inventory Monitoring and Modeling System (GIMMS NDVI3g) that offers considerable improvements over the previous GIMMS dataset and extends through 2011 (de Jong et al. 2013; Zhu et al. 2013). The dataset is constructed from Advanced Very High Resolution Radiometer (AVHRR) sensor data and provides 15-day maximum value composites at a spatial resolution of 8 km. The GIMMS NDVI3g dataset is considered to be the best dataset to use for the analysis of long-term trends and has been used numerous times for the detection of land degradation and greening (de Jong et al. 2013; Beck et al. 2011; Cook and Pau 2013). We divided the time series of the GIMMS NDVI data in three ten-year epochs: 1982–1991, 1991–2000 and 2000–2009 to maintain consistent statistical power in the trend analysis.

2.2 MODIS Nadir-BRDF Adjusted Reflectance and Land Surface Temperature Data

The MODIS instrument provides near-daily repeated coverage of the earth's surface, with 36 spectral bands and a swath width of approximately 2330 km. Seven bands are specifically designed for terrestrial remote sensing, with a spatial resolution of 250 m (bands 1–2) and 500 m (bands 3–7). Each MODIS swath is divided into 10 by 10° tiles that are numbered vertically and horizontally. For this study, we selected three MODIS products: the Nadir BRDF-Adjusted Reflectance (NBAR) dataset at two spatial resolutions, 500 m (MCD43A4v5) and 0.05° (5.6 km,

MCD43C4v5), and the Land Surface Temperature (LST)/Emissivity dataset with a spatial resolution of 1000 m (MOD11A2v5). We focus the analysis on three SIN tiles: h20v03, h21v03, and h21v04. We downloaded all available images obtained between January 2002 and December 2012. We calculated the Normalized Difference Vegetation Index (NDVI) based on the NBAR data and selected the day and night temperature data from the MOD11A2 dataset. We calculated growing degree-days (GDD) as follows:

$$GDD = \frac{Nighttime\ Temperature + Daytime\ Temperature}{2}$$

We accumulated 8-day GDD (AGDD) by simple summation commencing each January 1, when GDD exceeded the base temperature of 0 °C:

$$AGDD_t = AGDD_{t-1} + max(GDD_t, 0)$$

We chose a base temperature of 0 °C since this threshold is often used for high-latitude annual crops, such as wheat, and for perennial grasslands (Henebry 2013), since our study region is dominated by perennial grasslands and spring grains.

2.3 Landsat Data

We previously described our analyses for Chuvash Republic and Samara oblast (de Beurs and Ioffe 2013). Here we describe only the analyses for Stavropol kray, but the methodology that we apply is the same as for Chuvash Republic and Samara oblast. We used Landsat data from five different WRS-2 path/row scenes that covered Stavropol kray. Our time series consisted of nearly cloud-free Landsat 5 Thematic Mapper (TM) images for each tile between 2007 and 2011, with two exceptions: June 2000 in WRS-2 path 170/row 29 and August 2006 in path 172/row 29. Each path/row scene was represented by images from the summer months (June, July, August) occasionally supplemented by an image from May (p171r28) or September (p171r29), when there were not enough cloud-free summer images available (Table 1). Every image was atmospherically corrected with the

Table 1 Overview of Landsat Thematic Mapper images used in the land cover classification of Stavropol kray

P 170/R 29	P 171/R 28	P 171/R 29	P 172/R 28	P 172/R 29
02 JUN 2000	12 MAY 2007	29 JUL 2007	07 AUG 2007	04 AUG 2006
13 JUL 2009	31 JUL 2007	08 AUG 2010	25 JUN 2009	25 JUN 2009
01 AUG 2010	08 AUG 2010	26 JUL 2011	30 JUL 2010	11 JUL 2009
01 JUN 2011	28 SEP 2011	28 SEP 2011	31 AUG 2010	31 AUG 2010

P path, *R* row

ENVI FLAASH routine (Matthew et al. 2000), which is a first-principles atmospheric correction tool that incorporates the MODTRAN4 radiative transfer code to correct wavelengths in the visible through near-infrared and shortwave infrared regions. After correction, TM bands 1–5 and 7 for all files were stacked into a single file for each of the five path/row scenes.

3 Methods

3.1 Seasonal Kendall Trend Tests

Changes in variable time series have typically been summarized using the slope of a linear regression model, but many factors can degrade the reliability of the parameter estimates. Instead, we applied the Seasonal Kendall (SK) test corrected for first-order autocorrelation as a robust nonparametric alternative that relies on fewer statistical assumptions and is routinely used for analyses of climatological and hydrological time series (von Storch and Navarra 1999; de Beurs and Henebry 2004). The SK test is well equipped to pick up the kind of temporal changes that could result from shifts in above-ground biomass. We applied the nonparametric Seasonal Kendall (SK) trend test to both the AVHRR and MODIS time series to identify statistically significant increases and decreases in vegetation as measured by NDVI (de Beurs and Henebry 2004; de Beurs et al. 2009). We subsequently applied a median filter to remove single pixels with a significant trend. We interpreted areas with negative (or positive) test statistics and p-values below 0.01 as revealing significant multi-year trends in browning (or greening) of the vegetated land surface.

3.2 Cropland Probability and Cultivation Frequency

We used the FAO definition for arable land; specifically, all land with temporary crops, temporary meadows for mowing or pasture, market or kitchen gardens, or that was temporarily laid fallow (clean-fallow) was considered arable. Temporarily fallowed land is land set aside for one or more years before recultivation; that is, it is agricultural land to rest. Land abandoned as a result of shifting cultivation is excluded from this definition. The definition of agricultural lands includes arable lands and permanent pastures. Here we focus on arable lands only and do not consider permanent pastures or permanent crops.

It is difficult to distinguish between fallowed and abandoned fields based solely on remotely sensed imagery. This distinction is also difficult to obtain on the ground, since the distinction involves the intention of the crop farmers. While traveling around the Russian countryside, we saw several examples of fields that were considered arable, but were not actively cultivated. Some of these fields were set aside temporarily with the intent of using them again in the near future, while some of these fields were occasionally used to maintain them. We also found fields that had been abandoned, but were subsequently brought back into production. We learned that areas that were continuously cropped year after year or fields with standard rotation schedules were less likely to be abandoned in the future (Ioffe et al. 2004). For our purposes, we do not consider the technical distinction between fallowed fields and abandoned fields as critical. Instead, we want to determine if a field was successfully sown in the spring and then successfully harvested. Fields that were not successfully sown and subsequently harvested may result from weather impacts (drought, late frost, etc.). In contrast, fields intentionally fallowed are likely to be the result of management decisions.

We have previously described the methodology used to map the agricultural lands and cultivation frequency based on a combination of Landsat and MODIS imagery (de Beurs and Ioffe 2013). Below we summarize our approach to mapping agricultural lands and cultivation frequency as applied to Stavropol kray, which is more actively cropped than the other areas we investigated. See de Beurs and Ioffe (2013) for additional information.

3.2.1 Landsat Classification and Probabilistic Label Relaxation

We applied image segmentation to each image stack. We then generated a random sample of the segments (1 %) and displayed them within Google Earth to select training areas. We applied basic maximum likelihood classification to each stack of Landsat TM images (Table 1). We classified the images into the following five classes: water, urban, forest, grasslands, and croplands. After classification we used the class probability images generated by the maximum likelihood routine to apply probabilistic label relaxation (Richards and Jia 2006; Canty 2010).

3.2.2 MODIS Cropland Probability

Based on our extensive prior research using vegetation index time series to measure land surface phenology (LSP) (de Beurs and Henebry 2010; Henebry and Beurs 2013), we summarized the MODIS image time series into a set of annual metrics. The LSP metrics that we calculated included (1) the start of the growing season (SOS), (2) the length of the growing season (LOS), (3) the AGDD at SOS, (4) the NDVI peak timing measured in AGDD, (5) the NDVI peak height, and (6) the coefficient of determination (R^2) for convex quadratic LSP model.

For each pixel we calculated the average and standard deviation of the LSP metrics based on all years (2002–2012). We hypothesized that the standard deviation calculated for each phenological metric over the 11 years represented in our study (2002–2012) would be higher for croplands than for natural vegetation as a result of crop rotation. In addition, we anticipated a difference in the start of the growing season, the peak height, and the peak timing.

We randomly selected 5000 MODIS pixels to generate a logistic model linking the availability of cropland within a pixel (based on the Landsat classifications) with the LSP metrics. We used the following logistic form:

$$Probability\ of\ Cropland = \frac{1}{1 + e^{-(\beta_0 + \beta_1 x_1 \pm \cdots + \beta_k x_k)}}$$

where $x_1, \ldots, x_k$ are the LSP metrics, and $\beta_0, \ldots, \beta_k$ are the parameters to be fitted. When a particular grid cell was found to have more than 50 % cropland, we set the dependent variable to 1. In contrast, if it was found to have less than 50 % cropland, we set the variable to zero. We applied step-wise regression to find the best fitting model. The goodness-of-fit of the logistic regressions was assessed by the percentage of observations predicted correctly (PC), Cohen's Kappa, and the area under the curve (AUC) of the receiver-operating characteristic. Cohen's Kappa assesses the accuracy of location and PC describes the ratio of correctly predicted cells from the total number of cells. The AUC measures the performance of a model compared to a random model, while the cut-off threshold is varied from zero to one (Pontius and Schneider 2001).

To perform an independent validation of the final MODIS land cover classification, we selected 1000 random 500 m MODIS pixels across Stavropol kray. We displayed the pixel boundaries in Google Earth and visually determined the land cover class for each. We evaluated each MODIS pixel in Google Earth and labelled the percentages of land cover classes we saw on the high-resolution images. We kept track of the approximate image date provided by Google. We only kept the random pixels that (1) coincided with a high resolution image in Google Earth, (2) appeared homogeneous, and (3) had a Google image date that was less than five years from the date of the image we used for classification. This process resulted in 362 pixels suitable for validation.

3.2.3 Cultivation Frequency

To determine whether a pixel was successfully sown during a particular year, we applied a series of basic decision rules similar to de Beurs and Ioffe (2013). We identified the following rules and applied them to all pixels with a crop probability higher than 0.5:

If the LSP model fails, then label the area as "not cropped".
If the peak height > (mean peak height − standard deviation of peak heights)

AND (peak timing in AGDD < 1100), then label the area as "cropped".

Otherwise, label the area as "not cropped"

where both the mean peak height per pixel and the standard deviation of the peak heights per pixel are based on the peak heights estimated from the LSP model fitted for each year from 2002 through 2012. LSP Models typically fail for clean fallow regions, resulting in the correct identification of no crops. When a field is fallowed for a second consecutive year, we expect some weeds to start growing. However, typically, NDVI peaks will still be relatively low, and the peaks are reached slowly (for a larger number of growing degree days). In essence, we label a pixel as "cropped" when the peak NDVI of the growing season is not more than one standard deviation below normal, with the peak occurring before 1100 growing degree days are reached. We evaluated several cut-off degrees for AGDD and determined that a cut-off of 1100 generates the most accurate results.

3.3 Field Interviews

During three field trips to Russia in May 2010, June 2011, and September 2011, we collected official statistical yearbook data and updated rayon data collected previously (Ioffe et al. 2006; Pallot and Nefedova 2007). In addition, we designed our field interviews to improve our understanding of the economic and rural social situation on the ground (Ioffe et al. 2012). With this aim in mind, we asked all participants questions with respect to population dynamics, unemployment, subsidies and taxes, and their perception of drought and climate change. When appropriate, we asked to see farms and crops and asked about crop varieties and rotation schemes.

We visited typical settlements and enterprises in four to five selected subdistricts in each of the three selected study areas and interviewed rural administration heads, farm managers, and members of the local population. In total, we conducted twenty to twenty-five loosely structured interviews per region. Each interview lasted between 30 and 90 min. The interviews were typically attended by one to five respondents. We aimed at interviewing a cross-section of people with agricultural interests. Among the experts we interviewed were the head of agriculture of each selected subdistrict, the heads of at least three different farms, and the administration heads of the corresponding villages. In addition, we spoke with Samara's Ministers of Economics and Agriculture, the head of Samara's Land Use Committee, several faculty members of the department of geography at Chuvash Republic University, and agronomists and other agricultural specialists in the three areas. We also spoke with the heads of several agricultural companies and agricultural cooperatives. Among the farmers and administrators interviewed, there were people from Bashkir, Tatar, and Chuvash ethnicities.

Most of the statistical data, such as the total area sown to crops, are available at the rayon level. We have data for each crop type and farm type (household farms, registered private farms, or large scale commercial farms). Official statistics at the rayon level are typically skewed toward large-scale farming, and the farm data for important household productions are summarized into larger averages (Pallot and Nefedova 2007). Our fieldwork helped to create a more nuanced understanding of the ongoing population changes and their effects on croplands in Russia.

4 Results and Discussion

4.1 Broad Trends and Significant Changes

There have been widespread changes in the Normalized Difference Vegetation Index (NDVI) since 1982. In the first decade, we observed mainly positive vegetation trends or greening (Figs. 1 and 2), and this finding was consistent with the results of previous papers reporting the increased greening of the Northern Hemisphere during that period (Myneni et al. 1997; Tucker et al. 2001). The second decade (1991–2000) had fewer overall changes, but greening was still found to be

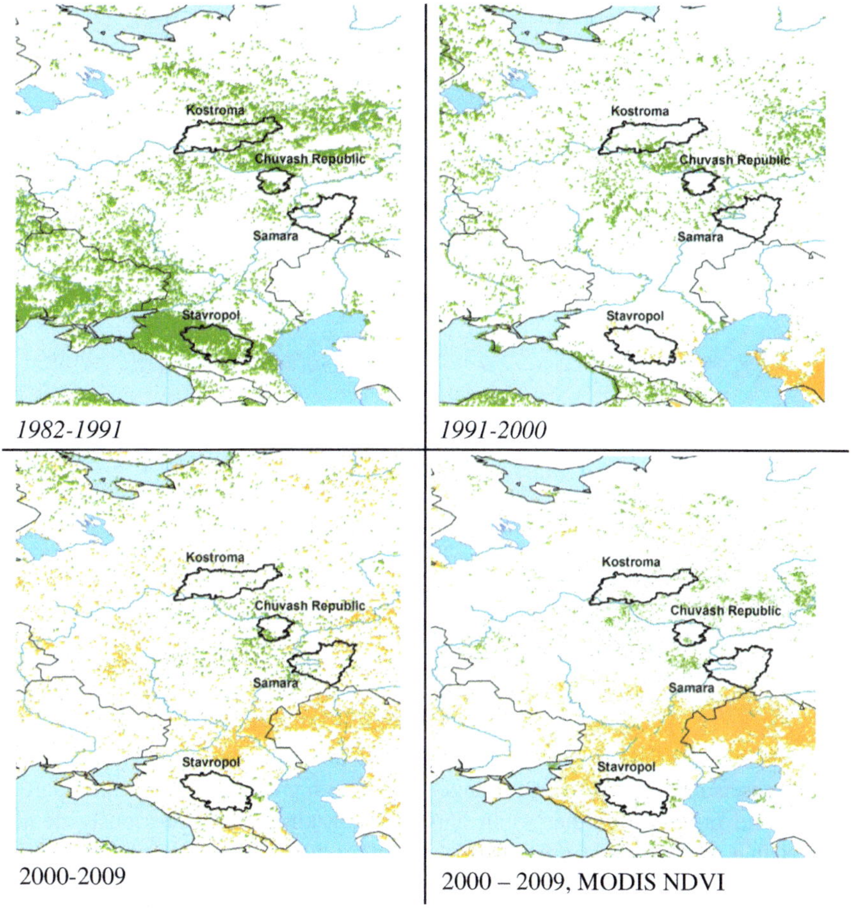

Fig. 1 Seasonal Kendall trend test for different 10-year periods between 1982 and 2009 based on AVHRR GIMMS data (*top row* and *lower left*) and MODIS data (*lower right*). *Green areas* reveal significant positive vegetation changes ($p < 0.01$), and *orange areas* reveal significant negative vegetation changes ($p < 0.01$)

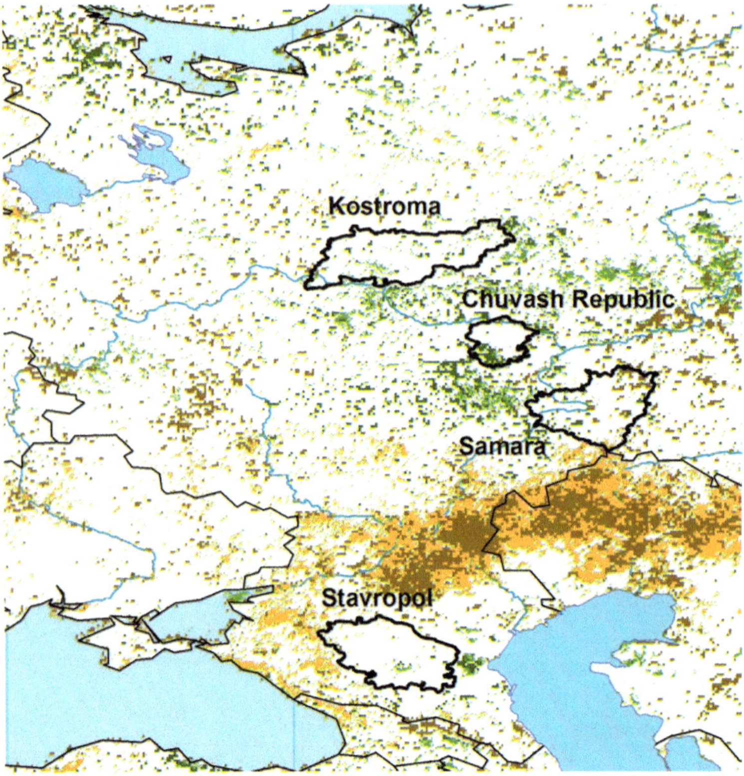

Fig. 2 Comparison of the AVHRR and MODIS trends from the same 10-year period (2000–2009). *Dark green areas* reveal significant positive vegetation change ($p < 0.01$) for AVHRR and *light green areas* reveal significant positive vegetation change ($p<0.01$) for MODIS. *Dark orange areas* reveal significant negative change ($p<0.01$) for AVHRR and *light orange* reveals significant negative change for MODIS

the dominant change for this region. The only area where browning (e.g., a negative NDVI trend) was visible was in southern Russia and Kazakhstan. The third decade (2000–2009) was found to have a greater mix of greening and browning, with most of the greening occurring in the central parts of Russia and most of the browning occurring farther south in Russia and Kazakhstan. Figure 2 compares the changes based on AVHRR NDVI and MODIS NDVI for the same time period (2000–2009). Both datasets reveal comparable change results, with greening in the central parts of Russia and browning farther south and into Kazakhstan. We previously demonstrated that most of this greening is consistent with large agricultural changes occurring in this region (de Beurs et al. 2009).

4.2 Cropland Probability and Cultivation Frequency

We previously calculated the cropland probability and intensity for the Samara oblast and Chuvash Republic (de Beurs and Ioffe 2013) based on data from 2002 through 2009. In this study, we have expanded the dataset for the Stavropol kray and calculate the cropland probability and cultivation frequency for this region based on the period 2002–2012. Table 2 provides the parameter estimates for Stavropol's cropland model, and Tables 3 and 4 provide the goodness of fit statistics and the confusion matrix. According to the model fits, 81.3 % of the croplands have been correctly identified. The independent Google Earth validation reveals that 85 % of the cropland pixels were correctly identified, with a Kappa

Table 2 Parameter coefficient estimates for the cropland model for Stavropol

Variable	Parameter estimate	S.E.
Constant	3.45	3.34
R^2_{adj} std. dev.	14.44	2.14
Peak timing	2.01E−3	2.85E−4
Peak timing std. dev.	−1.07E−3	2.76E−4
Peak height	4.58	0.60
Peak height std. dev.	38.26	1.99
LOS	−0.45	0.05
LOS std. dev.	0.58	0.06
AGDD at SOS	−3.41E−3	3.19E−4
AGDD at SOS std. dev.	7.15E−4	2.29E−4
y	−1.42E−6	6.34E−7

Table 3 Goodness-of-fit logistic regression assessment for the Stavropol kray cropland model

PC	0.813
AUC	0.865
Cohen's Kappa	0.625

PC percentage of the observations predicted correctly; *AUC* area under the curve of the receiver operating characteristic. Cohen's Kappa indicates that the model provides a significant (good) fit

Table 4 Confusion matrix for Stavropol kray

		GE reference			UAC	AC	K
		Cropland	Other	Σ			
MODIS logistic	Cropland	170	27	197	0.87		
	Other	26	139	165	0.84		
	Σ	196	166	362			
	PAC	0.87	0.84			0.85	0.71

GE Google Earth, *PAC* producer's accuracy, *UAC* users' accuracy, *AC* accuracy, *K* Kappa. 95 % confidence interval for Cohen's Kappa: (0.633–0.778), p-value < 0.01

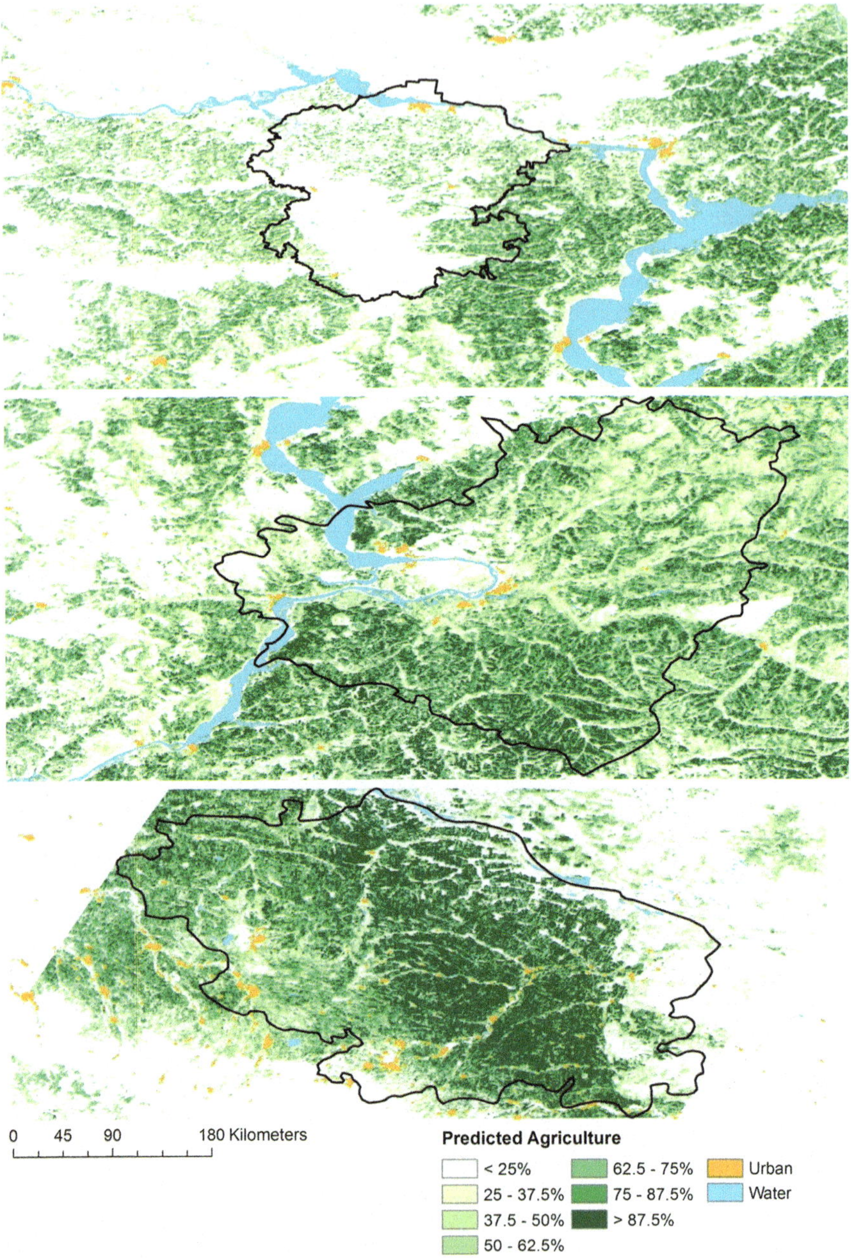

Fig. 3 Predicted agricultural lands in all Chuvash Republic (*top*), Samara oblast (*center*) and Stavropol krai (*bottom*)

coefficient of 0.71. The cropland model for Stavropol performs even better than the logistic cropland model we presented previously (de Beurs and Ioffe 2013) for areas located farther north.

Figure 3 provides the percentage of predicted agriculture for each pixel in the Chuvash Republic, Samara oblast, and Stavropol kray. We have not shown the results for the Kostroma oblast because there was so little agricultural activity to report. It is obvious from Fig. 3 that the amount of agricultural pixels increases dramatically for the southern oblasts with the least agricultural pixels in Chuvash and the most agricultural pixels in Stavropol. However, this picture is incomplete. We are much more interested in mapping cropland intensity because we think that it is important to understand not only where croplands are located, but also how many times in an 8- or 11-year period each field is successfully cropped (Fig. 4). Stavropol has substantially more croplands than Chuvash and Samara; however, most of the pixels are cropped approximately every other year. Only 0.5 % of the croplands in Stavropol are cropped continuously, while 81 % of the croplands in Chuvash are cropped every year. A little less than half of Samara's croplands (42 %) are cropped continuously. We believe that the low crop intensity and the change in the crop intensity can be interpreted with respect to the changing markets.

In recent years, planting decisions have become largely market-based in most of Russia and Ukraine (Lindeman 2004). As a result, crop production in Russia and other countries in the former Soviet Union has moved from growing mainly cereals to increased production of sunflower, soybean, and rapeseed. These three crops have increased in Russia from 5.6 to 11.9 million harvested hectares between 1990 and 2005 (European Bank for Reconstruction and Development 2008). Sunflower has become one of the most consistently profitable crops in Russia and Ukraine. The production costs of sunflower are relatively low, partly due to lower requirements for fertilizer and herbicide application (Lindeman 2004). The large leaves and tall stature of sunflowers makes them strong competitors for light against weeds, reducing the need for herbicides. However, sunflowers need more clean fallow time between plantings than small grains.

Crop rotation changes, including changes in the frequency of summer fallow and crop failures, are common in other global cereal belts as well (USDA Economic Research Service 2005). For example, in the early 1980s and 1990s, approximately half the cropland areas in Saskatchewan were set aside in summer fallow practices (Census Canada 1991). This pattern is comparable with that of the crop intensity documented in cereal growing regions in the southern Samara and northern Saratov oblasts.

4.3 Field Observations

This chapter has thus far discussed general patterns of agriculture in European Russia using case study areas. While it makes sense to discuss general Russian agricultural patterns in the context of satellite image analysis, there is wide variation

(a)

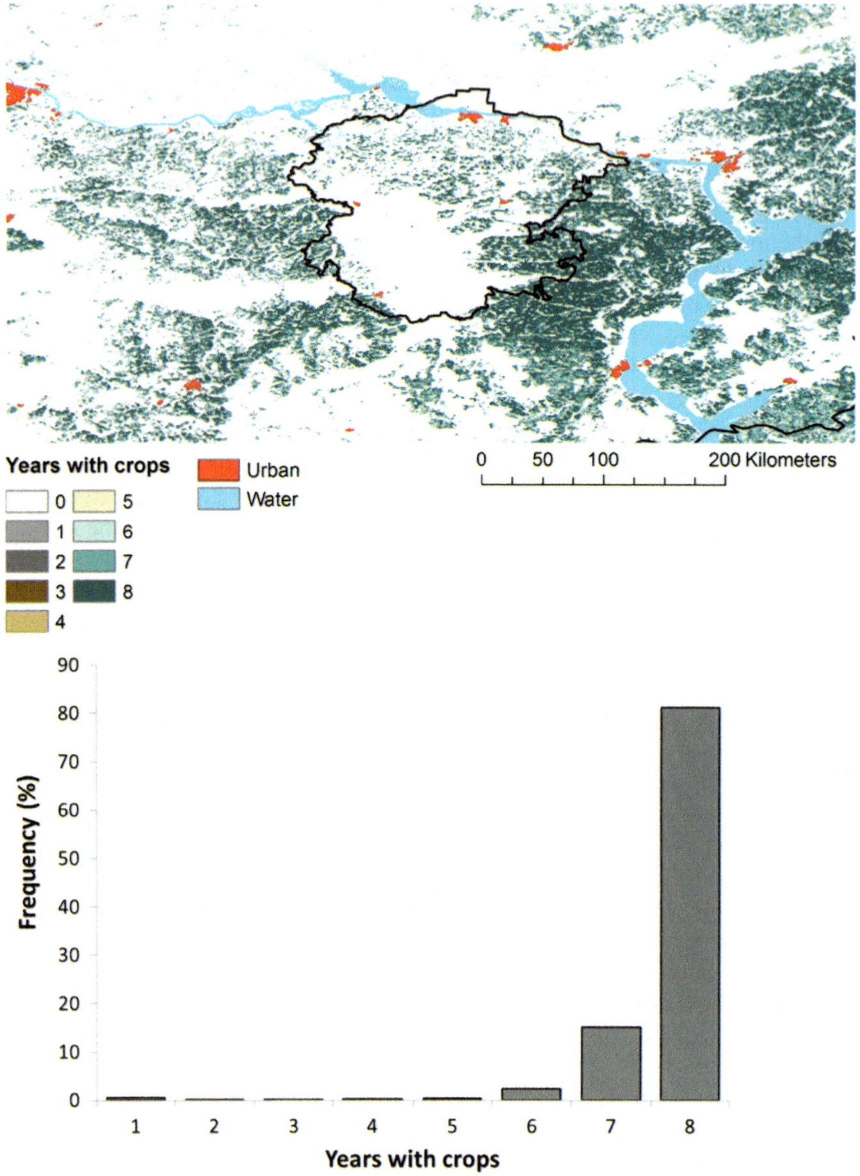

Fig. 4 **a** Chuvash Republic, 18.1 % of the oblast shows crop production at least once during the 8-year period investigated. Overall, 81 % of the crops are growing on continuously cropped fields. **b** Samara: 56.2 % of the oblast was found to have crop production at least once during the 8-year period investigated. 42 % of the crops were growing on continuously cropped fields. **c** Stavropol 68.5 % of the kray shows crop production at least once during the 11 year period investigated. Less than 0.5 % of the crops are grown on continuously cropped fields. Approximately 70 % of the crops are grown on fields that are cropped approximately every other year

(b)

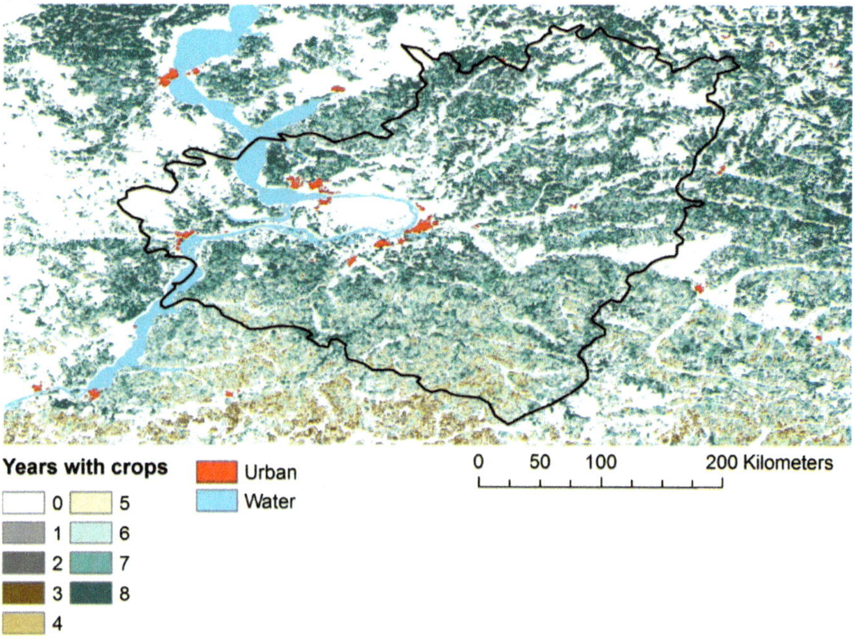

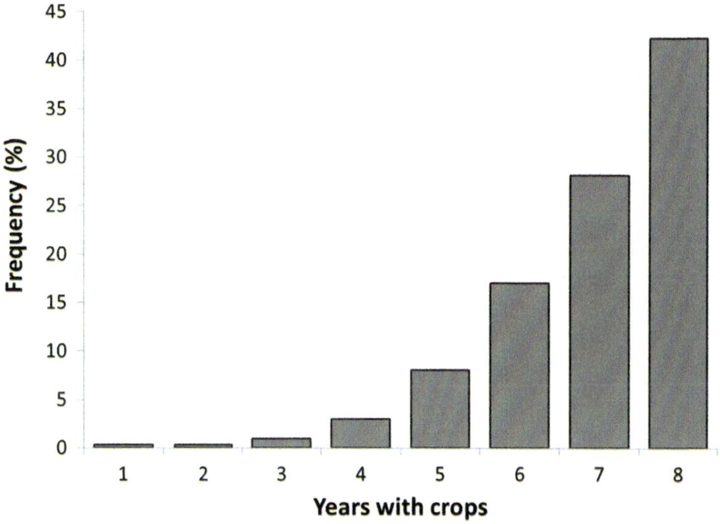

Fig. 4 (continued)

(c)

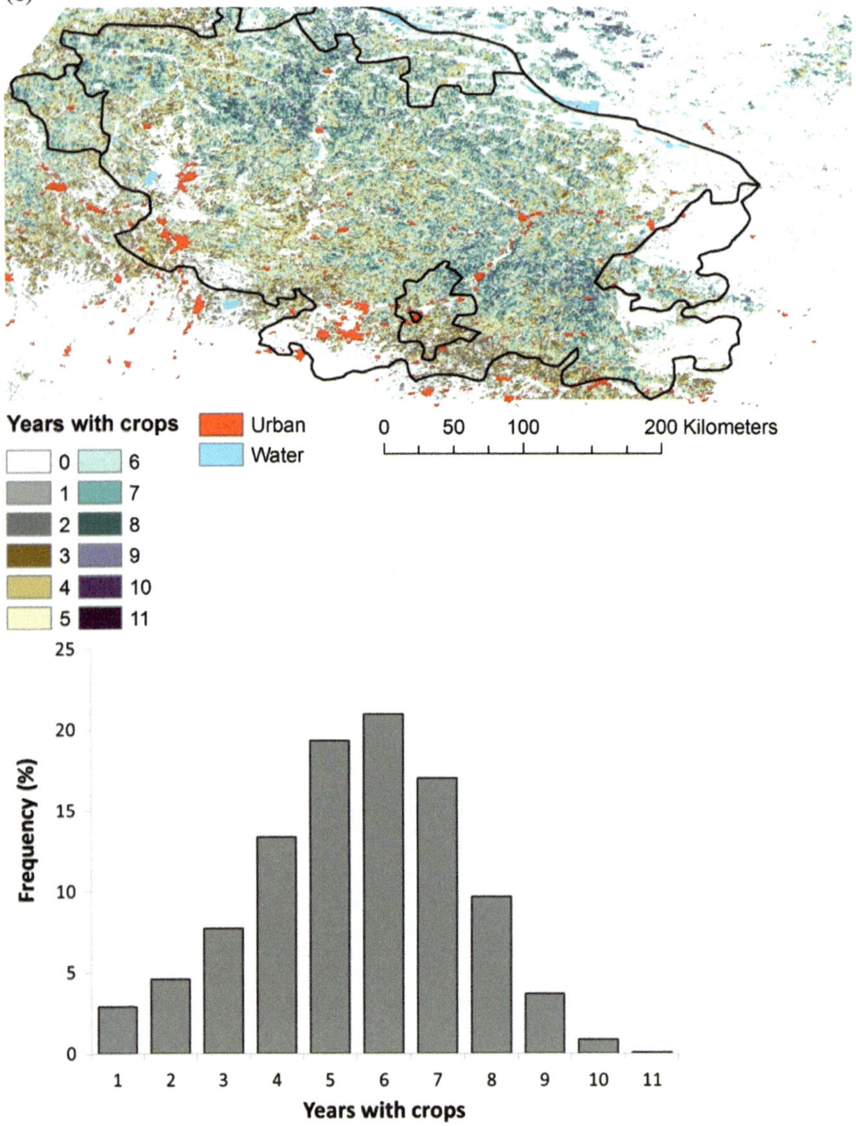

Fig. 4 (continued)

in European Russia in terms of the farm types, sizes, specialization, and productivity (Ioffe et al. 2012). We previously discussed the field work observations in detail (Ioffe et al. 2012, 2014). Figure 5 provides a few photographs from our field work. Here, we provide a basic summary of this work.

Fig. 5 *top left* Kostroma oblast, June 2010; *top right* Chuvash Republic, October 2011; *bottom left* Samara oblast, May 2010; *bottom right* Stavropol kray, June 2011

Kostroma's population size has declined by a factor of roughly three, and by 2010, the total area cultivated has shrunk to 25 % of the original 1950s levels. Furthermore, the total number of cattle in 2010 was approximately 10 % of the total in 1980 (Ioffe et al. 2012). Most hay meadows and pasture, as well as parts of arable lands, are no longer used. While traveling between sites, we found that many fields that had been sown with perennial grasses were no longer maintained. We found a strong link between the distance from urban areas and the lack of sown fields. We also found many rural villages that are either abandoned or shrinking rapidly, and this finding was consistent with our remotely sensed observations.

Compared to Kostroma oblast, the rural populations of Samara and Chuvash have been more stable. For example, Samara's population only declined by approximately 28 % between 1959 and 2010 (Ioffe et al. 2012). The social desert that can be found in Kostroma is not apparent in either Samara or Chuvash Republic. Moreover, significant pockets of local entrepreneurship exist, some of which are linked to ethnic minorities. In particular, we found re-cultivation of previously abandoned areas in Chuvash Republic. In Alatyr rayon, for example, where 41,000 ha were cultivated in 1990, all of this land had been abandoned by 2004. However, 22,000 ha were brought back into production by 2010, and another 16,000 ha were planned for production in 2012. Most re-cultivation is driven by agricultural firms owned by ethnic minorities. Remote sensing data cannot reveal why certain areas get re-cultivated, but re-cultivation is visible in these data and corresponds well with our observations on the ground.

5 Conclusions

The countryside in European Russia remains in flux and is not uniformly distributed. Our study demonstrates significant variability in agricultural development, from widespread abandonment to reinvigorated cultivation following abandonment, and from continuous cultivation to a range of fallow periods. Our satellite analysis demonstrates that it is important to track cultivation frequency in addition to locating where croplands are placed on the landscape. The frequency of cultivation is a critical management decision that affects soil carbon stocks (Lasco et al. 2006). Data on cultivation frequency provides insight into both the effects of drought and locations where the crop rotation schedules have been applied (de Beurs and Ioffe 2013). The manifold, cascading effects of declining rural populations can only be observed on the ground. Intensive field work has resulted in a nuanced understanding of the ongoing population changes and the effects on croplands in Russia (Ioffe et al. 2012, 2014).

References

Beck HE, McVicar TR, van Dijk AIJM, Schellekens J, de Jeu RAM, Bruijnzeel LA (2011) Global evaluation of four AVHRR–NDVI data sets: Intercomparison and assessment against Landsat imagery. Remote Sens Environ 115(10):2547–2563. doi:10.1016/j.rse.2011.05.012

Canty MJ (2010) Image analysis, classification and change detection in remote sensing, with algorithms for ENVI/IDL, 2nd edn. Taylor & Francis, CRC Press, Boca Raton

Census Canada (1991) Agricultural profile of Saskatchewan. Statistics Canada, Agriculture Division

Cook B, Pau S (2013) A global assessment of long-term greening and browning trends in pasture lands using the GIMMS LAI3g dataset. Remote Sens 5(5):2492–2512

de Beurs KM, Henebry GM (2004) Trend analysis of the Pathfinder AVHRR Land (PAL) NDVI data for the deserts of Central Asia. Geosci Remote Sens Lett 1(4):282–286

de Beurs KM, Henebry GM (2010) Spatio-temporal statistical methods for modelling land surface phenology. In: Phenological research: methods for environmental and climate change analysis, pp 177–208. doi:10.1007/978-90-481-3335-2_9

de Beurs KM, Ioffe G (2013) Use of Landsat and MODIS data to remotely estimate Russia's sown area. J Land Use Sci 1–25. doi:10.1080/1747423x.2013.798038

de Beurs KM, Wright CK, Henebry GM (2009) Dual scale trend analysis distinguishes climatic from anthropogenic effects on the vegetated land surface. Environ Res Lett 4(045012)

de Jong R, Verbesselt J, Zeileis A, Schaepman ME (2013) Shifts in global vegetation activity trends. Remote Sens 5(3):1117–1133

European Bank for Reconstruction and Development (2008) Fighting food inflation through sustainable investment. FAO

Henebry GM (2013) Phenologies of North American grasslands and grasses. In: Schwartz MD (ed) Phenology: an integrative environmental science. Springer Netherlands, pp 197–210. doi:10.1007/978-94-007-6925-0_11

Henebry GM, Beurs KM (2013) Remote sensing of land surface phenology: a prospectus. In: Schwartz MD (ed) Phenology: an integrative environmental science. Springer Netherlands, pp 385–411. doi:10.1007/978-94-007-6925-0_21

Ioffe G, Nefedova T, Zaslavsky I (2004) From spatial continuity to fragmentation: the case of Russian farming. Ann Assoc Am Geogr 94(4):913–943

Ioffe GV, Nefedova TiaGe, Zaslavsky I (2006) The end of peasantry?: the disintegration of rural Russia. University of Pittsburgh Pre

Ioffe G, Nefedova T, de Beurs KM (2012) Land abandonment in Russia. Eurasian Geogr Econ 53 (4):527–549

Ioffe G, Nefedova T, de Beurs KM (2014) Agrarian transformations in the Russian breadbasket: contemporary trends as manifest in Stravropol'. Post-Soviet Aff 30(6):441–463

Lasco RD, Ogle SM, Raison J, Verchot L, Wassman R, Yagi K, Bhattacharya S, Brenner JS, Partson Daka J, Gonzalez SP, Krug T, Li Y, Martino DL, McConkey BG, Smith P, Tyler SC, Zhakata W (2006) 2006 IPCC guidelines for national greenhouse gas inventories 4. Agriculture, forestry and other land use

Lindeman M (2004) Sunflowerseed production in Ukraine and Russia. Production estimates and Crop Assessment Division Foreign Agricultural Division FAS USDA. http://www.fas.usda.gov/pecad2/highlights/2004/06/ukr_sunseed/index.htm

Matthew MW, Adler-Golden SM, Berk A, Richtsmeier SC, Levine RY, Bernstein LS, Acharya PK, Anderson GP, Felde GW, Hoke MP, Ratkowski A, Burke H-H, Kaiser RD, Miller DP (2000) Status of atmospheric correction using a MODTRAN4-based algorithm. In: SPIE proceedings, Algorithms for multispectral, hyperspectral, and ultraspectral imagery VI, vol 4049, pp 199–207

Myneni RB, Keeling CD, Tucker CJ, Asrar G, Nemani RR (1997) Increased plant growth in the northern high latitudes from 1981 to 1991. Nature 386(6626):698–702

Pallot J, Nefedova T (2007) Russia's unknown agriculture: household production in Post-Soviet Russia. Oxford University Press, Oxford

Pontius RG, Schneider LC (2001) Land-cover change model validation by an ROC method for the Ipswich watershed, Massachusetts, USA. Agric Ecosyst Environ 85(1–3):239–248

Richards JA, Jia X (2006) Remote sensing digital analysis: an introduction, 4th edn. Springer, Berlin

Tucker CJ, Slayback DA, Pinzon JE, Los SO, Myneni RB, Taylor MG (2001) Higher northern latitude normalized difference vegetation index and growing season trends from 1982 to 1999. Int J Biometeorol 45(4):184–190

USDA Economic Research Service (2005) Land use, value, and management: major uses of land. http://www.ers.usda.gov/briefing/landuse/majorlandusechapter.htm

von Storch H, Navarra A (1999) Analysis of climate variability/applications of statistical techniques. Springer, Berlin

Zhu Z, Bi J, Pan Y, Ganguly S, Anav A, Xu L, Samanta A, Piao S, Nemani RR, Myneni RB (2013) Global data sets of vegetation leaf area index (LAI) 3g and Fraction of Photosynthetically Active Radiation (FPAR) 3g derived from Global Inventory Modeling and Mapping Studies (GIMMS) Normalized Difference Vegetation Index (NDVI3g) for the period 1981 to 2011. Remote Sens 5(2):927–948

Erratum to: Land Change in the Carpathian Region Before and After Major Institutional Changes

Catalina Munteanu, Volker Radeloff, Patrick Griffiths, Lubos Halada, Dominik Kaim, Jan Knorn, Jacek Kozak, Tobias Kuemmerle, Juraj Lieskovsky, Daniel Müller, Katarzyna Ostapowicz, Oleksandra Shandra and Premysl Stych

Erratum to: Chapter "Land Change in the Carpathian Region Before and After Major Institutional Changes" in: G. Gutman and V. Radeloff (eds.), *Land-Cover and Land-Use Changes in Eastern Europe after the Collapse of the Soviet Union in 1991*, DOI 10.1007/978-3-319-42638-9_4

The original version of Chapter 4 was inadvertently published with incorrect author name "Domink Kaim" instead of "Dominik Kaim". The erratum chapter and the book have been updated with the changes.

The updated original online version for this chapter can be found at
DOI 10.1007/978-3-319-42638-9_4

C. Munteanu (✉) · V. Radeloff
SILVIS Lab, Department of Forest and Wildlife Ecology, University of Wisconsin-Madison, 1630 Linden Drive, Madison, WI 53706, USA
e-mail: cmunteanu@wisc.edu

V. Radeloff
e-mail: radeloff@wisc.edu

P. Griffiths · J. Knorn · T. Kuemmerle · D. Müller
Geography Department, Humboldt-University Berlin, Unter den Linden 6, 10099 Berlin, Germany
e-mail: patrick.griffiths@geo.hu-berlin.de

J. Knorn
e-mail: jan.knorn@geo.hu-berlin.de

T. Kuemmerle
e-mail: tobias.kuemmerle@geo.hu-berlin.de

G. Gutman and V. Radeloff (eds.), *Land-Cover and Land-Use Changes in Eastern Europe after the Collapse of the Soviet Union in 1991*,
DOI 10.1007/978-3-319-42638-9_11

D. Müller
e-mail: mueller@iamo.de

L. Halada · J. Lieskovsky
Institute of Landscape Ecology, Slovak Academy of Sciences Bratislava, Branch Nitra, Akademická 2, 949 01 Nitra, Slovakia
e-mail: lubos.halada@savba.sk

J. Lieskovsky
e-mail: juraj.lieskovsky@savba.sk

D. Kaim · J. Kozak · K. Ostapowicz
Institute of Geography and Spatial Management, Jagiellonian University, Gronostajowa 7, 30-387 Kraków, Poland
e-mail: dkaim@gis.geo.uj.edu.pl

J. Kozak
e-mail: jkozak@gis.geo.uj.edu.pl

K. Ostapowicz
e-mail: kostapowicz@gis.geo.uj.edu.pl

T. Kuemmerle
Germany and Integrative Research Institute on Transformations in Human Environment Systems (IRI THESys), Humboldt-University Berlin, Unter den Linden 6, 10099 Berlin, Germany

D. Müller
Leibniz Institute of Agricultural Development in Transition Economies (IAMO), Halle (Saale), Germany

D. Müller
Integrative Research Institute on Transformations in Human Environment Systems (IRI THESys), Humboldt-University Berlin, Unter den Linden 6, 10099 Berlin, Germany

O. Shandra
Department of Physical Geography and Geoecology, Taras Shevchenko National University of Kyiv, Kiev, Ukraine
e-mail: alya.shandra@gmail.com

P. Stych
Faculty of Science, Charles University in Prague, Albertov 6, 128 43 Prague 2, Czech Republic
e-mail: premysl.stych@natur.cuni.cz

Index

G. Gutman and V. Radeloff (eds.), *Land-Cover and Land-Use Changes in Eastern Europe after the Collapse of the Soviet Union in 1991*,
DOI 10.1007/978-3-319-42638-9

The manufacturer's authorised representative in the EU is Springer Nature Customer Service Centre GmbH, Europaplatz 3, 69115 Heidelberg, Germany. If you have any concerns regarding our products, please contact ProductSafety@springernature.com

Printed and bound by CPI Group (UK) Ltd, Croydon, CR0 4YY
30/09/2026
02225571-0005